Elementary Theory & Application of Numerical Analysis

Revised Edition

Elementary Theory & Application of Numerical Analysis

Revised Edition

James E. Miller

Professor Emeritus
Mathematics & Computer Science
Transylvania University

David G. Moursund

Charles S. Duris

Dover Publications, Inc.
Mineola, New York

For My Students

Bibliographical Note

This Dover edition, first published in 2011, is a completely updated and revised version of the work originally published in 1967 by the McGraw-Hill Book Company, New York.

Library of Congress Cataloging-in-Publication Data

Miller, James E. (James Edward), 1937–
  Elementary theory and application of numerical analysis. —
[Completely updated and rev. version] / James E. Miller, David G. Moursund, Charles S. Duris.
       p. cm.
  Moursund's name appears first on the earlier edition.
  Includes bibliographical references and index.
  ISBN-13: 978-0-486-47906-4 (pbk.)
  ISBN-10: 0-486-47906-4 (pbk.)
  1. Numerical analysis.   I. Moursund, David G. II. Duris, Charles S.
III. Title.
  QA297.M67 2011
  518—dc22

                                                              2010045362

Manufactured in the United States by Courier Corporation
47906401
www.doverpublications.com

TABLE OF CONTENTS

PREFACE TO THE REVISED 2011 DOVER EDITION

This book is designed as a text for an undergraduate (junior-senior) level course in numerical analysis. A student that has an understanding of the calculus sequence has the necessary mathematics courses to succeed in this course. A course in differential equations and some computer programming will be helpful to the student but not essential. The book contains sufficient material for a one semester course. Some theory of the calculus is given, and the pure as well and the applied mathematics student can become further grounded in mathematical maturity as well as be challenged with algorithm development and computer programming. The language C++ is used to code the algorithms; however, the student can use the language of preference. Many of the problems in a course of this type lead to a great deal of computational effort. Because of this it will be to the student's benefit to review his/her skills in computer programming early in the course. To help with this effort a short course in C++ is given in the appendix.

Much of the material in the text is basically algorithmic in nature. In general an algorithm is stated or derived and then illustrated by examples. Then the underlying theory is presented and discussed. Theorems giving conditions under which the algorithm is applicable are stated and in some cases proof of the theorems are given. Frequent numerical examples for applying the just presented algorithm are given. Where appropriate, C++ programs are included to show how the algorithm can be implemented on a computer. In general the treatment of topics from numerical analysis is consistent with that ordinarily used in a modern graduate-level course in numerical analysis. Thus, a firm foundation is laid for future studies in this area.

The foremost goal of this book is to provide an introduction to modern numerical analysis. A relatively small number of basic concepts and techniques are considered. The emphasis is upon how and why each of these methods works. Underlying the entire development is a consideration of the error-analysis aspects of the various problems and algorithms discussed.

A secondary goal of the text is to review and solidify some of the basic concepts from elementary calculus and to help raise the overall level of mathematical maturity of the student. Repeated use is made of the **mean-value theorem, intermediate-value theorem** and **Taylor's** series. Considerable emphasis is placed upon theory and proofs, both in the text proper and in the exercises.

Throughout each chapter are found a large number of exercises. Answers to most of the exercises and short discussions on some of the theoretical questions are given at the end of the book.

A substantial portion of this text is derived from a 1967 book of the same name by David G. Moursund and Charles S. Duris. I used this text for many years in my numerical analysis course for junior and senior level students and found it to be sufficiently rigorous and able to cover the essential topics. When I found it no longer in print, I contacted David Moursund as Duris is now deceased. I was given permission to use the old text, however I thought best, and since this text is no longer in print and does such a good job of presenting the essential topics, I felt compelled to revise it and make it available. The programming in the text was with the language FORTRAN and has been replaced with the modern language C++.

I am grateful to my former colleague Dr. Tylene Garrett, Professor of Computer Science at Transylvania University for the technical review of the book. Finally, I am indebted to my wife Betty, for her understanding as I prepared the materials for this book.

James E. Miller
July, 2011

Solution of Equations by Fixed-Point Iteration

1-1 INTRODUCTION

It would seem appropriate to begin this text with a definition of **numerical analysis**. A useful, short definition of numerical analysis is as difficult to give as a useful, short definition of calculus. To know what calculus is, it is necessary to study the subject in some depth. The same is true of numerical analysis. In proceeding through this text, however, the reader might keep in mind the following definition due to Professor Preston C. Hammer: "Numerical analysis is the effective representation of anything by anything." That is, **the main goals of numerical analysis are to provide effective procedures for solving problems**. Because many numerical procedures can be effectively implemented on a computer, computers play an important role in modern-day numerical analysis.

The main usefulness of a digital computer lies in its ability to follow a repetitive set of directions accurately and rapidly. Included in this capability is the ability to modify the basic set of directions being followed during the execution of the computer program. By an iterative procedure we mean a process whereby a basic set of directions is followed repeatedly; modification of these directions in the course of the execution is allowed. In this chapter, some basic iterative procedures for solving functional equations are developed. The discussion is limited to functions of one variable. The application of iterative procedures to the solution of a system of simultaneous equations is given in Chapter 2. Further application of iterative procedures to the numerical solution of differential equations is given in Chapter 7.

This chapter also seeks to review and solidify some of the basic results from elementary calculus. Continuity, limit, the mean-value theorem, and graphing are among the topics considered. The reader may find it helpful to refer to one of the standard texts in elementary calculus.

1-2   EXAMPLE

We shall use one example to illustrate most of the ideas and results of this chapter. Suppose that one wishes to solve the equation $f(x) = 0$, where

$$f(x) = x - 1/2(x + A/x) \qquad (1-1)$$

Here $A > 0$ is assumed to be a given constant. The peculiar form of $f(x)$ will be explained in Sec. 1-6. Note that $f(x)$ is not defined for $x = 0$. It can be seen that the equation has two solutions, $x = \pm A^{1/2}$. Thus the problem is essentially that of computing $A^{1/2}$. The reader is undoubtedly familiar with one or more methods for accomplishing this. The procedure to be given here is easily programmed for a computer and is relatively fast.

Let $g(x) = 1/2(x + A/x)$ so that the equation $f(x) = 0$ is equivalent to the equation $x = g(x)$. Consider the following procedure. Let $p_0 > 0$ be a first approximation to $A^{1/2}$. Define p_1 by $p_1 = g(p_0)$. Define p_2 by $p_2 = g(p_1)$. In general,

$$p_0 > 0 \qquad \text{initial approximation}$$

$$p_n = \frac{1}{2}\left(p_{n-1} + \frac{A}{p_{n-1}} \right) \qquad n = 1,\ 2,... \qquad (1-2)$$

Because the terms of the sequence $\{p_n\}$ are formed by successively substituting into $g(x)$ the term most recently generated, Eq. (1-2) is known as a **method of successive substitution**. Let us apply this method to a particular example.

EXAMPLE 1-1.   Suppose that $A = 5$ and $p_0 = 2$. One measure of the accuracy of the approximation p_n to $5^{1/2}$ is the quantity $e_n = |p_n^2 - 5|$. A short table of $\{p_n\}$ and $\{e_n\}$ is given below.

$$p_0 = 2 \qquad\qquad e_0 = 1$$

$$p_1 = 9/4 \qquad\qquad e_1 = (1/4)^2$$

$$p_2 = 161/72 \qquad e_2 = (1/72)^2$$

$$p_3 = \frac{51,481}{23,184} \qquad e_3 = \left(\frac{1}{23,184} \right)$$

Notice the apparent rapid convergence of the sequence $\{e_n\}$ to zero. In the course of this chapter we shall develop sufficient theory to prove that the sequence $\{p_n\}$ of Eq. (1-2) converges to $A^{1/2}$.

EXERCISE 1.   Suppose that $A = 5$ and $p_0 = 3$ in Eq. (1-2). Compute p_1, p_2, p_3, and e_0, e_1, e_2, as in Example 1-1.

1-3   PRINCIPAL ALGORITHM

In this chapter we shall investigate one method for solving equations of the form $x = g(x)$. Various modifications of this method, along with methods of writing equations of the form $f(x) = 0$ into the form $x = g(x)$, will be considered.

Much of numerical analysis is devoted to the development and study of algorithms. By the term **algorithm** we shall mean a list of instructions specifying a sequence of operations to be used in solving a certain type of problem. Because the problems to be considered are mathematical, the operations used will be mathematical in nature.

An algorithm for approximating the square root of a positive number is given by Eq. (1-2). Ordinarily an algorithm will be accompanied by one or more theorems stating what problem the algorithm is designed to solve and when it works. The method of solving $x = g(x)$ which we shall investigate is that of **successive substitution**, or **fixed-point iteration**. The terminology fixed point arises from the fact that one seeks a point p which remains fixed under application of the function $g(x)$; that is, $p = g(p)$. The method of fixed-point iteration is defined by Algorithm 1-1.

ALGORITHM 1-1.   Let p_0 be an approximation to a solution to the equation $x = g(x)$. Generate the sequence $\{p_n\}$ recursively by the relation $p_n = g(p_{n-1})$, $n = 1, 2,\ldots$.

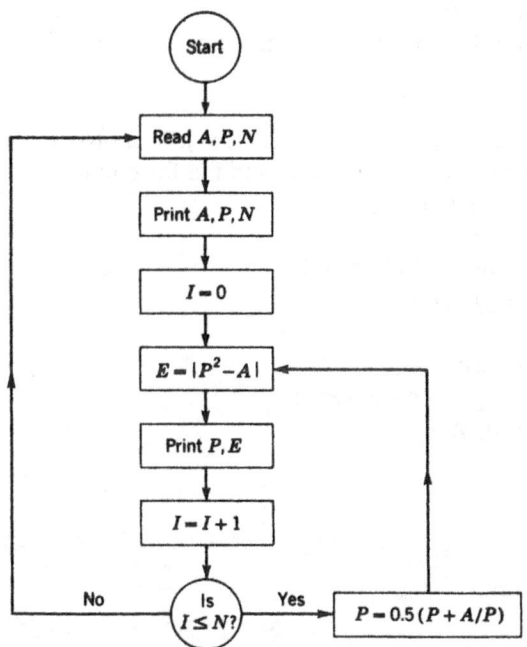

FIG. 1-1   Flow chart for square-root algorithm.

There are a number of ways to state an **algorithm**. One way is in words and mathematical symbols, as above. Another way is by means of a flow chart. Still another way is by the use of a specially designed algorithmic language. The computer language **ALGOL** was designed explicitly for

communication of algorithms. The reader is encouraged to code the algorithm in a language of choice and execute the code.

We shall give flow charts, and programs in some cases, and sample output
for many of the algorithms discussed in this text. To illustrate, suppose that
for arbitrary A, p_0, and n one wishes to compute $p_1, p_2, \ldots, p_n$ by the rule

$$p_i = \frac{1}{2}\left(p_{i-1} + \frac{A}{p_{i-1}} \right)$$

In addition, the quantities $|p_i^2 - A|$ are to be computed. A flow chart is given
in Fig. 1-1.

The flow chart shown in Fig. 1-1 provides for reading in the number A
along with an initial approximation p and the number n of iterations to be
performed. Printout is to include the data A, p, n and the elements of the sequences $\{p_n\}$ and $\{e_n\}$ as in Example 1-1.

Program 1-1. C++ program for square-root algorithm

```
#include <iostream>
#include <cmath>  // provides fabs
using namespace std;
// a  c++ program for square-root algorithm
int main(void)
{
  float a,p,e; // want square root of a where p is approximate value
  int n,i;  // n is number of iterations and i is for count
  cout<<"Input a, p and n:";
  cin>>a>>p>>n;
  cout<<a<<' '<<p<<' '<<n<<endl;
  for(i = 1; i <=n; ++i)
    {
    e = fabs( p*p - a);
    cout<<"    "<<p<<' '<<e<<endl;
    p = 0.5*(p + a/p);
    }
  return 0;
}


/* Sample Output

Input a, p and n:5 2 6
5 2 6
    2 1
    2.25 0.0625
    2.23611 0.000193138
    2.23607 1.46822e-007
    2.23607 1.46822e-007
    2.23607 1.46822e-007
```

Input a, p and n:10 3 5
10 3 5
   3 1
   3.16667 0.0277783
   3.16228 1.98451e-005
   3.16228 2.42537e-007
   3.16228 2.42537e-007

  */

EXAMPLE 1-2. The quadratic formula may be thought of as an algorithm for solving equations of the form $Ax^2 + Bx + C = 0$. Suppose that it is desired to solve a sequence of quadratic equations where it is known that $A \neq 0$ and $B^2 - 4AC \geq 0$. The necessary square root is to be approximated by using 10 iterations of Eq. (1-2) with $p_0 = 1$. (See Fig. 1-2.)

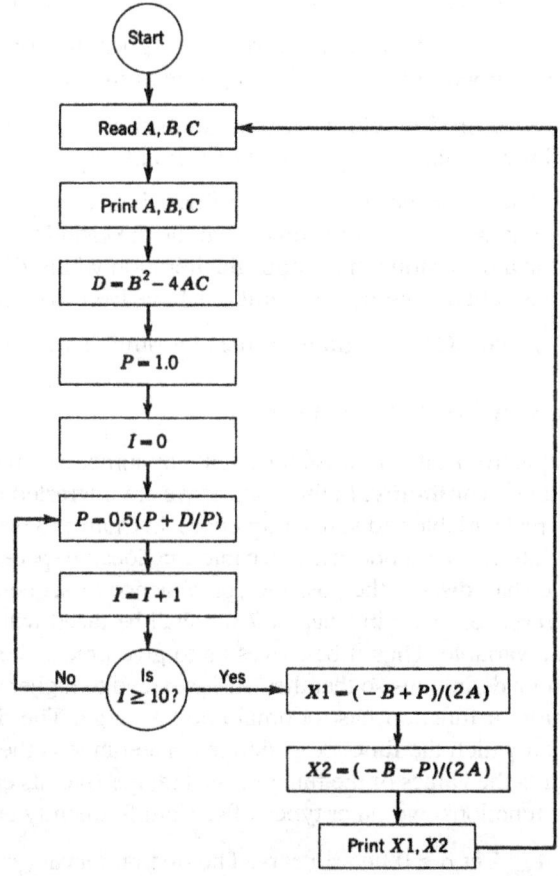

FIG. 1-2 Flow chart for solving a quadratic equation.

EXERCISE 2. Cramer's rule may be thought of as an algorithm for solving simultaneous linear equations. State the algorithm for the case of a system of two simultaneous equations, and state a theorem which tells when the algorithm works.

EXERCISE 3.   Let $\{x_1, x_2, \ldots, x_n\}$ be a given set of real numbers (not necessarily distinct). State in flow-chart form an algorithm to perform each of the following problems.

  (a) Determine the value of the largest element of the set.
  (b) Determine the value of the smallest element of the set.
  (c) Determine the subscript of the element of smallest subscript satisfying (a).
  (d) Determine the subscript of the element of largest subscript satisfying (a).

EXERCISE 4.   Let $\{x_1, x_2, \ldots, x_n\}$ be a given set of real numbers. State in flow-chart form an algorithm which reads the set of numbers, orders them from largest to smallest, and prints out the numbers in that order.

EXERCISE 5.   State in flow-chart form an algorithm for determining whether or not a given integer $n > 1$ is a prime number.

EXERCISE 6.   State in flow-chart form an algorithm for determining both the mean and the median of a given set of N numbers.

EXERCISE 7.   The algorithm given by the flow chart of Fig. 1-2 is not too useful as it stands. Add to it provisions to handle the case $D = B^2 - 4AC < 0$. In addition, add a provision which stops the iteration when $|P^2 - D| < 10^{-7}$ if this condition is satisfied before 10 iterations have been completed.

EXERCISE 8.   Write a C++ program for the algorithm given in Exercise 7.

1-4   REVIEW OF BASIC CONCEPTS

Certain concepts from calculus arise repeatedly in numerical analysis. Two of these are **limit** and **continuity**. In this chapter we are interested in real-valued functions of a real variable and in real sequences. In Chapter 3 we shall be considering functions from Euclidean n space into Euclidean n space, where $n > 1$; in addition we shall discuss the possible convergence of sequences of points in Euclidean n space. Again in Chapter 7 we shall be faced with functions of more than one variable. Thus it behooves us to give precise definitions here and to state these definitions so that they easily extend to higher dimensions.

By definition, a function has a domain and a range. The domain is the set of points on which the function is defined. The range is the set of points which constitute the values of the function as it ranges over its entire domain. In discussing functions two other types of sets are frequently encountered.

DEFINITION 1-1.   Let $\delta > 0$ be arbitrary. The open interval $(p - \delta, p + \delta)$ is called a **neighborhood of the point p**, with radius δ.

DEFINITION 1-2.   Let $\delta > 0$ be arbitrary. The set consisting of the two open intervals $(p - \delta, p)$ and $(p, p + \delta)$ is called a **deleted neighborhood of the point p**, with radius δ.

Algorithm 1-1 generates a sequence of points $p_1, p_2, p_3, \ldots$. In some cases this sequence will converge to a solution to the equation $x = g(x)$, while in other cases it will diverge. Definition 1-3 of the limit of a sequence will be used in developing the theory of fixed-point iterative procedures.

DEFINITION 1-3.   We write

$$\lim_{n\to\infty} p_n = p$$

to mean that, for each $\varepsilon > 0$, there is an integer N (which often depends on ε) such that $|p_n - p| < \varepsilon$ whenever $n > N$.

EXAMPLE 1-3.   Suppose that $p_n = n/(2n + 3)$. Then

$$\lim_{n\to\infty} p_n = 1/2$$

To prove this, suppose that $\varepsilon > 0$ is arbitrary. We consider

$$\left| \frac{n}{2n+3} - \frac{1}{2} \right|$$

for large n. This can be written as

$$\left| \frac{2n}{2(2n+3)} - \frac{2n+3}{2(2n+3)} \right| = \frac{3}{4n+6}$$

Observe that

$$\frac{3}{4n+6} < \frac{3}{4n}$$

for positive n. It is now a simple matter to select N such that $n > N$ implies that

$$\frac{3}{4n} < \varepsilon$$

Specifically, if $N \geq 3/(4\varepsilon)$ is an integer and $n > N$, then

$$\left| \frac{1n}{2n+3} - \frac{1}{2} \right| < \varepsilon$$

EXERCISE 9.   Prove from the definition that

$$\lim_{n\to\infty} \frac{1}{n} = 0 \quad \text{and} \quad \lim_{n\to\infty} 1 + \frac{2}{n} = 1$$

EXERCISE 10.   Prove from the definition that

$$\lim_{n\to\infty} \frac{2n}{4n+5} = \frac{1}{2} \quad \text{and} \quad \lim_{n\to\infty} \frac{3n+1}{n} = 3$$

Next we define the limit of a function at a point.

DEFINITION 1-4.   Suppose that $f(x)$ is defined on a domain D. Let p be a point such that every **deleted neighborhood of p** contains points of D. Then we write

$$\lim_{x \to p} f(x) = v$$

to mean that for each $\varepsilon > 0$ there exists a $\delta > 0$ (which often depends on ε) such that $|f(x) - v| < \varepsilon$ whenever $0 < |x - p| < \delta$ and $x \in D$.

Notice in Definition 1-4 that we are concerned only with points in a deleted neighborhood of the point p. The point p need not be in the domain of definition of the function f. If $p \in D$, then we can inquire into what is called the **continuity of $f(x)$** at p.

DEFINITION 1-5.   Suppose that $f(x)$ is defined on a domain D and $p \in D$. We say that $f(x)$ is continuous at p if

$$\lim_{x \to p} f(x) = f(p)$$

Definition 1-5 actually requires three things. In order to be continuous at a point p, the function $f(x)$ must be defined at p. Next, the function $f(x)$ must be such that

$$\lim_{x \to p} f(x)$$

exists. Finally, the value of the above limit must be $f(p)$.

EXAMPLE 1-4.   Using the appropriate definitions, we shall prove that $f(x) = 2x + 5$ is continuous at $x = 3$.

First we note that $f(3) = 11$. We must prove that

$$\lim_{x \to 3} (2x + 5)$$

exists and has the value 11. Let $\varepsilon > 0$ be arbitrary. We examine the quantity

$$|(2x + 5) - 11|$$

for x near 3. This quantity may be simplified and written in the form

$$|(2x + 5) - 11| = |2x - 6| = 2|x - 3|$$

This is to be made less than ε by making $|x - 3|$ less than some δ. It is now evident that if we set $\delta = \varepsilon/2$ then $|x - 3| < \delta$ implies that $2|x - 3| < \varepsilon$.

EXERCISE 11.

(a) Let $f(x) = 3x + 7$. Prove that $\lim_{x \to -1} f(x) = 4$ [that is, prove that $f(x)$ is continuous at $x = -1$],

(b) Prove that the above function is continuous for all x.

EXERCISE 12.

(a) Prove that the function $f(x) = x^2 + 1$ is continuous at the point $x = 2$.

(b) Prove that the above function is continuous for all $x \geq 1$.

The reader may be familiar with alternative definitions of continuity. Typical is the ε, δ definition.

EXAMPLE 1-5 Suppose that $f(x)$ is defined on D and is continuous at a point $p \in D$. Let $\varepsilon > 0$ be arbitrary. Show that there exists a $\delta > 0$ such that

$$\left| f(x) - f(p) \right| < \varepsilon$$

whenever $|x - p| < \delta$ and $x \in D$.

Solution From Definition 1-5 we know that

$$\lim_{x \to p} f(x) = f(p)$$

Using the definition of limit, for the above $\varepsilon > 0$, we know there exists a $\delta > 0$ such that

$$\left| f(x) - f(p) \right| < \varepsilon$$

whenever $|x - p| < \delta$ and $x \in D$. This completes the proof.

Frequently one uses a hypothesis that a function is **continuous** at each point of a given set. If $f(x)$ is continuous at each point of a set S, we say that $f(x)$ is continuous on S and write

$$f(x) \in C(S)$$

That is, $C(S)$ designates the class of **continuous functions** on S, and $f(x)$ belongs to $C(S)$. If the nth derivative of $f(x)$ is continuous on a set S, for some $n \geq 1$, we write

$$f(x) \in C^n(S)$$

$C^n(S)$ is the class of functions with n continuous derivatives on S.

Next, let us recall an important theorem from elementary calculus. The majority of functions we shall encounter are differentiable.

DEFINITION 1-6.   Suppose that $f(x)$ is defined on a domain D and $p \in D$. Then $f(x)$ is differentiable at p if

$$\lim_{x \to p} \frac{f(x) - f(p)}{x - p}$$

exists. The value of the limit, if it exists, is denoted by $f'(p)$.

THEOREM 1-4-1.   If $f(x)$ is differentiable at a point p, it is continuous at p. The proof of Theorem 1-4-1 is left as an exercise. The reader may find it helpful to refer to an elementary calculus text to refresh his memory on the details of a proof of this important result.

EXERCISE 13.   Prove Theorem 1-4-1.

EXERCISE 14.

(a) Prove that $f(x) = x^{1/2}$ is continuous on $[1,3]$.
(b) Is $f(x) = x^{1/2}$ continuous on $[0,1]$? Justify your assertion. (Hint: Consider the point $x = 0$ separately.)

EXERCISE 15.   Let R denote the real line, and let

$$f(x) = a_0 + a_1 x + a_2 x^2 + \cdots + a_m x^m$$

Prove that $f(x) \in C(R)$ and $f(x) \in C^n(R)$ for $n = 1, 2, \ldots$.

EXERCISE 16.   Suppose that $f(x) \in C^1(S)$ and $g(x) \in C^1(S)$. Prove that $f(x) \cdot g(x) \in C^1(S)$.

We now have enough definitions to prove an important result. **Algorithm 1-1** can be applied to any equation written in the form $x = g(x)$. Under certain conditions the sequence $\{p_i\}$ generated will converge. In such a case Theorem 1-4-2 is applicable.

THEOREM 1-4-2.   In Algorithm 1-1 suppose that $\lim_{n \to \infty} P_n = \rho$ and $g(x)$ is continuous at the point p. Then $p = g(p)$.

Proof.   Theorem 1-4-2 says that, if the sequence generated by the algorithm converges and the function $g(x)$ is continuous at the limit point, then the limit point is a solution to the equation $x = g(x)$.

Let $\varepsilon > 0$ be arbitrary. From the definition of continuity, applied at the point p, there exists a $\delta > 0$ such that if $|x - p| < \delta$ then $|g(x) - g(p)| < \varepsilon$. Next, apply the definition of limit to the sequence $\{p_n\}$. For the δ given above there exists an N such that $n > N$ implies that $|p_n - p| < \delta$. Thus, if $n > N$, it follows that $|g(p_n) - g(p)| < \varepsilon$. We have now proved that

$$\lim_{n \to \infty} g(p_n) = g(p)$$

To complete the proof, note that $g(p_n) = p_{n+1}$. Therefore,

$$\lim_{n \to \infty} p_{n+1} = g(p)$$

The hypotheses stated that $\lim_{n \to \infty} p_n = p$. Because $\lim_{n \to \infty} p_n = \lim_{n \to \infty} p_{n+1}$, the proof is complete.

The intermediate-value theorem states that if a function f is continuous on a closed interval $[a, b]$, then f takes on all values between $f(a)$ and $f(b)$ as x ranges over $[a, b]$. An equivalent form of this theorem is given below. This theorem is often useful in proving that an equation has a solution.

THEOREM 1-4-3.   Let $f(x) \in C([a, b])$, and suppose that $f(a) \cdot f(b) < 0$. That is, $f(a)$ and $f(b)$ are of opposite sign. Then there exists a point $p \in (a, b)$ such that $f(p) = 0$.

EXAMPLE 1-6.   Prove that the equation $x - \cos(x) = 0$ has a solution.

Solution.   Let $f(x) = x - \cos(x)$. Because $f'(x) = 1 + \sin(x)$ exists for all x, $f(x)$ is **continuous** for all x. We shall use Theorem 1-4-3. It is easy to see that $f(0) = -1$ and $f(\pi/2) = \pi/2$. Hence $f(x) = 0$ has a solution in $(0, \pi/2)$.

We could arrive at the same conclusion by considering the graph of $f(x)$. However, it is easier to consider two separate graphs

$$y = x \quad y = \cos x$$

Each point common to these two curves is a solution to the equation $x = \cos(x)$. Conversely, each solution to $x = \cos(x)$ is a point of intersection of the two curves. The graphs are given in Fig. 1-3.

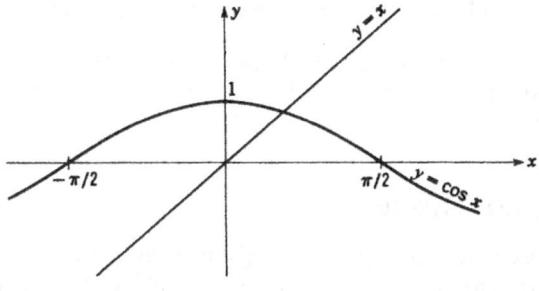

FIG. 1-3

EXERCISE 17.   Determine which of the equations below have at least one solution. For those which do, find an interval containing one of the solutions.

    (*a*) $x - e^{-x} = 0$
    (*b*) $x - e^x = 0$
    (*c*) $\sin x - 2 \cos x = 0$
    (*d*) $x - a^{-x} = 0$, for fixed $a \geq 1$

EXERCISE 18.   Let $P(x) = a_0 x^3 + a_1 x^2 + a_2 x + a_3$, where the a_i are given real numbers and $a_0 \neq 0$. Prove that $P(x) = 0$ has at least one real solution.

The final theorem to be given in this section is the **mean-value theorem** of differential calculus. It will be used repeatedly throughout this text. Geometrically, the theorem gives conditions under which a curve will have at least one tangent line parallel to its secant line for a given interval.

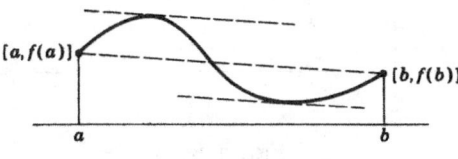

FIG. 1-4

In Fig. 1-4 there are two such points. A precise statement of the theorem is given below.

THEOREM 1-4-4.   Let $f(x) \in C([a, b])$, and suppose that $f(x)$ is differentiable on (a, b). Then there exists at least one point $t \in (a, b)$ such that

$$f(b) - f(a) = (b - a)f'(t)$$

EXERCISE 19.   Suppose that $f(x) \in C([a, b])$, that $f'(x)$ exists on (a, b), and that there exists a constant $k \ge 0$ such that $|f(x)| < k$ on (a, b). Prove that if $a \le x_1 < x_2 \le b$ then

$$|f(x_1) - f(x_2)| \le k(x_2 - x_1) \le k(b - a)$$

EXERCISE 20.   **(Rolle's Theorem).** Suppose that $a < b$, $f(x) \in C([a, b])$, $f'(x)$ exists on (a, b), and $f(a) = f(b)$. Prove that, for some point $t \in (a, b)$, $f'(t) = 0$.

EXERCISE 21.   Suppose that $f(x) \in C([a, b])$, $f'(x)$ exists on (a, b), and $f(x)$ has $n \ge 2$ distinct zeros in $[a, b]$. Prove that $f(x)$ has at least $n - 1$ distinct zeros in (a, b).

1-5   PRINCIPAL THEOREMS

The first theorem we shall prove in this section is an example of a fixed-point theorem. This theorem gives one set of conditions under which a function will have a fixed point. That is, it gives conditions under which $x = g(x)$ will have a solution.

THEOREM 1-5-1.   Suppose that $g(x) \in C([a, b])$ and has its range contained in the same interval $[a, b]$. Then the equation $x = g(x)$ has a solution in $[a, b]$.

Proof.   A picture of the situation under consideration is given in Fig. 1-5. If $g(a) = a$ or $g(b) = b$, then the assertion of Theorem 1-5-1 is true.

Otherwise, because $g(a) \ge a$ and $g(b) \le b$, it follows that $g(a) - a > 0$ and $g(b) - b < 0$. The function $g(x) - x$ is continuous, so by Theorem 1-4-3 there exists a $p \in (a, b)$ such that $g(p) - p = 0$.

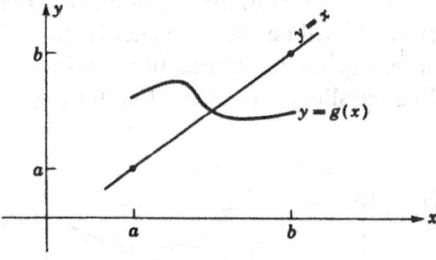

FIG. 1-5

An equation $x = g(x)$ satisfying the hypotheses of Theorem 1-5-1 may have more than one solution on $[a, b]$. A picture of such a situation is given in Fig. 1-6.

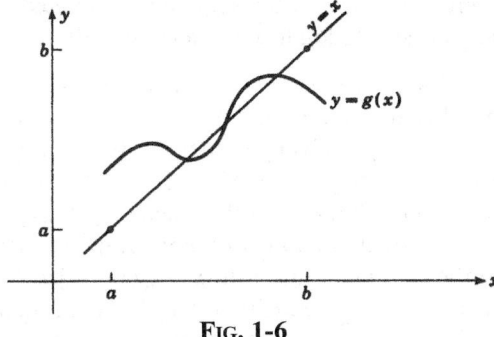

Fig. 1-6

Under certain circumstances one can prove that an equation has exactly one solution in an interval. That is, the solution is unique. A theorem attesting to this fact is called a **uniqueness** theorem. We shall give such a theorem for the problem at hand.

THEOREM 1-5-2.   Suppose that a function $g(x)$ satisfies the hypotheses of Theorem 1-5-1 and in addition is such that, for some constant $0 \le k < 1$, $|g'(x)| \le k$ for all $x \in (a, b)$. Then the equation $x = g(x)$ has exactly one solution in $[a, b]$.

Proof.   The hypothesis $|g'(x)| \le k$ means that the derivative of $g(x)$ exists and is uniformly bounded on (a, b). From Theorem 1-5-1 it follows that there is at least one solution to $x = g(x)$ in $[a, b]$. Suppose that there are two different solutions, p and q, in $[a, b]$.

Thus

$$p = g(p)$$
$$q = g(q)$$
$$p \ne q$$

The function $g(x)$ satisfies the hypotheses of the mean-value theorem on the interval with end points p and q. Hence there exists a point t between p and q such that

$$g(p) - g(q) = (p - q) g'(t)$$

Because $g(p) = p$ and $g(q) = q$, the left side of Eq. (1-3) may be simplified to give

$$p - q = (p - q) g'(t)$$

Division of both sides by $p - q$ gives the result $g'(t) = 1$. This is impossible, because $|g'(x)| \le k < 1$ for all $x \in (a, b)$.

Finally, we are ready to state and prove a theorem giving sufficient conditions for the convergence of the sequence generated by Algorithm 1-1 to a solution to $x = g(x)$. This is the fundamental result of this chapter.

THEOREM 1-5-3.   Let $g(x)$ be continuous on $[a, b]$ with range contained in $[a, b]$. Suppose that there exists a constant $k < 1$ such that $|g'(x)| < k$ for all $x \in (a, b)$. Then, if $p_0 \in [a, b]$, the sequence generated by Algorithm 1-1 converges to the unique solution to $x = g(x)$ which lies in $[a, b]$.

Proof.   We shall first prove that each of the points $p_1, p_2, \ldots$ lies in $[a, b]$. This is important, because the hypotheses give no information about $g(x)$ outside this interval. The proof is by induction. We are given that $p_0 \in [a, b]$. Since $p_1 = g(p_0)$ and the range of $g(x)$ is in $[a, b]$, it follows that $p_1 \in [a, b]$. In general, now, suppose that $p_n \in [a, b]$. Then $p_{n+1} = g(p_n)$, and hence p_{n+1} is also in $[a, b]$. Thus, by induction, all points of the sequence lie in $[a, b]$.

Let p denote the unique solution to $x = g(x)$ in $[a, b]$. It may happen that, for some value of n, $p_n = p$. If this does happen, then

$$p_{n+1} = g(p_n) = g(p) = p$$

and so $p_{n+1} = p$ also. By induction it then follows that

$$p = p_n = p_{n+1} = p_{n+2} = \cdots$$

It is obvious in this case that

$$\lim_{n \to \infty} p_n = p$$

If the above situation does not occur, then $p_n \neq p$ for each n. The hypotheses of the mean-value theorem are satisfied on the interval with end points p and p_n. Thus for each n there exists a point t_n between p and p_n such that

$$g(p_n) - g(p) = (p_n - p)g'(t_n)$$

This can be simplified to

$$p_{n+1} - p = (p_n - p)g'(t_n)$$

Now, by using the given bound k on the derivative, it follows that

$$|p_{n+1} - p| \leq k|p_n - p| \leq k(b-a) \quad \text{for } n = 0, 1, \ldots$$

In particular,

$$|p_1 - p| \leq k(b-a)$$
$$|p_2 - p| \leq k|p_1 - p| \leq k^2(b-a)$$
$$|p_3 - p| \leq k|p_2 - p| \leq k^3(b-a)$$
$$\cdots\cdots\cdots\cdots\cdots\cdots\cdots\cdots\cdots$$
$$|p_n - p| \leq k|p_{n-1} - p| \leq k^n(b-a)$$

Because $|k| < 1$, it follows that $\lim_{n \to \infty} k^n(b-a) = 0$. Thus

$$\lim_{n \to \infty} |p_n - p| = 0$$

That is, $\lim_{n \to \infty} p_n = p$.

In the course of the proof we have discovered another important result. A bound on the error $|p_n - p|$ can be given in terms of quantities given in the hypotheses of the theorem, and n.

COROLLARY 1-5-1. Under the hypotheses of Theorem 1-5-3 an error bound is given by

$$|p_n - p| < k^n(b - a).$$

EXAMPLE 1-7. Let us reconsider Example 1-1. Here $g(x) = .5(x + 5/x)$. We shall verify that the hypotheses of Theorem 1-5-3 are satisfied for the interval [2, 3].

$$g'(x) = .5\left(1 - \frac{5}{x^2}\right)$$

It is necessary to determine the maximum and minimum values of $g(x)$ on [2, 3] in order to verify that the range of the function, for this domain, is contained in the interval [2, 3]. The only zero of $g'(x)$ on this interval is at $x = 5^{1/2}$. Thus the relative extrema of $g(x)$ occur at $x = 2$, $x = 5^{1/2}$, $x = 3$. We have

$$g\left(5^{1/2}\right) = 5^{1/2}$$
$$g(2) = 9/4$$
$$g(3) = 7/3$$

Hence if $x \in [2, 3]$, then $g(x) \in [2, 3]$.

Next it is necessary to find the maximum value of $|g'(x)|$ on the interval

$$g''(x) = \frac{5}{x^3}$$

Because this is not zero on [2, 3], one need consider only the end points

$$g'(2) = -1/8$$
$$g'(3) = 2/9$$

Thus $|g'(x)| \leq k = 2/9$ on [2, 3]. It follows that if $p_0 \in [2, 3]$ the sequence $\{p_n\}$ generated by Algorithm 1-1 converges to $5^{1/2}$. In addition, a bound on the error is given by $|p_n - 5^{1/2}| \leq (2/9)^n$.

EXERCISE 22. Find an interval about the positive solution to the equation $x = .5(x + 12/x)$ which satisfies the hypotheses of Theorem 1-5-3.

EXERCISE 23. Repeat exercise 22 for the equation $x = x - 1/4(x^2 - 5)$.

EXERCISE 24. Repeat exercise 22 for the equation $x = e^{-x}$.

EXERCISE 25. Let $p_0 = 1/2$ in the iteration for the equation

$$x = \frac{1}{2}\left(x + \frac{5}{x}\right)$$

By computing the first few terms of the sequence $\{p_n\}$ one can verify, using the results of Example 1-7, that the sequence converges to $5^{1/2}$. Explain.

EXERCISE 26.   Let $g(x) = x - (x^2 - 5)$. Investigate the possible convergence of the sequence $\{p_n\}$ generated by Algorithm 1-1, for various starting values p_0 near $5^{1/2}$.

In many problems one knows that the equation $x = g(x)$ has a solution, and one possesses some additional information about $g(x)$ near the solution. For example, the equation $x = \frac{1}{2}(x + A/x)$, where $A > 0$, has solutions $\pm A^{1/2}$. The derivative of $1/2(x + A/x)$ at each solution is zero. In such a case, a theorem similar to Theorem 1-5-3, but with different hypotheses, is useful.

THEOREM 1-5-4.   Let $x - g(x)$ be an equation with a solution p. Suppose that there exist a $\delta > 0$ and a constant $k < 1$ such that $|g'(x)| \le k$ whenever $|x - p| \le \delta$. If $p_0 \in I = [p - \delta, p + \delta]$, then the sequence $\{p_n\}$ generated by Algorithm 1-1 converges to p.

Proof.   We shall prove that $g(x)$ satisfies the hypotheses of Theorem 1-5-3 on the interval I. If $x \in I$ and $x \ne p$, then the mean-value theorem is applicable to $g(x)$ on the interval with end points x, p. Thus, there exists a point t between x and p such that

$$g(x) - g(p) = (x - p)g'(t)$$

Because $|g'(t)| \le k < 1$, it follows that

$$\left|g(x) - g(p)\right| \le k\left|x - p\right| < \left|x - p\right| \le \delta$$

Finally because $g(p) = p$ we have proved that $g(x)$ is closer to the point p than is the point x. That is, the range of $g(x)$, for the domain I, is contained in I. The result then follows from Theorem 1-5-3.

The main applications of this theorem will be via the following corollary. This corollary is used in the discussion of the **Newton-Raphson** iteration in Sec. 1-6.

COROLLARY 1-5-2.   Let $x = g(x)$ be an equation with a solution p. If $g'(x)$ is continuous at the point p and $g'(p) = 0$, then the hypotheses, and hence the conclusion, of Theorem 1-5-4 are satisfied for some interval about p.

Proof.   Let ε satisfying $0 < \varepsilon < 1$ be arbitrary. From the continuity of $g'(x)$ at p there exists a $\delta > 0$ such that if $|x - p| < \delta$ then $|g'(x) - g'(p)| < \varepsilon$. Because $g'(p) = 0$, it follows that $|g'(x)| < \varepsilon$. Finally, let b satisfying $0 < b < \delta$ be arbitrary. Then with $k = \varepsilon$ the hypotheses of Theorem 1-5-4 are satisfied on the interval $I = [p - b, p + b]$.

EXAMPLE 1-8.   Use Corollary 1-5-2 to discuss the problem

$$x = \frac{1}{2}\left(x + \frac{5}{x}\right)$$

previously discussed in Example 1-7.

Solution.   The solutions to this equation are $x = \pm 5^{1/2}$. Setting

$$g(x) = \frac{1}{2}\left(x + \frac{5}{x}\right)$$

it is easily verified that $g'(\pm 5^{1/2}) = 0$. Since $g''(x)$ exists at these points, $g'(x)$ is continuous at $\pm 5^{1/2}$. Hence there is some interval about each solution for which the fixed-point iterative procedure will converge.

Results concerning the rate of convergence of a computational procedure are very important. Corollary 1-5-1 is an example of such a result. Under the hypotheses of Corollary 1-5-2 a high local rate of convergence is to be expected. That is, given any $k > 0$ there is an interval

$$I = [p - \delta, p + \delta]$$

about p in which $|g'(x)| \le k$. Thus no matter how small $k > 0$ is, if the iteration converges, then, for sufficiently large n, $|p_{n+1} - p| \le k |p_n - p|$. Roughly this says that, the closer one gets to the solution, the more rapidly one approaches the solution, Such a result is important if a highly accurate solution is desired. This is illustrated by the computations of Program 1-1.

Exercise 27.   Let $g(x) = 1/2(x + 5/x)$. Find the largest interval I with center at $5^{1/2}$ in which $|g'(x)| < 1/.2$.

Exercise 28.   Investigate the possible convergence of the sequence generated by Algorithm 1-1 for the equation $x = x - 1/6(x^2 - 9)$. Consider the solution -3 as well as 3.

1-6   Newton-Raphson Iteration

The **Newton-Raphson** procedure is the most commonly used method for converting an equation of the form $f(x) = 0$ into an equation of the form $x = g(x)$. The procedure is defined by the following algorithm.

ALGORITHM 1-2.   Given the function $f(x)$, set $g(x) = x - f(x)/f'(x)$. Let p_0 be an approximation to a solution of $f(x) = 0$. Generate the sequence $\{p_n\}$ recursively by the relation $p_n = g(p_{n-1})$, $n = 1, 2, \dots$.

Example 1-9.   Let $f(x) = x^2 - A$. From Algorithm 1-2 one can compute

$$g(x) = x - \frac{x^2 - A}{2x}$$

This simplifies to $g(x) = 1/2(x + A/x)$ so that the sequence $\{p_n\}$ generated when $p_0 > 0$ is the sequence considered in previous examples.

One of the derivations of Algorithm 1-2 is based upon Corollary 1-5-2. We shall discuss this derivation later. Given below is a sufficient set of conditions for convergence of $\{p_n\}$.

THEOREM 1-6-1.   Suppose that the equation $f(x) = 0$ has a zero at $x = p$, with $f'(p) \ne 0$ and $f''(x)$ continuous at $x = p$. Then there exists a $\delta > 0$ such that if $|p - p_0| \le \delta$ then the sequence given by Algorithm 1-2 converges to p.

Proof.

$$g(x) = x - \frac{f(x)}{f'(x)}$$

$$g'(x) = 1 - \frac{[f'(x)]^2 - f(x)f''(x)}{[f'(x)]^2}$$

Thus, $g'(p) = 0$ and $g'(x)$ is continuous at $x = p$. That is, all the hypotheses of Corollary 1-5-2 are satisfied.

EXERCISE 29.   Let $f(x) = 1/x - A$. Show that the **Newton-Raphson itera-tion formula** for solving $f(x) = 0$ is

$$p_{i+1} = p_i(2 - Ap_i)$$

This shows that one can find the inverse of a number by an iteration pro-cedure involving only multiplications and subtractions.

EXERCISE 30.   Find an interval I, about the solution to the equation $x = x(2 - Ax)$ obtained in Exercise 29, such that if $p_0 \in I$ the sequence $\{p_n\}$ converges. Assume that $A > 0$.

EXERCISE 31.   Let $f(x) = x^3 - A$. Find the Newton-Raphson form of the equation. For $A = 8$ compute several terms of the sequence $\{p_n\}$ when $p_0 = 1$ and when $p_0 = 3$.

EXERCISE 32.

(a) The equation $f(x) = x^3 - 5x^2 + 10x - 6$ has a real root $x = 1$. Find the Newton-Raphson iteration formula for this equation. Starting at $p_0 = 1.1$ and carrying all calculations to two decimal places, find p_1, p_2.

(b) Write a program which computes $p_1, p_2, \ldots, p_{10}$ for various p_0 in the problem of part (a).

One method of motivating and deriving the Newton-Raphson iteration formula is based upon consideration of equations of the form $x - g(x)$, where $g(x) = x - f(x)h(x)$. Notice that if $f(p) = 0$ then $p = g(p)$ provided that $h(p)$ is defined. That is, each solution to $f(x) = 0$ is a solution to $x = x - f(x)h(x)$. The function $h(x)$ is to be chosen so that Corollary 1-5-2 is applicable. That is, $h(x)$ must be such that if $f(p) = 0$ then $g'(p) = 0$.

$$g(x) = x - f(x)h(x)$$

$$g'(x) = 1 - f'(x)h(x) - f(x)h'(x)$$

Hence, if $f(p) = 0$, then $g'(p) = 1 - f'(p)h(p)$. If this is to be zero, then $h(p) = 1/f'(p)$. One way to achieve this, without knowing the point p, is to set $h(x) = 1/f'(x)$. The resulting equation is

$$x = x - \frac{f(x)}{f'(x)}$$

A geometric derivation of the Newton-Raphson formula is useful, as it suggests other possible iterative formulas.

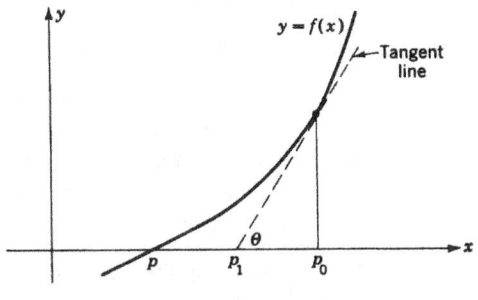

FIG. 1-7

Suppose that the graph of $y = f(x)$ near a zero p is somewhat similar to Fig. 1-7. If p_0 is an initial approximation to the zero p, then p_1 is determined by finding the zero of the tangent line to $y = f(x)$ at $(p_0, f(p_0))$.

This line is shown in dashes on the figure. From the geometry of the situation it follows that

$$\tan \theta = \frac{f(p_0)}{p_0 - p_1} = f'(p_0) \tag{1-4}$$

Equation (1-4) may be solved for p_1, giving

$$p_1 = p_0 - \frac{f(p_0)}{f'(p_0)}$$

Repeated application of this procedure corresponds to Algorithm 1-2. Thus, the algorithm makes use of successive tangents to the curve $y = f(x)$.

EXAMPLE 1-10.   Suppose that two different initial approximations, p_0 and p_1, to a zero of $f(x)$ are known and that the graph is somewhat similar to Fig. 1-8.

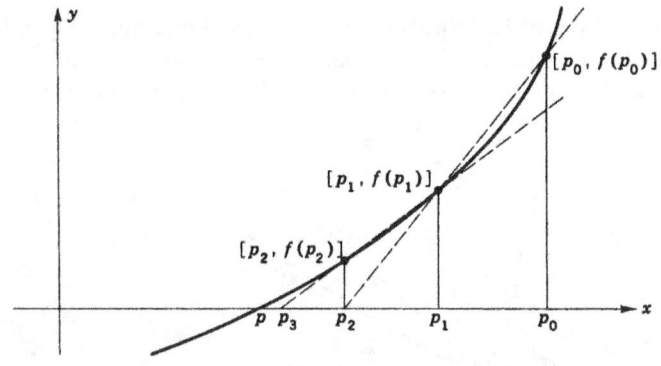

FIG. 1-8

The point p_2 is to be the zero of the line through the points $(p_0, f(p_0))$ and the point $(p_1, f(p_1))$. This line approximates the tangent line to $f(x)$ at

$(p_1, f(p_1))$. After p_2 is determined, the procedure may be repeated on the points p_1, p_2 to find p_3, and so on. This is called the **Secant Method**.

EXERCISE 33.   Derive the algorithm suggested by Example 1-10.

EXERCISE 34.   Write a program to apply the algorithm derived in Exercise 33 to equations of the form $x^2 - A = 0$, where $A > 0$ is given. Make a computational investigation of the convergence of this iteration for $A = 5$ and various starting values. Compare the rate of convergence with the Newton-Raphson iteration for the same problem.

EXERCISE 35.   Let $f(x) \in C[a, b]$, and suppose that $f(a) f(b) < 0$. Let $c = (a + b)/2$. Then at least one of the following holds:

$f(c) = 0$
$f(a) f(c) < 0$
$f(c) f(b) < 0$

(a) Prove the above assertion.
(b) Incorporate the above idea into an algorithm for the numerical calculation of a solution to $f(x) = 0$.

1-7   GRAPHICAL ANALYSIS

By now the reader should be fairly convinced that the **Newton-Raphson** iteration for determination of the square root of a number converges for all positive starting values. Such a result is not, however, immediately evident from any of the theorems proved so far. This particular problem is ideally suited to the method of graphical analysis. In general the method is useful for determining starting values for which convergence is to be expected for the iteration on $x = g(x)$ in cases where $g(x)$ is easily graphed. In addition, it gives considerable insight into the fixed-point iterative procedure.

The method of **graphical analysis** is begun by sketching the graphs of $y = x$ and $y = g(x)$ on one set of coordinate axes. The solutions to $x = g(x)$ correspond to intersections of the two curves. Figure 1-9 gives the graphs

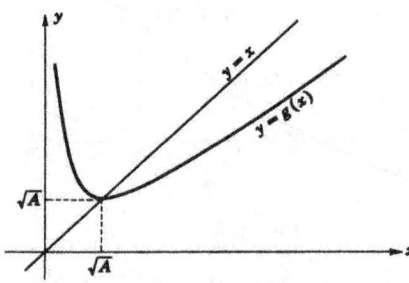

FIG. 1-9

for $x = 1/2(x + A/x)$ for $A > 0$ and $x > 0$. Notice that the scale is indicated by the quantity $A^{1/2}$ so that the graphs apply equally well to all $A > 0$.

The pertinent features of the graph are the vertical asymptote of $y = g(x)$ at $x = 0$, the intersection of $y = x$ and $y = g(x)$ at the horizontal tangent to $y = g(x)$, and the asymptote $y = x/2$ of $g(x)$ as x tends to infinity.

Now, by starting with an initial approximation $p_0 > 0$ to $A^{1/2}$, the value of p_1 can be determined graphically as follows:

1. Draw a vertical line from p_0 to the curve $y = g(x)$.
2. Draw a horizontal line from the intersection point in (1) to the curve $y = x$.
3. Draw a vertical line from the intersection point in (2) to the x axis. The point on the axis thus determined is p_1.

Repeated application of the process gives $p_2, p_3, \ldots$. Figure 1-10 shows the determination of p_1, p_2, where $p_0 > A^{1/2}$.

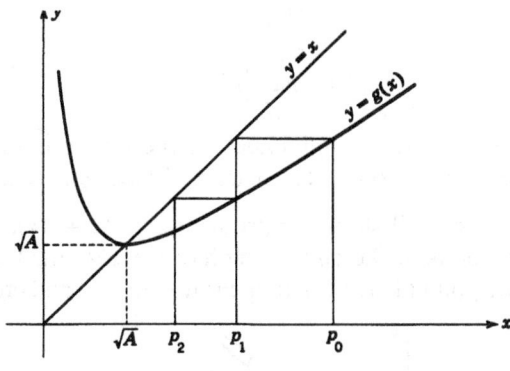

FIG. 1-10

Figure 1-11 shows the determination of p_1, p_2 for the same problem when $0 < p_0 < A^{1/2}$. Notice that it is evident from the graph that $p_1 > \sqrt{A}$.

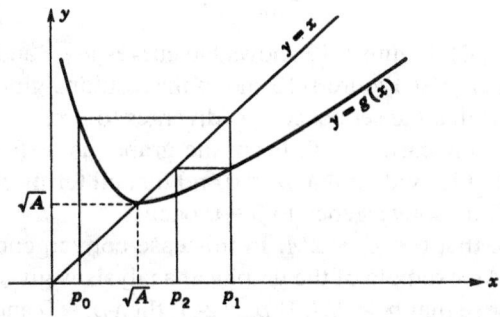

FIG. 1-11

From Figs. 1-10 and 1-11 one can argue that, if $A > 0$, $p_0 > 0$, and

$$p_{n+1} = .5\left(p_n + \frac{A}{p_n}\right)$$

then

$$\lim_{n \to \infty} p_n = A^{\frac{1}{2}}$$

If $p_0 \neq A^{1/2}$, then

$$p_1 > p_2 > p_3 > \cdots > A^{1/2}$$

A formal proof of the convergence can be given using the ideas suggested by the graphical analysis in conjunction with theorems previously proved. First one proves that $p_1 > A^{1/2}$. Next, assuming that we do not have the trivial case of $p_1 = A^{1/2}$, the interval $[A^{1/2}, p_1]$ is considered. On this interval, it can be shown that the hypotheses of Theorem 1-5-3 are satisfied with

$$g(x) = \frac{1}{2}\left(x + \frac{A}{x}\right)$$

$$k = \max_{[A^{1/2}, p_1]} |g'(x)| < 1/2$$

If we consider p_1 to be the initial guess to the solution, it follows that the sequence $\{p_n\}$ converges to $A^{1/2}$. The details of the proof are left as exercises.

EXAMPLE 1-11.   We shall study the problem $x = x(2 - Ax)$, where $A > 0$ is given. The first step is to sketch the graph of $y = x(2 - Ax)$. It is a parabola with vertex at the point $(1/A, 1/A)$. It opens downward and crosses the x axis

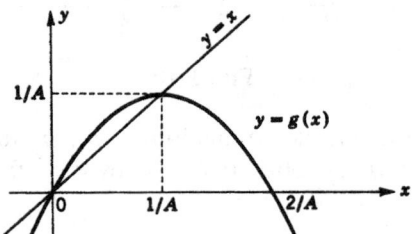

FIG. 1-12

at $(0, 0)$ and $(2/A, 0)$. Figure 1-12 shows the curves $y = x$ and $y = x(2 - Ax)$. First suppose that $p_0 < 0$. Figure 1-13 shows the resulting graphical analysis. The conclusion is that the sequence $\{p_n\}$ diverges to $-\infty$.

Next consider the case $p_0 = 0$. From the graph, as well as the formula $g(x) = x(2 - Ax)$, it is evident that $p_1 = 0$ and that all terms of the sequence $\{p_n\}$ are zero. Thus convergence to $p = 0$ occurs.

Next suppose that $0 < p_0 < 2/A$. In this case convergence to the value $p = 1/A$ occurs. An example of the graphical analysis is given in Fig. 1-14.

Finally, suppose that $p_0 \geq 2/A$. If $p_0 = 2/A$, then $p_1 = 0$ and the sequence converges to $p = 0$. If $p_2 > 2/A$, it is easily seen from the graph that $p_1 < 0$; then the previous discussion is applicable, and divergence to $-\infty$.

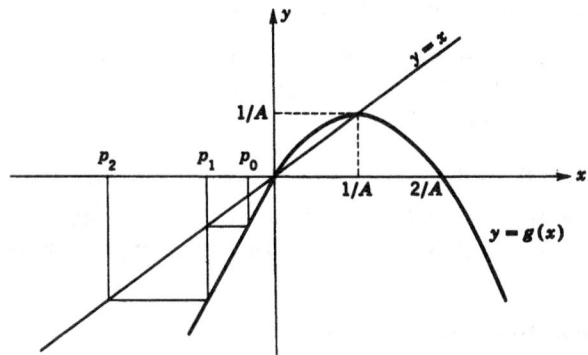

FIG. 1-13

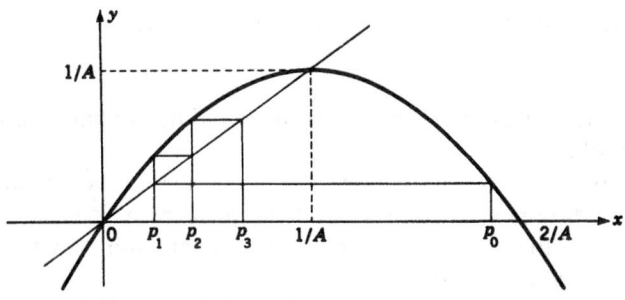

FIG. 1-14

EXERCISE 36. Do a graphical analysis for the iteration on

$$x = .5(x + A/x)$$

where $A > 0$ is given, for starting values $p_0 < 0$.

EXERCISE 37. Do a graphical analysis for the iteration on $x = x(2 - Ax)$, where $A < 0$ is given.

EXERCISE 38. Do a graphical analysis for the iteration on the equation $x = e^{-x}$.

EXERCISE 39. Suppose that $0 < p_0 < A^{1/2}$ in the iteration to solve $x = .5(x + A/x)$. Prove that $p_1 > A^{1/2}$. Find an expression for

$$\max_{[A^{1/2}, p_1]} |g'(x)|$$

and show that it is less than 1. Then complete the proof that, if $p_0 > 0$ and $p_{n+1} = g(p_n)$, then $\lim_{x \to \infty} p_n = A^{1/2}$.

EXERCISE 40. Determine the Newton-Raphson iteration formula for the equation $x^3 - A = 0$, where $A > 0$. Do a **graphical analysis** for this problem to determine the starting values for which the iteration will converge.

EXERCISE 41.   Do Exercise 40 for the case $A < 0$.

EXERCISE 42.   Determine the Newton-Raphson iteration formula for the equation $x^n - A = 0$, where $A > 0$ and $n \geq 2$ is an integer. Do a graphical analysis for the case $p_0 > 0$.

EXERCISE 43.   To find $A^{1/4}$ for $A > 0$, we could apply the procedure developed in Exercise 42 or find $\sqrt{\sqrt{A}}$. Write a program for each of these procedures. Try to determine which is the more efficient procedure.

1-8   ERROR BOUNDS

In solving any problem on a digital computer, it is highly desirable to obtain some measure of the accuracy of the computed results. Suppose, for example, that an exact solution to $x = g(x)$ is the number p, while an approximate solution is the number p_m. The quantity

$$r = |p - p_m|$$

is the absolute value of the error, and one would like to have a good bound on the size of r.

The problem is not one which admits an easy or universal solution. However, certain general ideas and techniques are worth discussing. We shall illustrate these by considering once again the Newton-Raphson iteration for calculating $A^{1/2}$.

Sometimes by a careful consideration of the quantities involved in a numerical calculation one can find an upper and a lower bound for the solution. Each iteration in the calculation of $A^{1/2}$ involves finding the average of the two quantities p_n and A/p_n. If $n > 1$ and $0 < p_0 \neq A^{1/2}$, then $p_n > A^{1/2}$ and $A/p_n < A^{1/2}$. Thus, p_{n+1} is the average of two quantities which bound the exact solution from above and below. That is,

$$p_{n+1} = .5\left(p_n + \frac{A}{p_n}\right)$$

Figure 1-15 indicates this graphically. From this discussion, it follows that for $n > 1$

$$p_{n+1} - A^{1/2} < .5(p_n - A^{1/2}) < .5\left(p_n + \frac{A}{p_n}\right) \tag{1-5}$$

Since p_n and A/p_n are available in the course of the computation of p_{n+1}, the result (1-5) is an effective and useful method for bounding the error.

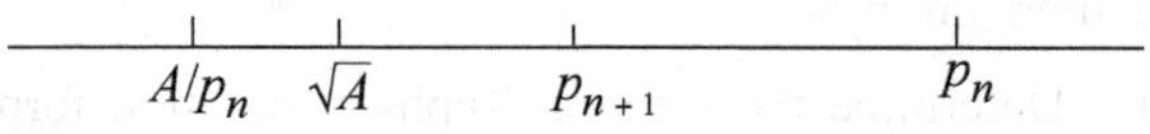

FIG. 1-15

We shall illustrate the various bounds obtained in this section by using the example $A = 5, p_0 = 20$. This very poor initial approximation provides a good picture of the convergence of p_n to $A^{1/2}$ and of the convergence of the error bounds to zero.

Example 1-12.   For $A = 5, p_0 = 20$, p_n and the bound on $|p_{n+1} - A^{1/2}|$ given by Eq. (1-5) are presented in Table 1-1.

Table 1-1

n	p_n	A/p_n	$.5|p_n - A/p_n|$
0	20.00000000	.2500000000	9.875000000
1	10.12500000	.4938271605	4.815586420
2	5.309413580	.9417234867	2.183844997
3	3.125568583	1.599708938	.7629298226
4	2.362638761	2.116277817	.1231804717
5	2.239458289	2.232682798	.003387745324
6	2.236070544	2.236065411	.000002566259
7	2.236067978	2.236067978	.0000000000

In Example 1-1 the quantity $e_n = |p_n^2 - A|$ was calculated and its magnitude used as an indication of the accuracy of the approximation p_n to $A^{1/2}$. Often one starts with an equation of the form $f(x) = 0$. A natural thing to do is to substitute the approximate solution into the original equation and determine the result. The quantity $e_n = |f(p_n)|$ may be a useful measure of the closeness of the point p_n to a point p such that $f(p) = 0$. Often however, the relationship between the magnitudes of $|f(p_n)|$ and $|p_n - p|$ is very difficult if not impossible to determine.

In the particular problem at hand, we have

$$e_n = |p_n^2 - A| = |p_n - A^{1/2}| \cdot |p_n + A^{1/2}|$$

Thus

$$r \equiv |p_n - A^{1/2}| = \frac{e_n}{|p_n + A^{1/2}|}$$

The difficulty with this result is that to find r one still needs to know the value of $A^{1/2}$. However, to find a bound on r when $p_n > 0$, we can use the result

$$r = \frac{e_n}{|p_n + A^{1/2}|} < \frac{e_n}{p_n} \tag{1-6}$$

We have previously seen that $p_n > A^{1/2}$ for $n > 1$. Thus another bound on r is given by

$$r = \frac{e_n}{\left|p_n + A^{1/2}\right|} \le \frac{e_n}{2A^{1/2}} = \frac{e_n A^{1/2}}{2A} \le \frac{e_n p_n}{2A} \tag{1-7}$$

Result (1-7) holds for $n \ge 1$. Notice that if p_n is close to $A^{1/2}$ then the quantity $e_n p_n / 2A$ is approximately $e_n / 2p_n$ so that the error bound given in Eq. (1-7) is about half the size of the bound given in Eq. (1-6).

EXAMPLE 1-13.   The table below gives p_n, e_n/p_n, and $e_n p_n / 2A$ for the iteration to determine $A^{1/2}$ when $A = 5$, $p_0 = 20$.

TABLE 1-2

n	p_n	e_n/p_n	$e_n p_n/2A$
0	20.00000000	19.75000000	790.0000000
1	10.12500000	9.631172840	98.73457031
2	5.309413580	4.367689993	12.31246243
3	3.125568583	1.525859645	1.490639596
4	2.362638761	.2463609435	.1375202040
5	2.239458289	.0067754906	.003398026066
6	2.236070944	.000005132525	.000002566268
7	2.236067978	.0000000000	.0000000000

A third approach to bounding the error is to consider the quantity

$$d_n \equiv \left|p_{n+1} - p_n\right|$$

Because $p_{n+1} = g(p_n)$, we are in actuality considering

$$d_n = \left|g(p_n) - p_n\right|$$

This in turn is merely the value of the quantity $|g(x) - x|$ at the point p_n.

Hence what we are actually doing is repeating the method just discussed, but with (ordinarily) a different form of the equation. In the square-root problem

$$d_n = \left|.5\left(p_n + \frac{A}{p_n}\right) - p_n\right| = \frac{\left|A - p_n^2\right|}{2\left|p_n\right|}$$

Using the same techniques as before, we get

$$r = \frac{2p_n d_n}{\left|p_n + A^{1/2}\right|} < \frac{2p_n d_n}{p_n} = 2d_n \tag{1-8}$$

If $n \ge 1$, then as before we get

$$r = \frac{2p_n d_n}{\left|p_n + A^{1/2}\right|} \le \frac{2p_n d_n}{2A^{1/2}} = \frac{p_n d_n A^{1/2}}{A} \le \frac{p_n^2 d_n}{A} \tag{1-9}$$

Again, if p_n is close enough to $A^{1/2}$, then the error bound provided by Eq. (1-9) is about half the size of the error bound given by Eq. (1-8).

EXAMPLE 1-14.   For the problem of Example 1-13, p_n, $2d_n$, and $p_n^2 d_n/A$ are given below. Notice that to find d_1 we need p_0 and p_1, so d_0 is not defined.

TABLE 1-3

n	p_n	$2d_n$	$p_n^2 d_n/A$
1	10.12500000	19.75000000	202.4683594
2	5.309413580	9.631172840	27.15015350
3	3.125568583	4.367689993	4.266874523
4	2.362638761	1.525859645	.8517443013
5	2.239458289	.2463609434	.1235542857
6	2.236070544	.006775490707	.0003387753129
7	2.236067978	.0000051325187	.0000025662593
8	2.236067978	.0000000000	.0000000000

A word of caution is in order. In addition to fixed-point iterative procedures, many other numerical procedures produce a sequence of approximations $p_1, p_2, \ldots$ to a desired result p. It can easily happen that the quantity $d_n \equiv |p_{n+1} - p_n|$ becomes quite small even though $|p_n - p|$ is not small.

EXAMPLE 1-15.   Consider the sequence defined by $p_1 = 1$,

$$p_{n+1} = p_n + \frac{1}{n+1}$$

$n = 1, 2, \ldots$. This is called the **harmonic sequence**. It is easily seen that $p_n = 1 + \frac{1}{2} + \frac{1}{4} + \ldots + 1/n$. It is known (and, indeed, the proof is not difficult) that $\lim_{n \to \infty} p_n = \infty$. However,

$$d_n \equiv |p_{n+1} - p_n| = \frac{1}{n+1}$$

Thus $\lim_{n \to \infty} d_n = 0$. That is, the magnitude of d_n gives no indication of the nearness of p_n to the limit of the sequence $\{p_n\}$. A more detailed discussion of this and other, related problems will be given in Chapter 5.

As a final example, we shall give another derivation of the **Newton-Raphson** iterative procedure. Our goal here is to find an approximate bound on the error in an approximation p_n to a root p of an equation $f(x) = 0$.

Suppose that $f(x) = 0$ has a root p and that $p_n = p + \varepsilon$, where $|\varepsilon| > 0$. By use of the mean-value theorem, we can write

$$f(p+\varepsilon) - f(p) = [(p+\varepsilon) - p] f'(t)$$

where t is between p and $p + \varepsilon$. Then, because $f(p) = 0$, this becomes

$$f(p+\varepsilon) = \varepsilon f'(t)$$

If $f'(x)$ is continuous in a neighborhood of p and $|\varepsilon|$ is sufficiently small, then $f'(t)$ will be approximately the same as $f'(p + \varepsilon)$. This is denoted by

$$f'(t) \approx f'(p+\varepsilon)$$

The symbol $\approx$ is read "approximately equal to." Thus, under the above conditions

$$f(p+\varepsilon) \approx \varepsilon f''(p+\varepsilon)$$

Finally, if $f'(x) \neq 0$ in the neighborhood of p under consideration, we can solve for ε to get

$$\varepsilon \approx \frac{f(p+\varepsilon)}{f'(p+\varepsilon)} = \frac{f(p_n)}{f'(p_n)}$$

Since $\varepsilon = p_n - p$, we have

$$p \approx p_n - \frac{f(p_n)}{f'(p_n)}$$

Once again the Newton-Raphson procedure is suggested.

The point to be made here is that the quantity

$$\left| \frac{f(p_n)}{f'(p_n)} \right|$$

which in turn is just $|p_{n+1} - p_n|$ in a Newton-Raphson iteration, is a good measure of the size of $|p_n - p|$ provided that $|p_n - p|$ is sufficiently small and certain other conditions are met. This furnishes support for using the magnitude of the quantity

$$d_n \equiv |p_{n+1} - p_n|$$

in a convergence criterion for a Newton-Raphson iterative procedure.

EXAMPLE 1-16.   A numerical procedure was used to find a root of the equation $f(x) = x^3 - 200x^2 - x + 200 = 0$. For a certain n the results $x_n = 200.00053$ and $f(x_n) \approx 21.2$ were obtained. Estimate the error $|x_n - p|$, where it is assumed that p is a root of $f(x) = 0$ near x_n.

Solution.

$$f'(x_n) \approx 40,000$$

$$\frac{f(x_n)}{f'(x_n)} \approx .00053$$

Thus one would expect that the error is approximately .00053 and that a better approximation to the solution is

$$x_{n+1} = x_n - \frac{f(x_n)}{f'(x_n)} \approx 200.00000$$

EXERCISE 44.   Equations of the form $x = a^{-x}$ are to be solved for various $a > 1$ by the iteration

$$p_0 = 1 \qquad p_{n+1} = a^{-p_n} \qquad n = 0, 1, \ldots$$

Derive an error bound for the computation. (Hint: Consider the graphical analysis for this problem.)

EXERCISE 45.   The equation $x^3 - A = 0$, $A > 0$, is to be solved by Newton-Raphson iteration for various A. Derive an error bound for this problem.

EXERCISE 46.

(a) Suppose that $P(x) = x^3 + a_1 x^2 + a_2 x + a_3$ has three real distinct roots. It can be proved (Stiefel, p. 102) that if a Newton-Raphson iteration is applied starting to the right of the largest root, convergence to the largest root will occur. Write a program to read in a_1, a_2, a_3, and find the largest root of such an equation.

(b) Incorporate into the above program a procedure for dividing out the factor containing the largest root and then finding the remaining two roots by use of the quadratic formula.

EXERCISE 47.   The equations

$$x^2 - A = 0 \qquad x - \frac{A}{x} = 0 \qquad 1 - \frac{A}{x^2} = 0$$

all have $x = \pm A^{1/2}$ for their solution. We have investigated the application of the Newton-Raphson algorithm to the first of these equations. Make a similar investigation for the other two equations.

Solution of Linear Equations and Matrix Computations

2-1 INTRODUCTION

A very large class of problems in science and engineering have as their final computational form a **system of equations** of the form

$$
\begin{aligned}
a_{11}x_1 + a_{12}x_2 + \cdots + a_{1n}x_n &= b_1 \\
a_{21}x_1 + a_{22}x_2 + \cdots + a_{2n}x_n &= b_2 \\
&\cdots\cdots\cdots\cdots\cdots\cdots\cdots \\
a_{n1}x_1 + a_{n2}x_2 + \cdots + a_{nn}x_n &= b_n
\end{aligned}
\tag{2-1}
$$

The a_{ij}'s and b_i's are known **real numbers,** and the x_i's are **unknown** real numbers to be determined so that each of the n equations in Eq. (2-1) is satisfied by these x_i's. For example,

$$
\begin{aligned}
2x_1 + 3x_2 &= 1 \\
x_1 - x_2 &= -2
\end{aligned}
$$

has the form (2-1) for $n = 2$.

The equations in Eq (2-1) form what is called a **system of linear equations**. In particular Eq. (2-1) is a system of n linear equations in n unknowns.

This chapter is devoted to the study of algorithms for determining the unknown numbers $x_1, x_2, \ldots, x_n$ in Eq (2-1) and to related computational problems. Most algorithms developed in this chapter are easily implemented on a computer.

EXAMPLE 2-1. Let $f(x)$ be a given function defined on the distinct points $x_0, x_1, \ldots, x_n$. It is desired to find a polynomial

$$
P(x) = b_0 + b_1 x + b_2 x^2 + \cdots + b_n x^n
\tag{2-2}
$$

such that

$$P(x_i) = f(x_i) \qquad i = 0, 1, \ldots, n \tag{2-3}$$

Set up the problem as a system of simultaneous linear equations.

Solution. Here the unknown quantities are $b_0, b_1, \ldots, b_n$. By using Eq. (2-2), Eqs. (2-3) become

$$b_0 + b_1 x_1 + b_2 x_1^2 + \cdots + b_n x_1^n = f(x_1) \quad i = 0, \ldots, n$$

This system can be rewritten in the form (2-1).

$$1b_0 + x_0 b_1 + x_0^2 b_2 + \cdots + x_0^n b_n = f(x_0)$$
$$1b_0 + x_1 b_1 + x_1^2 b_2 + \cdots + x_1^n b_n = f(x_1)$$
$$\cdots\cdots\cdots\cdots\cdots\cdots\cdots\cdots\cdots\cdots\cdots\cdots$$
$$1b_0 + x_n b_1 + x_n^2 b_2 + \cdots + x_n^n b_n = f(x_n)$$

This is a system of $n + 1$ simultaneous linear equations in $n + 1$ unknowns.

2-2   MATRICES

To begin with, it is necessary to define what is meant by the word matrix and to define and discuss certain basic operations on **matrices**.

DEFINITION 2-1. An $n \times m$ **matrix** is a collection of elements arranged to form a rectangular table having n rows and m columns. An $n \times m$ matrix is said to have dimension $n \times m$.

The elements of a matrix may be real or complex numbers, functions, other **matrices**, etc. We shall be interested only in matrices of real numbers. An example of a 4×3 matrix is given below.

$$\begin{bmatrix} 1 & -1 & 2 \\ 5 & 3 & 1 \\ 4 & 2 & 1 \\ 7 & -5 & 4 \end{bmatrix}$$

Notations used for an $n \times m$ matrix are

$$A = \begin{bmatrix} a_{11} & a_{12} & \cdots & a_{1m} \\ a_{21} & a_{22} & \cdots & a_{2m} \\ \cdots & \cdots & \cdots & \cdots \\ a_{n1} & a_{n2} & \cdots & a_{nm} \end{bmatrix}$$

or $A = (a_{ij})n \times m$. Whenever no confusion can occur, we shall drop the $n \times m$ and use the notation $A = (a_{ij})$. The number a_{ij} is called the (i, j)th element of the matrix A. It appears in the ith row, jth column. As would be expected, we shall say that two matrices are equal if and only if they have the same number of rows and columns and the corresponding elements are equal.

Matrices are closely related to vectors. It is sometimes convenient, for example, to consider the rows or columns of a matrix as vectors.

Definition 2-2. A matrix consisting of just one row is called a **row matrix**; a matrix consisting of just one column is called a **column matrix**.

There are several basic operations which can be performed on or between matrices. We shall study scalar multiplication, addition and subtraction of matrices, and multiplication of matrices.

Definition 2-3. The product of a matrix $A = (a_{ij})$ by a number t is the matrix $tA = (ta_{ij})$. That is, each element of A is multiplied by t. The quantity t is often called a **scalar**, and the process of multiplying A by t is called **scalar multiplication**.

Definition 2-4. The **sum** of two matrices $A = (a_{ij})n \times m$ and $B = (b_{ij})n \times m$ is defined only when the matrices are of the same dimension and is given by $A + B = (a_{ij} + b_{ij})n \times m$. That is, the corresponding elements are added. The **difference** of two matrices of the same dimension is the matrix of differences of the corresponding elements.

Example 2-2. Let $t = -3$,

$$A = \begin{bmatrix} 2 & -1.5 \\ 3 & 7 \end{bmatrix} \quad B = \begin{bmatrix} 1 & -1 \\ 2 & -1 \end{bmatrix}$$

Then

$$A + B = \begin{bmatrix} 2+1 & -1.5-1 \\ 3+2 & 7-1 \end{bmatrix} = \begin{bmatrix} 3 & -2.5 \\ 5 & 6 \end{bmatrix}$$

$$A - B = \begin{bmatrix} 1 & -.5 \\ 1 & 8 \end{bmatrix}$$

and

$$tA = \begin{bmatrix} -6 & 4.5 \\ -9 & -21 \end{bmatrix}$$

The reader will want to implement the above flow chart in the language of his/her choice. Most languages **BASIC, FORTRAN, C, C++, Pascal,** etc allow for arrays of multiple dimensions and makes implementation of the above relatively easy.

Example 2-3. Draw a **flow chart** (see Fig. 2-1), and write a program to add $A = (a_{ij})_{n \times m}$ and $B = (b_{ij})_{n \times m}$ and store the results in $C = (c_{ij})_{n \times m}$. Assume that n, m do not exceed 10.

Solution. Let A, B, and C be the variable names of three two-dimensional arrays each capable of holding a 10×10 matrix. Let n and m be variable names for the storage of the number of rows and columns, respectively, to be used. The program is illustrated below using the language C++. Implement in the language of your choice and execute with examples.

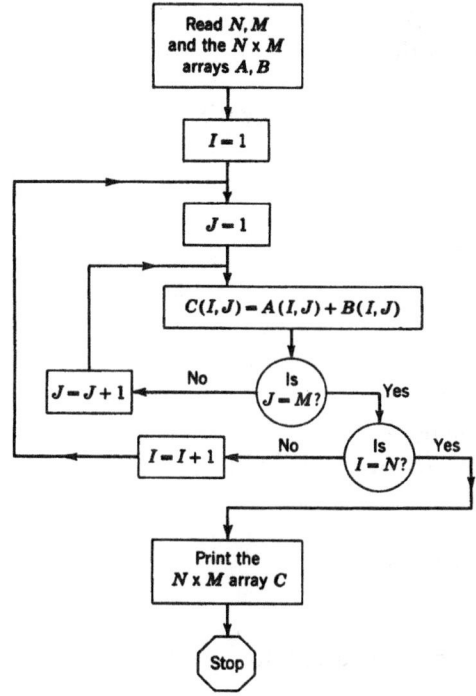

FIG 2-1   Flow chart for matrix addition.

Program 2-1. C++ program for matrix addition

```
#include <iostream>
using namespace std;
// a program for matrix addition
int main(void)

{
float a[11][11],b[11][11],c[11][11]; // add a and b and store in c
int n,m;  // size of matrices to add
int i,j;
cout<<"Input the size of matrices:";
cin>>n>>m;
cout<<"Input matrix a row-by-row"<<endl;
for(i = 1; i<=n; ++i)
 for(j = 1; j<=m; ++j)
  cin>>a[i][j];
cout<<"Input matrix b row-by-row"<<endl;
for(i = 1; i<=n; ++i)
 for(j = 1; j<=m; ++j)
  cin>>b[i][j];
// now to add the two matrices
for(i = 1; i<=n; ++i)
```

```
for(j = 1; j<=m; ++j)
  c[i][j] = a[i][j] + b[i][j];
// output c the sum of a and b
cout<<"The sum of matrix a and matrix b follows"<<endl;
for(i = 1; i<=n; ++i)
{
for(j = 1; j<=m; ++j)
cout<<c[i][j]<<' ';
cout<<endl;
}
return 0;
}
```

Input the size of matrices:2 2
Input matrix a row-by-row
2 3
4 5
Input matrix b row-by-row
1 2
3 4
The sum of matrix a and matrix b follows
3 5
7 9

EXERCISE 1. Let

$$A = \begin{bmatrix} 2 & 4 \\ -1 & 3 \\ 0 & -2 \end{bmatrix} \quad B = \begin{bmatrix} 3 & 6 \\ 0 & -2 \\ 1 & 6 \end{bmatrix} \quad C = \begin{bmatrix} 1 & 2 & 0 \\ 1 & 0 & 3 \end{bmatrix}$$

Determine which of the quantities below are defined. Calculate those which are defined.

$$A+B \quad A+C \quad B-A \quad 3A+2B$$
$$B+A \quad C-B \quad C-C \quad 3A-2B$$

EXERCISE 2. Find two matrices A, B each 2×2, such that

$$3A+2B = \begin{bmatrix} 0 & 0 \\ 0 & 0 \end{bmatrix} \quad \text{and} \quad A-B = \begin{bmatrix} 1 & 2 \\ 1 & 3 \end{bmatrix}$$

The operation of multiplying two **matrices** is more involved than the previous operations discussed. It is an extension of the idea of the inner product of vectors. The reader may recall that the inner product of two vectors $X = (x_1, x_2, \ldots, x_n)$, $Y = (y_1, y_2, \ldots, y_n)$ is a scalar given by

$$(X, Y) = x_1 y_1 + \cdots + x_n y_n$$

For example,

if $X = (2, 3, -1)$ and $Y = (0, 3, 5)$, then
$$(X, Y) = (2)(0) + (3)(3) + (-1)(5) = 4$$

The product AB of two matrices A and B is defined only when the rows of A are the same length (contain the same number of elements) as the columns of B. The product is then a matrix C whose (i, j)th element is the inner product of the ith row of A with the jth column of B. A more formal definition is given below.

DEFINITION 2-5. The product AB of the two matrices $A = (a_{ij})_{n \times m}$ and $B = (b_{ij})_{n \times k}$ is defined to be the matrix $C = (c_{ij})_{n \times k}$, where

$$c_{ij} = \sum_{s=1}^{m} a_{is} b_{sj}$$

EXAMPLE 2-4. Let

$$A = \begin{bmatrix} 1 & 0 & -2 \\ 0 & 3 & .5 \\ 2 & -.5 & 1 \end{bmatrix} \qquad B = \begin{bmatrix} 2 \\ -1 \\ 0 \end{bmatrix}$$

Find AB and AA.

Solution. Because A is 3×3 and B is 3×1, the product AB is defined and will be a 3×1 matrix.

$$AB = \begin{bmatrix} (1)(2) & + & (0)(-1) & + & (-2)(0) \\ (0)(2) & + & (3)(-1) & + & (.5)(0) \\ (2)(2) & + & (-.5)(-1) & + & (1)(0) \end{bmatrix} = \begin{bmatrix} 2 \\ -3 \\ 4.5 \end{bmatrix}$$

$$AA = \begin{bmatrix} -3 & 1 & -4 \\ 1 & 8.75 & 2 \\ 4 & -2 & -3.25 \end{bmatrix}$$

It should be noted that it would not make sense to try to compute BA in the example given above. For the product to be defined, the number of columns in the first matrix must be equal to the number of rows in the second matrix.

The definition of matrix multiplication allows a simple representation for a system of simultaneous linear equations. Considering the system of equations in Eq. (2-1), we note that if we put

$$A_{n \times n} = \begin{bmatrix} a_{11} & a_{12} & \cdots & a_{1n} \\ \cdots & \cdots & \cdots & \cdots \\ a_{n1} & a_{n2} & \cdots & a_{nn} \end{bmatrix} \qquad X = \begin{bmatrix} x_1 \\ x_2 \\ \cdots \\ x_n \end{bmatrix} \qquad B = \begin{bmatrix} b_1 \\ b_2 \\ \cdots \\ b_n \end{bmatrix}$$

then we may write Eq. (2-1) in the form

$$\begin{bmatrix} a_{11} & a_{12} & \cdots & a_{1n} \\ \cdots & \cdots & \cdots & \cdots \\ a_{n1} & a_{n2} & \cdots & a_{nn} \end{bmatrix} \begin{bmatrix} x_1 \\ x_2 \\ \cdots \\ x_n \end{bmatrix} = \begin{bmatrix} b_1 \\ b_2 \\ \cdots \\ b_n \end{bmatrix}$$

To see this, notice that the product on the left is an $n \times 1$ matrix. For this product to be the same as the matrix B on the right, the corresponding elements must be equal. Each equation of Eq. (2-1) corresponds to a statement of equality of two elements in the column matrices.

In shorthand notation the above is just $AX = B$. The matrix notation for a system of linear equations is quite useful in the study of simultaneous linear equations.

EXERCISE 3. Let

$$A = \begin{bmatrix} 1 & 2 & 3 \\ 0 & 3 & -1 \end{bmatrix} \qquad B = \begin{bmatrix} 1 & 2 \\ 0 & -2 \\ 1 & 0 \end{bmatrix}$$

Find AB, BA.

EXERCISE 4. Write out in the form (2-1) the linear system $AX = B$, where

$$A = \begin{bmatrix} -1 & 2 & 1.5 & 3 \\ 2 & 4 & 1.5 & 2 \\ 8 & 2.5 & 0 & 1 \\ -3 & -3.5 & 0 & 3 \end{bmatrix} \qquad B = \begin{bmatrix} 0 \\ 2 \\ 1.5 \\ -2.5 \end{bmatrix}$$

EXERCISE 5. Let

$$A = \begin{bmatrix} 1 & 0 & 0 \\ 0 & 1 & 0 \\ 0 & 0 & 1 \end{bmatrix} \qquad B = \begin{bmatrix} 1 & -2 & 13 \\ 5 & -5 & 6 \\ 7 & 18 & -9 \end{bmatrix}$$

Find AB and BA.

If a and b are two real numbers, we know that $ab = ba$. This need not be the case for matrices. Even if A and B are matrices such that AB and BA are both defined, it still need not be the case that $AB = BA$.

EXAMPLE 2-5. Give an example of two 2×2 matrices A and B such that $AB \neq BA$.

Solution. Let

$$A = \begin{bmatrix} 1 & -1 \\ 1 & 1 \end{bmatrix} \quad \text{and} \quad B = \begin{bmatrix} 2 & 1 \\ 1 & 0 \end{bmatrix}$$

Then

$$AB = \begin{bmatrix} 1 & -1 \\ 1 & 1 \end{bmatrix}\begin{bmatrix} 2 & 1 \\ 1 & 0 \end{bmatrix} = \begin{bmatrix} 1 & 1 \\ 3 & 1 \end{bmatrix}$$

and

$$BA = \begin{bmatrix} 2 & 1 \\ 1 & 0 \end{bmatrix}\begin{bmatrix} 1 & -1 \\ 1 & 1 \end{bmatrix} = \begin{bmatrix} 3 & -1 \\ 1 & -1 \end{bmatrix}$$

Thus $AB \neq BA$.

It sometimes happens, however, for two particular matrices A and B, that $AB = BA$. If this is so, then A and B are said to **commute**, or to satisfy the **commutative law** for multiplication. One very simple non-trivial $n \times n$ matrix commutes with all $n \times n$ matrices. It is the identity matrix.

DEFINITION 2-6. The identity matrix $I_n = (a_{ij})$ $n \times n$ is defined by

$$a_{ij} = \begin{cases} 1 & \text{if } i = j \\ 0 & \text{if } i \neq j \end{cases}$$

EXAMPLE 2-6.

$$I_2 = \begin{bmatrix} 1 & 0 \\ 0 & 1 \end{bmatrix}$$

$$I_4 = \begin{bmatrix} 1 & 0 & 0 & 0 \\ 0 & 1 & 0 & 0 \\ 0 & 0 & 1 & 0 \\ 0 & 0 & 0 & 1 \end{bmatrix}$$

It is convenient to have a shorthand notation for the elements of the identity matrix. The Kronecker δ serves this purpose.

DEFINITION 2-7.

$$\delta_{ij} = \begin{cases} 1 & i = j \\ 0 & i \neq j \end{cases}$$

Thus one often uses the notation

$$I_n = \left(\delta_{ij} \right)_{n \times n}$$

If the value of n is clear then one writes $I = (\delta_{ij})$.

THEOREM 2-2-1. If B is an $n \times m$ matrix, then $I_n B = B$. If C is an $m \times n$ matrix, then $CI_n = C$.

Proof.

$$I_n B = \left(\sum_{s=1}^{n} \delta_{is} b_{sj} \right)_{n \times m} = \left(b_{ij} \right)_{n \times m} = B$$

This is because $\delta_{is} b_{sj} = 0$ when $s \neq i$. Similarly

$$CI_n = \left(\sum_{S=1}^{n} c_{is} \delta_{sj} \right)_{m \times n} = \left(c_{ij} \right)_{m \times n} = C$$

An immediate consequence of this theorem is that I_n commutes with any $n \times n$ matrix A. That is, $AI_n = I_n A = A$.

The operation of matrix multiplication has several properties which we shall use in later proofs. These are:

1. If $A = \left(a_{ij}\right)_{n \times m}$, $B = \left(b_{ij}\right)_{m \times k}$, and $C = \left(c_{ij}\right)_{k \times r}$, then

$$(AB)C = A(BC)$$

This is an associative law for matrix multiplication.

2. $a(AB) = (aA)B = A(aB)$ where a is a scalar.
3. If A and B are $m \times k$ and C is $k \times r$, then $(A + B)C = AC + BC$. This is a distributive law for matrix multiplication.
4. If A and B are $m \times k$ and C is $n \times m$, then $C(A + B) = CA + CB$. This is also a distributive law for matrix multiplication.

We shall prove property 1. The proofs of (2), (3), and (4) are left as exercises.

Proof of (1).   The main idea needed for the proof is that of interchanging the order of summation in a double summation. Suppose that one is given numbers g_{ij} for $1 < i < n$, $1 < j < m$. Then the expressions

$$\sum_{i=1}^{n} \left(\sum_{j=1}^{n} q_{ij} \right) \equiv \sum_{i=1}^{n} \sum_{j=1}^{m} q_{ij}$$

$$\sum_{j=1}^{m} \left(\sum_{i=1}^{n} q_{ij} \right) \equiv \sum_{j=1}^{m} \sum_{i=1}^{n} q_{ij}$$

each denotes the sum of all elements of the matrix $(q_{ij})_{n \times m}$; hence

$$\sum_{i=1}^{n} \sum_{j=1}^{m} q_{ij} \equiv \sum_{j=1}^{m} \sum_{i=1}^{n} q_{ij}$$

To begin the proof, note that

$$BC = \left(\sum_{s=1}^{k} b_{is} c_{sj} \right)_{m \times r}$$

Hence

$$A(BC) = \left[\sum_{t=1}^{m} a_{it} \left(\sum_{s=1}^{k} b_{ts} c_{sj} \right) \right]_{n \times r}$$

$$= \left(\sum_{t=1}^{m} \sum_{s=1}^{k} a_{it} b_{ts} c_{sj} \right)_{n \times r}$$

By interchanging the order summation this can be written as

$$A(BC) = \left(\sum_{s=1}^{k} \sum_{t=1}^{m} a_{it} b_{ts} c_{sj} \right)_{n \times r} \tag{2-4}$$

Using the same techniques, we have

$$AB = \left(\sum_{t=1}^{m} a_{it} b_{tj} \right)_{n \times k}$$

$$(AB)C = \left[\sum_{s=1}^{k} \left(\sum_{t=1}^{k} a_{it} b_{ts} \right) c_{sj} \right]_{n \times r} \qquad (2\text{-}5)$$

$$= \left(\sum_{s=1}^{k} \sum_{t=1}^{m} a_{it} b_{ts} c_{sj} \right)_{n \times r}$$

Because Eqs. (2-4) and (2-5) are identical, the proof is complete.

EXERCISE 6. Write out a careful proof for property 2.

EXERCISE 7. Write out a careful proof for property 3.

EXERCISE 8. Write out a careful proof for property 4.

EXERCISE 9. Prove, using properties 1 to 4, that

(a) $D(A+B+C) = DA + DB + DC$

(b) $C(dA+eB) = dCA + eCB$

(c) $(A+B)(C+D) = AC + AD + BC + BD$

Here A, B, C, D are all $n \times n$ matrices.

The computer program implementation of **matrix multiplication** is rather simple. The example below illustrates the point.

EXAMPLE 2-7. Suppose that $A = (a_{ij})_{n \times m}$, $B = (b_{ij})_{m \times k}$ are stored in a computer and the product AB is to be computed and stored in C. A program fragment for this purpose is given below:

Solution.

```
// program matrix multiply
    float a[10][10], b[10][10], c[10][10], t
    int i,j,n,k,m
    .
    .
    .
    for(i = 1; i<=n; ++i)
    {
    for(j = 1; j<=k; ++j)
    {
    t = 0.;
    for (l=1; l<=m; ++l)
     t = t + a[i][l] * b[l][j];
    c[i][j] = t;
    }
    .
    .
    }
```

EXAMPLE 10. Let

$$A = \begin{bmatrix} .1 & .2 \\ -3 & 0 \end{bmatrix} \qquad B = \begin{bmatrix} 1.1 \\ -.7 \end{bmatrix} \qquad X_0 = \begin{bmatrix} .5 \\ -.5 \end{bmatrix}$$

Compute
$$X_1 = AX_0 + B$$
$$X_2 = AX_1 + B$$
$$X_3 = AX_2 + B$$

[Compute X_3 with the vector $\begin{bmatrix} 1 \\ -1 \end{bmatrix}$, the solution to $(I-A)X = B$. In Chapter 3 an algorithm for solving a linear system of equations will the based on this procedure.]

EXERCISE 11. Write a program which reads in an $N \times N$ matrix A and an integer $M \geq 1$ and computes

$$C = I + A + A^2 + A^3 + \cdots + A^M$$

2-3   UPPER-TRIANGULAR SYSTEM

We now return to the main goal of this chapter, that of giving an algorithm for solving a linear system (2-1). Certain systems of linear equations are easily solved. For example, if the matrix of coefficients for a system of n equations is I_n, then the task is trivial. In this section we shall show how to solve a system of linear equations which has a triangular matrix of coefficients. In the next section we shall give an algorithm for reducing an arbitrary system (2-1) into an equivalent system with a triangular matrix of coefficients.

DEFINITION 2-8. An $n \times n$ system of linear equations is called **upper-triangular** if it has the form

$$
\begin{aligned}
a_{11}x_1 + a_{12}x_2 + a_{12}x_3 + &\cdots + a_{1n}x_n = b_1 \\
a_{22}x_2 + a_{23}x_3 + &\cdots + a_{2n}x_n = b_2 \\
a_{33}x_3 + &\cdots + a_{3n}x_n = b_3 \\
&\cdots\cdots\cdots\cdots\cdots\cdots \\
&\quad a_{nn}x_n = b_n
\end{aligned}
\tag{2-6}
$$

In the matrix form this is $AX = B$, where

$$A = \begin{bmatrix} a_{11} & a_{12} & \cdots & a_{1n} \\ 0 & a_{22} & \cdots & a_{2n} \\ 0 & 0 & \cdots & a_{3n} \\ \cdots & \cdots & \cdots & \cdots \\ 0 & 0 & \cdots & a_{nn} \end{bmatrix} \qquad X = \begin{bmatrix} x_1 \\ x_2 \\ \cdots \\ x_n \end{bmatrix} \qquad B = \begin{bmatrix} b_1 \\ b_2 \\ \cdots \\ b_n \end{bmatrix}$$

This matrix A is called upper-triangular. In any $n \times n$ matrix A the elements $a_{11}, a_{22}, a_{33}, \ldots, a_{nn}$ are called the **diagonal** elements of A.

In general, if all the elements below the diagonal in an $n \times n$ matrix are zero, the matrix is called upper-triangular.

An upper-triangular system of n simultaneous linear equations in n unknowns is easily solved if the diagonal elements are all nonzero. Before giving an algorithm for the process, we illustrate by a simple example.

EXAMPLE 2-8. Solve the system:

$$
\begin{aligned}
x_1 + 2x_2 - 4x_3 - x_4 &= 2 \\
3x_2 + 7x_3 - 3x_4 &= 3 \\
-2x_3 + 5x_4 &= -1 \\
3x_4 &= 5
\end{aligned}
$$

Solution. The fourth equation is easily solved for x_4. Upon substitution of the result in the third equation, the third equation in turn is easily solved to give x_3. Continuation of the process gives x_2 and x_1.

$$
x_4 = 5/3
$$
$$
x_3 = \frac{-1 - 5x_4}{-2} = \frac{-1 - (5)(5/3)}{-2} = \frac{14}{3}
$$
$$
x_2 = \frac{3 + 3x_4 - 7x_3}{3} = \frac{-74}{9}
$$
$$
x_1 = 2 + x_4 + 4x_3 - 2x_2 = 349/9
$$

This example leads us to the following algorithm for solving upper-triangular systems of linear equations. This algorithm is commonly called **solution by back substitution**.

ALGORITHM 2-1. (Back substitution). Given the upper-triangular linear system

$$
\begin{aligned}
a_{11}x_1 + a_{12}x_2 + \cdots + a_{1n}x_n &= b_1 \\
a_{22}x_2 + \cdots + a_{2n}x_n &= b_2 \\
\cdots\cdots\cdots\cdots\cdots\cdots\cdots \\
a_{nn}x_n &= b_n
\end{aligned}
\qquad (2\text{-}7)
$$

where $a_{ij} \neq 0$ for $i = 1, 2, \ldots, n$.

1. Put $x_n = b_n / a_{nn}$.
2. Put $i = 1$.

3. $x_{n-1} = \left(b_{n-1} - \displaystyle\sum_{j=0}^{i-1} a_{n-i,\,x-j} x_{n-j} \right) \Big/ a_{n-i,\,n-i}.$

4. Increase i by 1.
5. If i has reached n, go on to (6). If i is less than n, go back to (3).
6. The algorithm is completed.

Note that this algorithm fails to work if $a_{ii} = 0$ for any i. This is clear because in either step 1 or step 3 a division by 0 would occur. The numbers $x_1, x_2 \ldots x_n$ can be written in the vector form

$$X = \begin{bmatrix} x_1 \\ x_2 \\ \cdot\quad\cdot\quad\cdot \\ x_n \end{bmatrix}$$

This vector is called the solution or the **solution vector** for the linear system.

THEOREM 2-3-1. If $a_{ii} \neq 0$ for $i = 1, 2, \ldots, n$ in Eqs. (2-7), then Algorithm 2-1 produces a solution which is the only solution $x_1, x_2, \ldots, x_n$ to Eqs. (2-7).

(If a problem in mathematics has only one solution, we say this solution is the **unique** solution.)

Proof. The fact that Algorithm 2-1 produces a solution is seen from steps 1 and 3 of the algorithm. That is, the method of constructing $x_n, x_{n-1}, \ldots, x_1$ ensures that each of the equations is satisfied. We must show that the solution produced by the algorithm is the only solution to the system.

Suppose that $x_1, x_2, \ldots, x_n$ is a solution and that $x'_1, x'_2, \ldots, x'_n$ is also a solution. Then $a_{nn}x_n = b_n$, and $a_{nn}x'_n = b_n$. Thus

$$a_{nn}x_n - a_{nn}x'_n = 0$$

giving $a_{nn}(x_n - x'_n) = 0$. Since $a_{nn} \neq 0$, this means that $x_n - x'_n = 0$. Hence $x_n = x'_n$.

Next $a_{n-1,n-1}x_{n-1} + a_{n-1,n}x_n = b_{n-1}$, and $a_{n-1, n-1}x'_{n-1} + a_{n-1, n}x'_n = b_{n-1}$. Thus

$$a_{n-1,n-1}x_{n-1} + a_{n-1,n}x_n = a_{n-1,n-1}x'_{n-1} + a_{n-1,n}x'_n$$

giving

$$a_{n-1,n-1}(x_{n-1} - x'_{n-1}) = 0$$

Because $a_{n-1,n-1} \neq 0$, this means that $x_{n-1} = x'_{n-1}$.
We may continue this process. In general

$$a_{n-i,n-1}x_{n-i} + \sum_{j=n-i+1}^{n} a_{n-i,j}x_j = a_{n-i,n-i}x'_{n-i} + \sum_{j=n-i+1}^{n} a_{n-i,j}x'_j$$

Hence, because $x_j = x'_j$ for $j = n - i + 1, \ldots, n$, we get

$$a_{n-i,n-i}x_{n-i} = a_{n-i,n-i}x'_{n-i}$$

Thus $x_{n-i} = x'_{n-i}$ because $a_{n-i, n-i} \neq 0$.

The fact that $a_{ii} \neq 0$ for $i = 1, 2, \ldots, n$ was used in showing that a solution could be constructed by Algorithm 2-1 and also in concluding that the solution to Eqs. (2-7) is unique. If one or more of the a_{ii}'s in Eqs. (2-7) equal zero, then either (1) no solution to Eqs. (2-7) exists or (2) infinitely many solutions to Eqs. (2-7) exist.

The following examples illustrate these points.

EXAMPLE 2-9. Discuss the system below:

$$
\begin{aligned}
2x_1 + x_2 - x_3 + x_4 &= -3 \\
0x_2 - 2x_3 + 3x_4 &= 1 \\
x_3 + x_4 &= 3 \\
x_4 &= 1
\end{aligned}
$$

Solution. This example has no solution. If it had, then from the last two equations we would have $x_4 = 1$ and $x_3 = 2$. But this is an impossible requirement, because $-2x_3 + 3x_4 = 1$ is required by the second equation, and $(-2)(2) + (3)(1) = -1$.

From this we see that, if an upper-triangular system of equations has a zero diagonal element, then the equation having the zero diagonal element must be satisfied by numbers already found from the lower equations. A slight change in the example gives a system with an infinite number of solutions.

EXAMPLE 2-10. Solve the system below:

$$
\begin{aligned}
2x_1 + x_2 - x_3 + x_4 &= -3 \\
0x_2 - 2x_3 + 3x_4 &= -1 \\
x_3 + x_4 &= 3 \\
x_4 &= 1
\end{aligned}
$$

Solution. In this case the second equation is satisfied by $x_3 = 2$ and $x_4 = 1$. The first equation may now be solved for x_1.

$$
x_1 = \frac{-3 - x_4 + x_3 - x_2}{2}
$$

or

$$
x_1 = \frac{-3 - 1 + 2 - x_2}{2} = \frac{-2 - x_2}{2}
$$

Because the second equation has zero as the coefficient of x_2, it did not restrict x_2 in any way. Hence for any real number t a solution to the system is given by

$$
\begin{aligned}
x_1 &= \frac{-2 - t}{2} \\
x_2 &= t \\
x_3 &= 2 \\
x_4 &= 1
\end{aligned}
$$

EXERCISE 12. Solve the following upper-triangular system of equations by back substitution:

$$
\begin{aligned}
5x_1 - 2x_2 + 6x_3 + x_4 + 2x_5 &= 4 \\
-4x_2 + 7x_3 - x_4 - x_5 &= -3 \\
3x_3 + x_4 + x_5 &= 7 \\
x_4 - x_5 &= 10 \\
2x_5 &= 14
\end{aligned}
$$

EXERCISE 13. Suppose that the third equation in Exercise 12 were changed to $0x_3 + x_4 + x_5 = 7$. What would the 7 on the right-hand side have to be changed to for the new system to have a solution?

EXERCISE 14. Write out the following as a system of linear equations, and solve this system:

$$
\begin{bmatrix}
2 & -4 & 0 & 1 \\
0 & -1 & 2 & 1 \\
0 & 0 & 7 & -2 \\
0 & 0 & 0 & 2
\end{bmatrix}
\begin{bmatrix}
x_1 \\
x_2 \\
x_3 \\
x_4
\end{bmatrix}
=
\begin{bmatrix}
3 \\
7 \\
-3 \\
10
\end{bmatrix}
$$

EXERCISE 15. A system is said to be **lower-triangular** if all the elements above the diagonal are zero. The general notation is

$$
\begin{aligned}
a_{11}x_1 &= b_1 \\
a_{21}x_1 + a_{22}x_2 &= b_2 \\
a_{31}x_1 + a_{32}x_2 + a_{33}x_3 &= b_3 \\
&\cdots\cdots\cdots \\
a_{n1}x_1 + a_{n2}x_2 + \cdots + a_{nn}x_n &= b_n
\end{aligned}
$$

Assuming that $a_{ii} \neq 0$, $i = 1, \ldots, n$, devise an algorithm for solving a lower-triangular system.

EXERCISE 16. Can a linear system be both upper-triangular and lower-triangular at the same time? Explain.

EXERCISE 17. Let $B = (b_{ii})_{n \times 1}$ be a given column matrix. What is the solution to the system $I_n X = B$?

EXERCISE 18. Prove that the product of two upper-triangular $n \times n$ matrices is upper-triangular.

EXERCISE 19. Draw a flow chart for the back-substitution algorithm (Algorithm 2-1).

EXERCISE 20. Write a program for solving an upper-triangular system of linear equations with nonzero diagonal elements.

2-4   REDUCTION OF A LINEAR SYSTEM TO UPPER-TRIANGULAR FORM

Two systems of equations are said to be **equivalent** if and only if their solutions are the same. It is not too difficult to show that every system of n linear equations in n unknowns can be put into an equivalent upper-triangular form (see Exercise 28). In this section we shall establish an algorithm which will put any system having a unique solution into an equivalent upper-triangular form with 1s down the diagonal. The solution to the original system can then be obtained by **back substitution** on the upper-triangular system.

From his previous studies in algebra the reader is probably aware that certain transformations on an equation, such as squaring each side, may introduce extraneous roots. That is, the equation which results need not be equivalent to the original. Certain other transformations, such as multiplying both sides of an equation by a nonzero constant, give an equivalent equation. In dealing with a linear system we shall employ three types of transformations, each of which yields an equivalent system. These are

1. Interchange the order of two equations.
2. Add a multiple of one equation to another equation.
3. Multiply an equation by a nonzero constant.

We shall prove that transformation 2 produces an **equivalent system**. The proofs for (1) and (3) are left as exercises.

Proof. Suppose that the original system is

$$
\begin{aligned}
a_{11}x_1 + a_{12}x_2 + \;\cdot\;\cdot\;\cdot\; + a_{1n}x_n &= b_n \\
a_{21}x_1 + a_{22}x_2 + \;\cdot\;\cdot\;\cdot\; + a_{2n}x_n &= b_2 \\
&\cdots\cdots\cdots \\
a_{n1}x_1 + a_{n2}x_2 + \;\cdot\;\cdot\;\cdot\; + a_{nn}x_n &= b_n
\end{aligned}
\tag{2-8}
$$

Let t be a constant, and suppose that t times the ith equation is added to the jth equation, where $i \neq j$. The result is a system

$$
\begin{aligned}
a_{11}x_1 + a_{12}x_2 + \;\cdot\;\cdot\;\cdot\; + a_{1n}x_n &= b_1 \\
&\cdots\cdots\cdots \\
a_{i1}x_1 + a_{i2}x_2 + \;\cdot\;\cdot\;\cdot\; + a_{in}x_n &= b_i \\
&\cdots\cdots\cdots \\
(ta_{i1} + a_{j1})x_1 + (ta_{i2} + a_{j2})x_2 + \;\cdot\;\cdot\;\cdot\; + (ta_{in} + a_{jn})x_n &= tb_i + b_j \\
&\cdots\cdots\cdots \\
a_{n1}x_1 + a_{n2}x_2 + \;\cdot\;\cdot\;\cdot\; + a_{nn}x_n &= b_n
\end{aligned}
\tag{2-9}
$$

We must show that if $(u_1, u_2, \ldots, u_n)$ is a solution to Eqs. (2-8) then it is a solution to Eqs. (2-9), and vice versa. If $(u_1, u_2, \ldots, u_n)$ is a solution to Eqs. (2-8), then

$$
\sum_{i=1}^{n} a_{ki}u_i = b_k \qquad \text{for } k = 1, 2, \ldots, n
$$

Hence all but the jth equation in Eqs. (2-9) are automatically satisfied. Because

$$t\left(\sum_{l=1}^{n} a_{il}u_{l}\right) = \sum_{l=1}^{n} ta_{il}u_{l} = tb_{i}$$

and

$$\sum_{l=1}^{n} a_{jl}u_{l} = b_{j}$$

it follows that

$$\sum_{l=1}^{n} ta_{il}u_{l} + \sum_{l=1}^{n} a_{jl}u_{l} = tb_{i} + b_{j}$$

This can written as

$$\sum_{l=1}^{n} (ta_{il} + a_{jl})u_{l} = tb_{i} + b_{j}$$

so that the jth equation of Eqs. (2-9) is also satisfied.

Suppose on the other hand that $(u_{1}, u_{2}, \ldots, u_{n})$ is a solution to Eqs.(2-9). Then

$$\sum_{l=1}^{n} a_{kl}u_{l} = b_{k} \quad \text{for } k = 1, \ldots, n; \quad k \neq j$$

$$\sum_{l=1}^{n} (ta_{il} + a_{jl})u_{l} = tb_{i} + b_{j}$$

Thus all but the jth equation of Eqs. (2-8) are automatically satisfied. In particular, multiplication of the ith equation by $-t$ gives

$$-t\sum_{l=1}^{n} a_{il}u_{l} = \sum_{l=1}^{n} -ta_{il}u_{l} = -tb_{i}$$

Thus, by addition

$$\sum_{l=1}^{n} (ta_{il} + a_{jl})u_{l} + \sum_{l=1}^{n} -ta_{il}u_{l} = tb_{i} + b_{j} - tb_{i}$$

Upon simplification this becomes

$$\sum_{l=1}^{n} (ta_{il} + a_{jl} - ta_{il})u_{l} = \sum_{l=1}^{n} a_{jl}u_{l} = b_{j}$$

That is, the jth equation of Eqs. (2-8)is satisfied.

EXERCISE 21. Prove that transformation 1 produces an equivalent system when applied to a system of linear equations.

EXERCISE 22. Do the same thing for transformation 3.

EXERCISE 23. Show how transformation 1 can be carried out by a sequence of transformations of types 2 and 3.

EXERCISE 24. Explain why it is necessary that the constant in transformation 3 be nonzero if an equivalent system is to be produced.

The example below illustrates how a linear system can be upper-triangulated by use of the three basic transformations.

EXAMPLE 2-11. Upper-triangulate the system below:

$$\begin{aligned} x_2 + 2x_3 &= 1 \\ x_1 + x_2 + x_3 &= 0 \\ 2x_1 - x_2 &= 5 \end{aligned}$$

Solution.

(*a*) Interchange the first and third equations.

$$\begin{aligned} 2x_1 - x_2 &= 5 \\ x_1 + x_2 + x_3 &= 0 \\ x_2 + 2x_3 &= 1 \end{aligned}$$

(*b*) Multiply the first equation by 1/2.

$$\begin{aligned} x_1 - 1/2\,x_2 &= 5/2 \\ x_1 + x_2 + x_3 &= 0 \\ x_2 + 2x_3 &= 1 \end{aligned}$$

(*c*) Add -1 times the first equation to the second equation.

$$\begin{aligned} x_1 - 1/2\,x_2 &= 5/2 \\ 3/2\,x_2 + x_3 &= -5/2 \\ x_2 + 2x_3 &= 1 \end{aligned}$$

(*d*) Multiply the second equation by $2/3$.

$$\begin{aligned} x_1 - 1/2\,x_2 &= 5/2 \\ x_2 + 2/3\,x_3 &= -5/3 \\ x_2 + 2x_3 &= 1 \end{aligned}$$

(*e*) Add -1 times the second equation to the third equation.

$$\begin{aligned} x_1 - 1/2\,x_2 &= 5/2 \\ x_2 + 2/3\,x_3 &= -5/3 \\ 4/3\,x_3 &= 8/3 \end{aligned}$$

(*f*) Multiply the third equation by $3/4$

$$\begin{aligned} x_1 - 1/2\,x_2 &= 5/2 \\ x_2 + 2/3\,x_3 &= -5/3 \\ x_3 &= 2 \end{aligned}$$

Note that in this procedure we have done more than just upper-triangulate the original system; we have made the diagonal elements $a_{ii} = 1$.

As mentioned previously, while it is always possible to upper-triangulate a system of linear equations, it is not always possible to upper-triangulate a system and have the diagonal elements $a_{ii} = 1$. This is because a system of equations having no solutions or infinitely many solutions must necessarily upper-triangulate into a form having one or more of the diagonal elements equal to zero. However, a system of equations having a unique solution must necessarily upper-triangulate into a system having no zero elements on the diagonal, and hence the diagonal elements can be made to be 1 by use of transformation 3.

We shall now discuss an algorithm which transforms the $n \times n$ system

$$a_{11}x_1 + a_{12}x_2 + \cdots + a_{1n}x_n = b_1$$
$$\dotfill \tag{2-10}$$
$$a_{n1}x_1 + a_{n2}x_2 + \cdots + a_{nn}x_n = b_n$$

into an equivalent system of the form

$$x_1 + c_{12}x_2 + \cdots + c_{1n}x_n = d_1$$
$$c_{22}x_2 + \cdots + c_{2n}x_n = d_2$$
$$\dotfill \tag{2-11}$$
$$c_{n2}x_2 + \cdots + c_{nn}x_n = d_n$$

It should be evident that the same algorithm could then be applied to the $(n-1) \times (n-1)$ system formed when the first equation of Eqs. (2-11) is deleted. Following this idea recursively will give an algorithm for the upper triangulation of a linear system.

Basically the algorithm must consist of two parts. The first part is a search for an equation with a nonzero leading coefficient. Once such an equation is found, if it is not the first equation, then it is interchanged with the first equation.

To complete the algorithm, appropriate multiples of the (new) first equation are subtracted from the other equations to produce the form (2-11). All this will be possible if and only if at least one of the coefficients $a_{ii}, a_{21}, \ldots, a_{n1}$ is nonzero.

As usual we have in mind implementation of our algorithm in C++; thus it is convenient to make a slight change of notation. The right-hand-side elements $b_1, b_2, \ldots, b_n$ are stored in the $(n+1)$st column of matrix A. Thus the system under consideration is

$$a_{11}x_1 + a_{12}x_2 + \cdots + a_{1n}x_n = a_{1,n+1}$$
$$\dotfill$$
$$a_{n1}x_1 + a_{n2}x_2 + \cdots + a_{nn}x_n = a_{n,n+1}$$

Figure 2-2 presents the desired algorithm. It has several additional features which are discussed in the next paragraph.

In this particular formulation of the algorithm a leading coefficient of largest possible absolute value is selected. If this element is zero, it follows that the algorithm cannot be completed. The element selected, provided that it is nonzero, plays a very important role in the remainder of the algorithm. It is called the **pivot element.** It is clear that, as long as a nonzero pivot element can be found, the algorithm can be completed. However, considerable experimental and theoretical evidence suggests that better computational results are achieved on the average if a pivot element of largest absolute value is used. This assumes that the calculations are carried out in floating-point arithmetic. (See Chapter 5.)

As previously mentioned, the idea of Fig. 2-2 can be used repeatedly to upper-triangulate a linear system. Once the linear system is upper-triangulated, the system can be solved by the back-substitution algorithm. The result is a general-purpose algorithm for solving a linear system (see Exercise 27).

EXERCISE 25. Solve the following system by upper triangulation and **back substitution**:

$$x_1 + x_2 + 2x_3 = 4$$
$$4x_1 - 3x_2 + x_3 = 2$$
$$x_1 - 2x_2 + 4x_3 = -2$$

EXERCISE 26. Solve the following system by upper triangulation and **back substitution**:

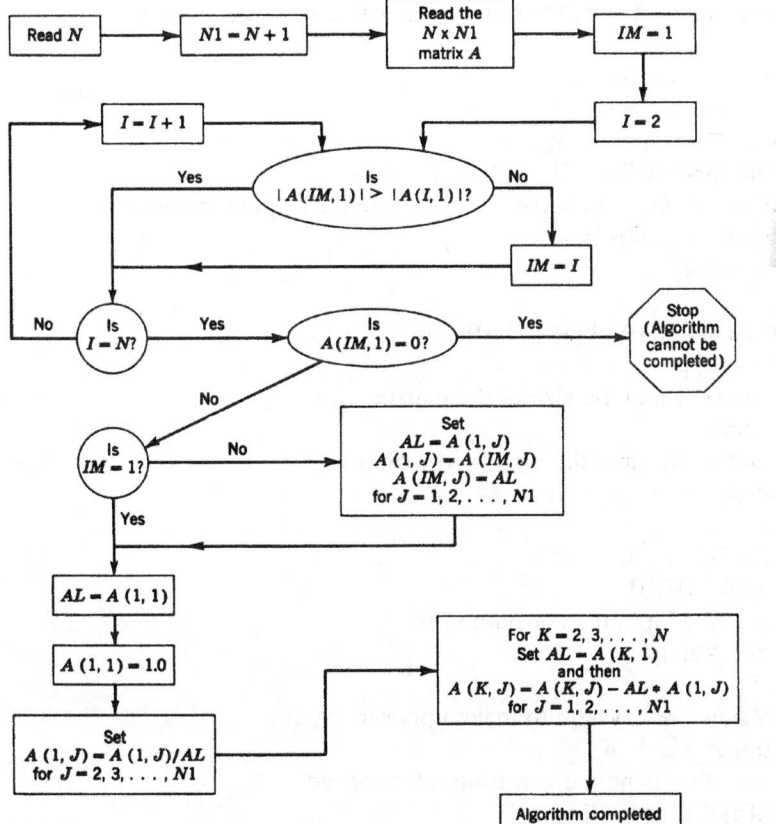

FIG. 2-2   Flow chart for one step in upper-triangulation algorithm.

$$2x_1 + x_2 - 4x_3 + x_4 = 0$$
$$-4x_1 + 3x_2 + 5x_3 - 2x_4 = 10$$
$$x_1 - x_2 + x_3 - x_4 = 3$$
$$x_1 + 3x_2 - 3x_3 + 2x_4 = -1$$

EXERCISE 27. If a linear system $AX = B$ is being solved by upper triangulation, the value of the determinant of A may easily be found at the same time. The value of the determinant is the product of the pivotal elements used in the upper triangulation algorithm, times -1 raised to the power k where k is the number of row interchanges which occured in using the algorithm. Write a computer program to read in n, $A_{n \times n}$, $B_{n \times 1}$, then solve $AX = B$, and find the determinant of A.

EXERCISE 28 Explain how to modify the upper triangulation described in Sec. 2.4 so that it will upper-triangulate an arbitrary matrix A.

The following program takes a linear system and solves by the upper triangulation method. A feature is added to use the absolute maximum element in a column to zero out all elements below.

Program 2–2 A C++ program to solve a linear system

```
#include <iostream>
#include <cmath>
using namespace std;
void upper_t(float [][10], float [], int);
// C++ Program to solve a linear system by upper triangulate
// and back substitute
 int main()
 {
 float a[10][10], b[10],x[10],sum;
 int n, i, j;
 cout<<"Input the size of the matrix: ";
 cin>>n;
 cout<<"\n Input the matrix row by row \n";
 for(i = 1; i<=n; ++i)
  {
 for(j=1; j<=n; ++j)
  cin>>a[i][j];
 cout<<"Input the constant term  ";
 cin>>b[i];
  }
 // now pass system to make upper triangular
 upper_t(a, b, n);
 // system is now upper triangular - solve
 x[n] = b[n]/a[n][n];
```

```
for(i = 1; i < =n; ++i)
  {
  sum = 0;
  for(j=0; j<= i-1; ++j)
   sum = sum + a[n-i][n-j]*x[n-j];
  x[n-i]= (b[n-i] - sum) / a[n-i][n-i];
  }
cout<<"The solution is:"<<endl;
for(i = 1; i <=n; ++i)
  cout<< i<<"   "<< x[i]<<endl;

return 0;
}

void upper_t(float a[][10], float b[], int n )
{
int i,j,is,c,k   ;
int ii,jj;
float as,bs,larg, fact;
for(i = 1; i<=n; ++i)
  {
  for(j = 1; j<=n; ++j)
   cout<< a[i][j]<<' ';
   cout<< b[i] <<'\n';
  }
cout<<endl;
// Now to make upper triangular
for(c = 1; c < n; ++c)     // goes across columns
  {
   larg = fabs(a[c][c]);
   is = c;
   for(j=c+1; j<=n; ++j)
    if(fabs(a[j][c]) > larg)
     {
      is = j; // saves row with max element in column i
      larg = fabs(a[j][c]);
     }
   // now interchange row i with row is if necessary
   if( is != c)
     {
      for(j = c; j<=n; ++j)
       {
       as = a[c][j];
       a[c][j] = a[is][j];
       a[is][j] = as;
```

```
        }
      bs = b[c];
      b[c] = b[is];
      b[is] = bs;
      }

  for(k = c+1; k<= n; ++k) // zeroes out all below in col c
  {
  fact = a[k][c]/a[c][c];
  for(j= c; j <= n; ++j)
  {
    a[k][j] = a[k][j] -  fact * a[c][j];
  }
  b[k] = b[k] -  fact * b[c];

  }
  } //end of c loop
  cout<<"End of function \n";
  cout<<endl;
  }
```

2-5   THE INVERSE OF A SQUARE MATRIX

We now know how to solve a system $AX = B$ provided that a **unique solution** X exists. Such a system is similar notationally to a single equation $ax = b$. This single equation can be solved as follows:

$$ax = b \quad a \neq 0$$
$$a^{-1}ax = a^{-1}b$$
$$x = a^{-1}b$$

In this section we shall show that a similar procedure is applicable to $AX = B$. We want to know how to find a matrix A^{-1} which has the property that the unique solution to $AX = B$ is $X = A^{-1}B$.

Before investigating this topic, we need to know when $AX = B$ has a unique solution. From Sec. 2-4 it follows that $AX = B$ has a unique solution if and only if $AX = B$ can be put into upper-triangular form with 1s down the diagonal. By careful consideration of the upper-triangulation algorithm we can see that, if upper triangulation can be accomplished with 1s down the diagonal for one vector B, it can be accomplished for any vector B. Hence the question as to whether a system $AX = B$ can be upper-triangulated with 1s down the diagonal depends only on the matrix A.

DEFINITION 2-9. If the $n \times n$ matrix A is such that the system $AX = B$ can be upper-triangulated with 1s down the diagonal we say that A is **nonsingular**.

Definition 2-10. If a matrix C exists such that $X = CB$ is the unique solution to $AX = B$ for any vector B, then C is called an **inverse** of A. The notation most commonly used to designate an inverse of a matrix A is A^{-1}.

THEOREM 2-5-1. If A is a nonsingular $n \times n$ matrix, then there exists a unique inverse matrix A^{-1}.

A proof of this theorem will be given by using Lemma 2-5-1 below. First it should be pointed out that we could have defined the inverse matrix of A to be the matrix A^{-1} such that $A^{-1}A = I$. The reader may be familiar with this definition and the fact that $A^{-1}A = AA^{-1} = I$. This definition was not used because we wish to give a constructive procedure for finding A^{-1} which is more suitable to our needs (see Exercise 31).

Lemma 2-5-1. Let $Y_1, Y_2, \ldots, Y_n$ and $B_1, B_2, \ldots, B_n$ be column vectors such that $AY_i = B_i$, for $i = 1, 2, \ldots, n$, and let $c_1, c_2, \ldots, c_n$ be arbitrary real numbers. Then

$$A(c_1 Y_1 + c_2 Y_2 + \cdots + c_n Y_n) = c_1 B_1 + c_2 B_2 + \cdots + c_n B_n$$

Proof. By an extention of property 4 (page 40) we know that

$$A(c_1 Y_1 + c_2 Y_2 + \cdots + c_n Y_n) = A(c_1 Y_1) + A(c_2 Y_2) + \cdots + A(c_n Y_n)$$

By property 2 we know that $A(c_2 Y_1) = c_2 AY_1$. Thus

$$A(c_1 Y_1 + c_2 Y_2 + \cdots + c_n Y_n) = c_1 AY_1 + c_2 AY_2 + \cdots + c_n AY_n$$

Finally, because $AY_i = B_{ij}$, we get

$$A(c_1 Y_1 + c_2 Y_2 + \cdots + c_n Y_n) = c_1 B_1 + c_2 B_2 + \cdots + c_n B_n$$

Proof of Theorem 2-5-1. we begin by defining certain special vectors, called *base* vectors. These are given by

$$E_i = \begin{bmatrix} 0 \\ \cdots \\ 0 \\ 1 \\ 0 \\ \cdots \\ 0 \end{bmatrix} \leftarrow i\text{th row} \qquad (2\text{-}12)$$

where the 1 occurs in the ith position. For example, if $n = 3$, then

$$E_1 = \begin{bmatrix} 1 \\ 0 \\ 0 \end{bmatrix} \quad E_2 = \begin{bmatrix} 0 \\ 1 \\ 0 \end{bmatrix} \quad E_3 = \begin{bmatrix} 0 \\ 0 \\ 1 \end{bmatrix}$$

Any vector

$$B = \begin{bmatrix} b_1 \\ b_2 \\ \cdots \\ b_n \end{bmatrix}$$

can be written in the form

$$B = b_1 E_1 + b_2 E_2 + \cdots + b_n E_n \qquad (2\text{-}13)$$

For example, if $b_1 = 5$, $b_2 = -3$, $b_3 = 7$, then

$$\begin{bmatrix} 5 \\ -3 \\ 7 \end{bmatrix} = 5 \begin{bmatrix} 1 \\ 0 \\ 0 \end{bmatrix} - 3 \begin{bmatrix} 0 \\ 1 \\ 0 \end{bmatrix} + 7 \begin{bmatrix} 0 \\ 0 \\ 1 \end{bmatrix} \qquad (2\text{-}14)$$

Thus the system $AX = B$ can be written as

$$AX = b_1 E_1 + b_2 E_2 + \cdots + b_n E_n \qquad (2\text{-}15)$$

Now let $Y_1, Y_2, \ldots, Y_n$ be the unique solutions to

$$AY_1 = E_1 \quad AY_2 = E_2 \quad \cdots \quad AY_n = E_n$$

respectively. Then by Lemma 2-5-1

$$A(b_1 Y_1 + b_2 Y_2 + \cdots + b_n Y_n) = b_1 E_1 + b_2 E_2 + \cdots + B_n E_n = B$$

Thus $X = b_1 Y_1 + b_2 y_2 + \cdots + b_n Y_n$ is the unique solution to

$$AX = B$$

Let

$$Y_i = \begin{bmatrix} y_{1i} \\ y_{2i} \\ \cdots \\ y_{ni} \end{bmatrix}$$

Then

$$X = b_1 \begin{bmatrix} y_{11} \\ y_{21} \\ \cdots \\ y_{n1} \end{bmatrix} + b_2 \begin{bmatrix} y_{12} \\ y_{22} \\ \cdots \\ y_{n2} \end{bmatrix} + \cdots + b_n \begin{bmatrix} y_{1n} \\ y_{2n} \\ \cdots \\ y_{nn} \end{bmatrix}$$

This can be written in matrix-multiplication form as

$$X = \begin{bmatrix} y_{11} & y_{12} & \cdots & y_{1n} \\ y_{21} & y_{22} & \cdots & y_{2n} \\ \cdots & \cdots & \cdots & \cdots \\ y_{n1} & y_{n2} & \cdots & y_{nn} \end{bmatrix} \begin{bmatrix} b_1 \\ b_2 \\ \cdots \\ b_n \end{bmatrix}$$

Because we are assuming that A is nonsingular, each of the Y_i's exists and is unique. Thus from our definition

$$A^{-1} = \begin{bmatrix} y_{11} & y_{12} & \cdots & y_{1n} \\ y_{21} & y_{22} & \cdots & y_{2n} \\ \cdots & \cdots & \cdots & \cdots \\ y_{n1} & y_{n2} & \cdots & y_{nn} \end{bmatrix} \qquad (2\text{-}16)$$

is an inverse of A.

Finally we shall prove that the matrix (2-16) is the unique inverse matrix of A. Suppose that

$$
C = \begin{bmatrix} c_{11} & c_{12} & \cdots & c_{1n} \\ \cdots & \cdots & \cdots & \cdots \\ c_{n1} & c_{n2} & \cdots & c_{nn} \end{bmatrix}
$$

also has the property that $X = CB$ is the unique solution to $AX = B$ for every vector B. Then, because

$$
Y_i = \begin{bmatrix} y_{1i} \\ y_{2i} \\ \cdots \\ y_{ni} \end{bmatrix}
$$

is the unique solution to $AX = E_{ij}$, we get

$$
\begin{bmatrix} y_{1i} \\ y_{2i} \\ \cdots \\ \cdots \\ y_{ni} \end{bmatrix} = \begin{bmatrix} c_{11} & c_{12} & \cdots & c_{1n} \\ c_{21} & c_{22} & \cdots & c_{2n} \\ \cdots & \cdots & \cdots & \cdots \\ \cdots & \cdots & \cdots & \cdots \\ c_{n1} & c_{n2} & \cdots & c_{nn} \end{bmatrix} \begin{bmatrix} 0 \\ \cdots \\ 1 \\ \cdots \\ 0 \end{bmatrix} \leftarrow i\text{th position}
$$

or

$$
\begin{bmatrix} y_{1i} \\ y_{2i} \\ \cdots \\ y_{ni} \end{bmatrix} = \begin{bmatrix} c_{1i} \\ c_{2i} \\ \cdots \\ c_{ni} \end{bmatrix}
$$

Hence the ith column of C is equal to the ith column of the A^{-1} in Eq. (2-16). Since i can equal $1, 2, \ldots, n$, we conclude that each column of C agrees with the corresponding column of A^{-1}. Thus $C = A^{-1}$.

From the proof we see that the inverse of the matrix A may be determined by solving the n systems of linear equations $AY_i = E_i$ for $i = 1, 2, \ldots, n$. The following example will indicate that the upper-triangulation algorithm need be carried out only once to solve all n systems of linear equations.

EXAMPLE 2-12. Find A^{-1} for the 3×3 matrix

$$
A = \begin{bmatrix} 1 & 2 & 1 \\ 3 & 1 & -1 \\ 1 & 1 & 1 \end{bmatrix}
$$

Solution. We want to find

$$
\begin{bmatrix} c_{1i} \\ c_{2i} \\ c_{3i} \end{bmatrix} \quad \text{for } i = 1, 2, 3
$$

such that

$$\begin{bmatrix} 1 & 2 & 1 \\ 3 & 1 & -1 \\ 1 & 1 & 1 \end{bmatrix}\begin{bmatrix} c_{1i} \\ c_{2i} \\ c_{3i} \end{bmatrix} = \left\{\begin{bmatrix} 1 \\ 0 \\ 0 \end{bmatrix} \text{ for } i=1 \quad \begin{bmatrix} 0 \\ 1 \\ 0 \end{bmatrix} \text{ for } i=2 \quad \begin{bmatrix} 0 \\ 0 \\ 1 \end{bmatrix} \text{ for } i=3\right\}$$

The steps for obtaining the inverse are now shown.

Step 1. Reorder equations for maximum pivot element.

$$\begin{bmatrix} 3 & 1 & -1 \\ 1 & 2 & 1 \\ 1 & 1 & 1 \end{bmatrix}\begin{bmatrix} c_{1i} \\ c_{2i} \\ c_{3i} \end{bmatrix} = \left\{\begin{bmatrix} 0 \\ 1 \\ 0 \end{bmatrix} \begin{bmatrix} 1 \\ 0 \\ 0 \end{bmatrix} \begin{bmatrix} 0 \\ 0 \\ 1 \end{bmatrix}\right\}$$

Step 2. Complete the algorithm of Fig. 2-2.

$$\begin{bmatrix} 1 & 1/3 & -1/3 \\ 0 & 5/3 & 4/3 \\ 0 & 2/3 & 4/3 \end{bmatrix}\begin{bmatrix} c_{1i} \\ c_{2i} \\ c_{3i} \end{bmatrix} = \left\{\begin{bmatrix} 0 \\ 1 \\ 0 \end{bmatrix} \begin{bmatrix} 1/3 \\ -1/3 \\ -1/3 \end{bmatrix} \begin{bmatrix} 0 \\ 0 \\ 1 \end{bmatrix}\right\}$$

Step 3. Carry out the algorithm of Fig. 2-2 on the lower 2×2 systems.

$$\begin{bmatrix} 1 & 1/3 & -1/3 \\ 0 & 1 & 4/5 \\ 0 & 0 & 4/5 \end{bmatrix}\begin{bmatrix} c_{1i} \\ c_{2i} \\ c_{3i} \end{bmatrix} = \left\{\begin{bmatrix} 0 \\ 3/5 \\ -2/5 \end{bmatrix} \begin{bmatrix} 1/3 \\ -1/5 \\ -1/5 \end{bmatrix} \begin{bmatrix} 0 \\ 0 \\ 1 \end{bmatrix}\right\}$$

Step 4 Multiply the botton equations by 5/4.

$$\begin{bmatrix} 1 & 1/3 & -1/3 \\ 0 & 1 & 4/5 \\ 0 & 0 & 1 \end{bmatrix}\begin{bmatrix} c_{1i} \\ c_{2i} \\ c_{3i} \end{bmatrix} = \left\{\begin{bmatrix} 0 \\ 3/5 \\ -1/2 \end{bmatrix} \begin{bmatrix} 1/3 \\ -1/5 \\ -1/4 \end{bmatrix} \begin{bmatrix} 0 \\ 0 \\ 5/4 \end{bmatrix}\right\}$$

Hence we have by back substitution

$$\begin{bmatrix} c_{11} \\ c_{21} \\ c_{31} \end{bmatrix} = \begin{bmatrix} -1/2 \\ 1 \\ -1/2 \end{bmatrix} \quad \begin{bmatrix} c_{12} \\ c_{22} \\ c_{32} \end{bmatrix} = \begin{bmatrix} 1/4 \\ 0 \\ -1/4 \end{bmatrix} \quad \begin{bmatrix} c_{13} \\ c_{23} \\ c_{33} \end{bmatrix} = \begin{bmatrix} 3/4 \\ -1 \\ 5/4 \end{bmatrix}$$

Thus

$$A^{-1} = \begin{bmatrix} -1/2 & 1/4 & 3/4 \\ 1 & 0 & -1 \\ -1/2 & -1/4 & 5/4 \end{bmatrix}$$

EXAMPLE 2-13. Using the result of Example of 2-12, solve the linear system $AX = B$, where

$$A = \begin{bmatrix} 1 & 2 & 1 \\ 3 & 1 & -1 \\ 1 & 1 & 1 \end{bmatrix} \quad B = \begin{bmatrix} 3 \\ -11 \\ 3 \end{bmatrix}$$

Solution. The unique solution is $X = A^{-1}B$. Thus

$$X = \begin{bmatrix} -1/2 & 1/4 & 3/4 \\ 1 & 0 & -1 \\ -1/2 & -1/4 & 5/4 \end{bmatrix} \begin{bmatrix} 3 \\ -11 \\ 3 \end{bmatrix} = \begin{bmatrix} -2 \\ 0 \\ 5 \end{bmatrix}$$

EXERCISE 29.

(a) Determine the inverse of the 3 × 3 matrix

$$A = \begin{bmatrix} 1 & 2 & -1 \\ 0 & 1 & 10 \\ -7 & 3 & -9 \end{bmatrix}$$

(b) Use the inverse determined in (*a*) to solve the system

$$\begin{bmatrix} 1 & 2 & -1 \\ 0 & 1 & 10 \\ -7 & 3 & -9 \end{bmatrix} \begin{bmatrix} x_1 \\ x_2 \\ x_3 \end{bmatrix} = \begin{bmatrix} 2 \\ 1 \\ -1 \end{bmatrix}$$

EXERCISE 30.

(a) Determine the inverse of the 4 × 4 matrix

$$\begin{bmatrix} 1 & 2 & -1 & 1 \\ -1 & 0 & 2 & 5 \\ 3 & 1 & 7 & 1 \\ 0 & 2 & -1 & 1 \end{bmatrix}$$

(b) Use the inverse determined in (*a*) to solve the system

$$\begin{bmatrix} 1 & 2 & -1 & 1 \\ -1 & 0 & 2 & 5 \\ 3 & 1 & 7 & 1 \\ 0 & 2 & -1 & 1 \end{bmatrix} \begin{bmatrix} x_1 \\ x_2 \\ x_3 \\ x_4 \end{bmatrix} = \begin{bmatrix} 1 \\ 0 \\ -12 \\ 3 \end{bmatrix}$$

EXERCISE 31. Show that if A is nonsingular and A^{-1} is defined by Definition 2-10 then $A^{-1}A = I$. Hint: Let $A_1, A_2, \ldots, A_n$ denote the columns of A, and let $E_1, E_2, \ldots, E_n$ denote the columns of I. First show that $AE_i = A_i$, so that $E_i = A^{-1}A_i$.

2-6   A BETTER ALGORITHM FOR INVERTING A MATRIX

The procedure described in Sec. 2-5 for finding the inverse of a matrix is somewhat cumbersome to apply. In this section we shall give an improved version of that procedure.

DEFINITION 2-11. If $A = (a_{ij})_{n \times n}$ is such that $a_{ij} = 0$ when $i \neq j$, then A is **diagonal**. If the matrix of a linear system of equations is diagonal, the system is said to be diagonal.

Examples of 3 × 3 diagonal matrices include

$$\begin{bmatrix} 1 & 0 & 0 \\ 0 & 1 & 0 \\ 0 & 0 & 1 \end{bmatrix} \quad \begin{bmatrix} -5 & 0 & 0 \\ 0 & 13 & 0 \\ 0 & 0 & 2 \end{bmatrix} \quad \begin{bmatrix} 0 & 0 & 0 \\ 0 & 1 & 0 \\ 0 & 0 & -5 \end{bmatrix}$$

By only a slight modification of the algorithm for upper triangulation one can get an **algorithm for diagonalizing** a linear system. Such an algorithm is quite convenient in the problem of finding the inverse of a matrix. In particular, if a diagonalization procedure uses only the transformations previously listed and reduces A to the identity matrix, then the same procedure applied to the identity matrix will produce A^{-1}. This is because the back-substitution part of the procedure for solving a linear system is not needed in this case. The example below illustrates the basic ideas involved.

EXAMPLE 2-14. Find A^{-1}, where

$$A = \begin{bmatrix} 3 & 1 & -1 \\ 1 & 2 & 1 \\ 1 & 1 & 1 \end{bmatrix}$$

Solution. Start with a matrix

$$\begin{bmatrix} 3 & 1 & -1 & \cdot & 1 & 0 & 0 \\ 1 & 2 & 1 & \cdot & 0 & 1 & 0 \\ 1 & 1 & 1 & \cdot & 0 & 0 & 1 \end{bmatrix}$$

Step 1.

$$\begin{bmatrix} 1 & 1/3 & -1/3 & \cdot & 1/3 & 0 & 0 \\ 1 & 2 & 1 & \cdot & 0 & 1 & 0 \\ 1 & 1 & 1 & \cdot & 0 & 0 & 1 \end{bmatrix}$$

Step 2.

$$\begin{bmatrix} 1 & 1/3 & -1/3 & \cdot & 1/3 & 0 & 0 \\ 0 & 5/3 & 4/3 & \cdot & -1/3 & 1 & 0 \\ 0 & 2/3 & 4/3 & \cdot & -1/3 & 0 & 1 \end{bmatrix}$$

Step 3.

$$\begin{bmatrix} 1 & 1/3 & -1/3 & \cdot & 1/3 & 0 & 0 \\ 0 & 1 & 4/5 & \cdot & -1/5 & 3/5 & 0 \\ 0 & 2/3 & 4/3 & \cdot & -1/3 & 0 & 1 \end{bmatrix}$$

Step 4.

$$\begin{bmatrix} 1 & 0 & -3/5 & \cdot & 2/5 & -1/5 & 0 \\ 0 & 1 & 4/5 & \cdot & -1/5 & 3/5 & 0 \\ 0 & 0 & 4/5 & \cdot & -1/5 & -2/5 & 1 \end{bmatrix}$$

Step 5.

$$\begin{bmatrix} 1 & 0 & -3/5 & \cdot & 2/5 & -1/5 & 0 \\ 0 & 1 & 4/5 & \cdot & -1/5 & 3/5 & 0 \\ 0 & 0 & 1 & \cdot & -1/4 & -1/2 & 5/4 \end{bmatrix}$$

Step 6.

$$\begin{bmatrix} 1 & 0 & 0 & \cdot & 1/4 & -1/2 & 3/4 \\ 0 & 1 & 0 & \cdot & 0 & 1 & -1 \\ 0 & 0 & 1 & \cdot & -1/4 & -1/2 & 5/4 \end{bmatrix}$$

Thus

$$A^{-1} = \begin{bmatrix} 1/4 & -1/2 & 3/4 \\ 0 & 1 & -1 \\ -1/4 & -1/2 & 5/4 \end{bmatrix}$$

Exercise 32. Repeat Exercise 29*a*, using the algorithm of this section.

Exercise 33. Repeat Exercise 30*a*, using the algorithm of this section.

Exercise 34. Let

$$H_n = \begin{bmatrix} 1 & 1/2 & 1/3 & \cdots & \dfrac{1}{n} \\ 1/2 & 1/3 & 1/4 & \cdots & \dfrac{1}{n+1} \\ 1/3 & 1/4 & 1/5 & \cdots & \dfrac{1}{n+2} \\ \cdots & \cdots & \cdots & \cdots & \cdots \\ \dfrac{1}{n} & \dfrac{1}{n+1} & \dfrac{1}{n+2} & \cdots & \dfrac{1}{2n-1} \end{bmatrix}$$

This is known as the **Hilbert matrix of order n**. Using a computer, find H_n^{-1}, $H_n H_n^{-1}$, and $H_n^{-1} H_n$ for $n = 2, 3, \ldots, m$. (Here m should depend upon the speed of the computer you have available.) Notice the rapid decrease in accuracy as n increases.

2-7 Error Aspects

Each of the algorithms described in this chapter involves only a finite number of arithmetic operations. If the matrix A is nonsingular and all calculations are performed exactly, then, using these algorithms, one can find the exact solution to a system $AX = B$ and one can find the exact elements of A^{-1}. In some problems it is desirable (or necessary) to expend the effort needed to perform the necessary calculations exactly. This can be done, for example, if the coefficients are given as rational numbers. In such a case a program could be written to solve the problem in rational (integer) arithmetic. In general this results in a greatly increased running time and a greatly increased storage-locations requirement. More generally, however, such problems are solved by using floating-point arithmetic on a computer. Usually this means that in the course of the computation round-off errors will occur. Because of this the algorithms used may produce results which differ considerably from the exact solution. A discussion of floating-point arithmetic and several examples involving matrices are given in Chapter 5.

In general the problem of finding error bounds for an approximate solution to a linear system, or for an approximate inverse of a matrix, is beyond the scope of this text. However, we can give a brief idea of some of the difficulties involved. Since our matrix-inversion algorithm is based upon solving linear systems of equations, we shall restrict our attention to the problem of solving a linear system.

Given an $n \times n$ system $AX = B$ with exact solution P and an approximate solution P_0, the problem is to find bounds on the magnitudes of the components of $P - P_0$. A natural approach is to calculate the column vector $AP_0 - B = E$. If $E = 0$, then P_0 must be the solution to the system. If $P \neq P_0$, then $E \neq 0$. Then using

$$AP = B$$
$$AP_0 = B + E$$

we get

$$A(P - P_0) = -E$$
$$P - P_0 = -A^{-1}E \tag{2-17}$$

From Eq. (2-17) the relation between $P - P_0$ and E can now be seen. The ith component of $P - P_0$ is formed by taking the negative of the ith row of A^{-1} times the column matrix E. If A^{-1} has elements of large magnitude, then the components of $P - P_0$ may have considerably larger magnitude than the magnitudes of the components of E. This will be illustrated in Example (2-15) and Exercise 36 below.

The main difficulty with Eq. (2-17) is that it involves the (usually) un-known matrix A^{-1}. Ordinarily one would hope to compute an error bound using a computer. The quantity $-A^{-1}E$ is the solution to the linear system

$$AX = -E \tag{2-18}$$

Thus, not only must the quantity $AP_0 - B = E$ be computed, but the system (2-18) must be solved. If errors occurred in solving the original linear systems $AX = B$, then one would certainly expect similar errors to occur in calculating E and solving $AX = -E$. Thus the error expression which results would be suspect in much the same manner as the solution to the original system.

EXAMPLE 2-15 Consider the system $AX = B$, where

$$A = \begin{bmatrix} 1 & 1/2 & 1/3 & 1/4 \\ 1/2 & 1/3 & 1/4 & 1/5 \\ 1/3 & 1/4 & 1/5 & 1/6 \\ 1/4 & 1/5 & 1/6 & 1/7 \end{bmatrix}$$

Suppose that an approximate solution P_0 obtained such that

$$E \equiv AP_0 - B = \begin{bmatrix} e_1 \\ e_2 \\ e_3 \\ e_4 \end{bmatrix}$$

where $|e_i| \leq 10^{-5}$ ($i = 1, 2, 3, 4$). Let P denote the exact solution and

$$P - P_0 = \begin{bmatrix} c_1 \\ c_2 \\ c_3 \\ c_4 \end{bmatrix}$$

Find bounds on $|c_i|$.

Solution. The matrix A is the Hilbert matrix of order 4. The magnitudes of the elements of the inverse of Hilbert matrices grow very rapidly with n. For the case $n = 4$

$$A^{-1} = \begin{bmatrix} 16 & -120 & 240 & -140 \\ -120 & 1,200 & -2,700 & 1,680 \\ 240 & -2,700 & 6,480 & -4,200 \\ -140 & 1,680 & -4,200 & 2,800 \end{bmatrix}$$

Thus

$$-A^{-1}E = \begin{bmatrix} -16e_1 & +120e_2 & -240e_3 & +140e_4 \\ 120e_1 & -1,200e_2 & +2,700e_3 & -1,680e_4 \\ -240e_1 & +2,700e_2 & -6,480e_3 & +4,200e_4 \\ 140e_1 & -1,680e_2 & +4,200e_3 & -2,800e_4 \end{bmatrix}$$

From this it follows that

$$|c_1| \leq 5.16 \times 10^{-3}$$
$$|c_2| \leq 5.7 \times 10^{-2}$$
$$|c_3| \leq 1.362 \times 10^{-1}$$
$$|c_4| \leq 8.82 \times 10^{-2}$$

Moreover, the maximum possible errors may actually occur. For example, if

$$E = 10^{-5} \begin{bmatrix} -1.0 \\ 1.0 \\ -1.0 \\ 1.0 \end{bmatrix}$$

then

$$P - P_0 = \begin{bmatrix} 5.16 & \times & 10^{-3} \\ -5.7 & \times & 10^{-2} \\ 1.362 & \times & 10^{-1} \\ -8.82 & \times & 10^{-2} \end{bmatrix}$$

EXERCISE 35. Determine the uniform bound on the $|e_i|$, $i = 1, \ldots, 4$ needed in Example 2-15 to guarantee $|c_i| \leq 10^{-4}$.

EXERCISE 36. Let $A = [1/(i + j - 1)]_{5 \times 5}$ be the Hilbert matrix of order 5. Then

$$A^{-1} = \begin{bmatrix} 52 & -300 & 1,050 & -1,400 & -630 \\ -300 & 4,800 & -18,900 & 26,880 & -12,600 \\ 1,050 & -18,900 & 79,380 & -117,600 & 56,700 \\ -1,400 & 26,880 & -117,600 & 179,200 & -88,200 \\ 630 & -12,600 & 56,700 & -88,200 & 44,100 \end{bmatrix}$$

(a) Repeat Example 2-15 for this matrix.
(b) Repeat Exercise 35 for this matrix.

EXERCISE 37 Repeat Exercise 36 for the matrix

$$A = \begin{bmatrix} 1 & 2 & 1 \\ 3 & 1 & -1 \\ 1 & 1 & 1 \end{bmatrix}$$

given in Example 2-12.

2.8   ESTIMATION OF COMPUTATION TIMES

It is often desirable to be able to estimate the time it will take to solve a system of linear equations or invert a matrix. For a given algorithm one can estimate (or count) the greatest possible number of additions, subtractions, multiplications, divisions, branching operations, etc., which must be performed by the computer. Then, by using the add, subtract, multiply, etc., times of the particular computer to be used, an upper bound for the computation time may be computed.

In actual practice a rough approximation is often employed. To illustrate, consider Program 2-3. It is evident that it is fairly complicated to count carefully the largest possible number of operations which may be involved in solving a problem. The indexing, double subscripting, etc., would all need to be taken into account. However, it is not too difficult to observe that, each time through the main loop for ($ip = 1$; $ip <= N$, ++ ip), somewhere between half and all the elements of the matrix A are operated upon. Because there are $2N^2$ such elements ($2N$ columns and N rows) we conclude that the main FOR loop will involve operations on between N^3 and $2N^3$

elements. Thus the total number of operations involved will be bounded by some constant times N^3.

In Ralston and Wilf (p. 55) the following approximation for the time for **matrix inversion** is discussed:

$$\text{Time} \approx \pm \mu N^3$$

Here N is the order of the system, μ the multiplication time for the computer, and 'α' is a constant dependent upon the particular computer program under consideration.

Iterative Solution of Systems of Equations

3-1 INTRODUCTION

In this chapter some of the ideas and results from Chapter 1 are extended to systems of n equations in n unknowns. Both **linear** and **nonlinear** systems of equations are considered. This first section gives examples of some of the algorithms we consider.

Many applied and theoretical problems in mathematics can be reduced to a system of simultaneous functional equations. We shall consider a system of n equations in n unknowns expressed in the form

$$f_1(x_1, x_2, \ldots, x_n) = 0$$
$$f_2(x_1, x_2, \ldots, x_n) = 0$$
$$\ldots\ldots\ldots\ldots\ldots\ldots\ldots\ldots$$
$$f_n(x_1, x_2, \ldots, x_n) = 0$$

(3-1)

The systems of linear equations studied in Chapter 2 serve as examples. A system $AX = B$ can be written in the form

$$a_{11}x_1 + a_{12}x_2 + \cdots + a_{1n}x_n - b_1 = 0$$
$$a_{21}x_1 + a_{22}x_2 + \cdots + a_{2n}x_n - b_2 = 0$$
$$\ldots\ldots\ldots\ldots\ldots\ldots\ldots\ldots\ldots\ldots\ldots$$
$$a_{n1}x_1 + a_{n2}x_2 + \cdots + a_{nn}x_n - b_n = 0$$

(3-2)

In the notation of Eqs. (3-1), then,

$$f_1(x_1, \ldots, x_n) = a_{11}x_1 + a_{12}x_2 + \cdots + a_{1n}x_n - b_1$$
$$\ldots\ldots\ldots\ldots\ldots\ldots\ldots\ldots\ldots\ldots\ldots$$
$$f_n(x_1, \ldots, x_n) = a_{n1}x_1 + a_{n2}x_2 + \cdots + a_{nn}x_n - b_n$$

An **n-tuple** of real numbers $(x_1, x_2, \ldots, x_n)$ is considered both as a point in Euclidean n space and as an n vector. That is, Euclidean n space, abbreviated E^n, may be thought of as all n-tuples of real numbers. We shall use capital letters $X = (x_1, x_2, \ldots, x_n)$, $Y = (y_1, y_2, \ldots, y_n)$ to denote points in E^n. The symbol 0 is used for the point $(0, 0, \ldots, 0)$, which we identify with the zero vector. The system (3-1) is symbolized by

$$F(X) = 0$$

Systems of nonlinear equations are often quite difficult to solve. Because of the diversity of such problems, it is not even possible to give a usable procedure for determining whether or not a general system has a solution. Each problem has to be considered individually. It will often be the case that both the existence of one or more solutions and some knowledge of their location will come from a knowledge of the origin of the problem. Thus, the person who is working with a particular problem and needs the numerical solution may well be best qualified to try to find a solution.

In general we shall assume that the problems, $F(X) = 0$, we consider have a finite number of solutions in a particular region under consideration.

Certain of the theorems we shall discuss give conditions under which a system $F(X) = 0$ will have a unique solution or a unique solution in a particular region. For convenience most of the examples we shall consider have $n = 2$. However, the techniques developed are applicable for any size n.

EXAMPLE 3-1. Give an example of a system of two simultaneous equations in two unknowns which has three distinct solutions.

Solution. There are, of course, many possible systems of this kind. One such is

$$f_1(x_1, x_2) \equiv x_1^2 + x_2^2 - 1 = 0$$
$$f_2(x_1, x_2) \equiv 2x_1^2 - x_2 - 1 = 0$$

A graph of the relation $f_1(x_1, x_2) = 0$, $f_2(x_1, x_2) = 0$ is given in Fig. 3-1.

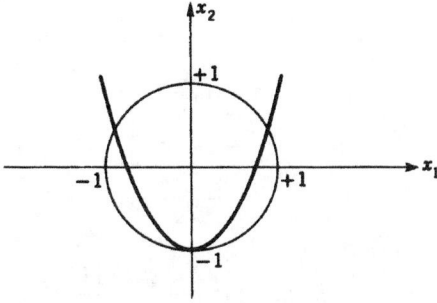

FIG. 3-1

EXERCISE 1. Give an example of a system of two simultaneous equations in two unknowns which has exactly four distinct solutions.

EXERCISE 2. Graph the relations $f_1(x_1, x_2) = 0$, $f_2(x_1, x_2) = 0$, and find all solutions to to the system where

$$f_1(x_1, x_2) = x_1^2 - x_2^2 - x_1$$
$$f_2(x_1, x_2) = x_1 - x_2 - 1/2$$

EXERCISE 3. Repeat Exercise 2 using

$$f_1(x_1, x_2) = x_1 - x_2$$
$$f_2(x_1, x_2) = x_1 \sin x_1 - x_2$$

The method of **fixed-point iteration** for a system of equations is a simple extension of the same method for a single equation. We consider a system of the form $X = G(X)$. That is,

$$x_1 = g_1(x_1, x_2, \ldots, x_n)$$
$$x_2 = g_2(x_1, x_2, \ldots, x_n)$$
$$\ldots\ldots\ldots\ldots\ldots\ldots\ldots\ldots\ldots \tag{3-3}$$
$$x_n = g_n(x_1, x_2, \ldots, x_n)$$

One of the problems we shall discuss in this chapter is that of writing a given system $F(X) = 0$ into an equivalent form $X = G(X)$. Then the algorithm given below may be applied.

ALGORITHM 3-1. Let $P_0 = (p_{10}, p_{20}, \ldots, p_{n0})$ be a first approximation to a solution of a system $X = G(X)$. Generate the sequence $\{P_m\}$ recursively by the relation

$$P_m = G(P_{m-1}) \qquad m = 1, 2, \ldots$$

We shall investigate the possible convergence of the sequence $\{P_m\}$ to a solution of $X = G(X)$.

EXAMPLE 3-2. Apply Algorithm 3 – 1 to the system given below, with $P_0 = \{0, 0\}$.

$$x_1 = \frac{x_1}{4} + \frac{x_2}{8} + \frac{11}{8}$$
$$x_2 = \frac{-x_1}{8} + \frac{x_2}{4} + 1$$

Here

$$g_1(x_1, x_2) = \frac{x_1}{4} + \frac{x_2}{8} + \frac{11}{8}$$
$$g_2(x_1, x_2) = \frac{-x_1}{8} + \frac{x_2}{4} + 1$$

The exact solution is given by $x_1 = 2$, $x_2 = 1$. The iteration gives

$$P_1 = [g_1(0,0), g_2(0,0)] = (11/8, 1)$$

$$P_2 = [g_1(11/8, 1), g_2(11/8, 1)] = (59/32, 69/64)$$

$$P_2 = [g_1(59/32, 69/64), g_2(59/32, 69/64)] = \left(\frac{1,009}{512}, \frac{133}{128}\right)$$

$$P_4 = \left(\frac{4,091}{2,048}, \frac{4,151}{4,096}\right)$$

Because we know the exact solution, we can compute the errors in the approximations provided by P_1, P_2, P_3, P_4. The values in Table 3-1 have been rounded to three decimal places.

TABLE 3-1

i	$|p_{1i} - 2|$	$|p_{2i} - 1|$
0	2.000	1.000
1	.875	.000
2	.094	.078
3	.006	.039
4	.002	.011

EXERCISE 4. Calculate P_1, P_2, P_3 in Example 3-2, starting with $P_0 = (1, 1)$. Compare the results with the exact solution.

EXERCISE 5. Calculate P_1, P_2, P_3 in Example 3-2, starting with $P_0 = (2, 1)$.

EXERCISE 6. Suppose that the values of the quantities

$$a_{11} \quad a_{12}$$
$$a_{21} \quad a_{22}$$
$$b_1 \quad b_2$$

are read as input and let

$$A = \begin{bmatrix} a_{11} & a_{12} \\ a_{21} & a_{22} \end{bmatrix} \qquad B = \begin{bmatrix} b_1 \\ b_2 \end{bmatrix}$$

Write a program to input A and B then calculate and output

$$X_1 \ X_2 \ X_3 \ \ldots, X_{20}$$

where

$$X_0 = 0 \quad \text{and} \quad X_{i+1} = AX_i + B$$

for $i = 0, 1, 2, \ldots, 19$
Test the program on the problem of Example 3–2.

EXERCISE 7. Show that the system

$$x_1^2 - x_2^2 - x_1 = 0$$
$$x_1 - 4x_2 \quad\quad = 0$$

is equivalent to the system

$$x_1 = x_1^2 - x_2^2$$
$$x_2 = \frac{x_1 - x_2}{3}$$

Then find P_1, P_2, P_3 by starting with $P_0 = (1/4, -1/4)$ and applying Algorithm 3-1. To what point does the iteration seem to be converging?

Algorithm 3-1 describes a procedure where, given

$$P_0 = (p_{10}, p_{20}, \ldots, p_{n0})$$

we compute $P_m = (p_{1m}, p_{2m}, \ldots, p_{nm})$ by $P_m = G(P_{m-1})$. Thus

$$p_{1,m} = g_1(p_{1,m-1}, p_{2,m-1}, \ldots, p_{n,m-1})$$
$$p_{2,m} = g_2(p_{1,m-1}, p_{2,m-1}, \ldots, p_{n,m-1})$$
$$\cdots\cdots\cdots\cdots\cdots\cdots\cdots\cdots\cdots\cdots\cdots$$
$$p_{n,m} = g_n(p_{1,m-1}, p_{2,m-1}, \ldots, p_{n,m-1})$$

Normally one would expect that $p_{1m}, p_{2m}, \ldots, p_{nm}$ would be calculated sequentially. Thus, at the time p_{2m} is being calculated, the value of p_{1m} is known. If the iterative procedure is converging, one would expect that p_{1m} is closer to the first component of the solution vector than is $p_{1,m-1}$. Thus it would seem natural to compute p_{2m} by the formula

$$p_{2m} = g_2(p_{1m}, p_{2,m-1}, p_{3,m-1}, \ldots, p_{n,m-1})$$

Repetition of this reasoning leads to the following formulas for $P_m = (p_{1m}, p_{2m}, \ldots, p_{nm})$:

$$p_{1,m} = g_1(p_{1,m-1}, p_{2,m-1}, \ldots, p_{n,m-1})$$
$$p_{2,m} = g_2(p_{1,m}, p_{2,m-1}, \ldots, p_{n,m-1})$$
$$p_{3,m} = g_3(p_{1,m}, p_{2,m}, p_{3,m-1}, \ldots, p_{n,m-1})$$
$$\cdots\cdots\cdots\cdots\cdots\cdots\cdots\cdots\cdots\cdots\cdots$$
$$p_{n,m} = g_n(p_{1,m}, p_{2,m}, p_{3,m}, \ldots, p_{n-1,m}, p_{n,m-1})$$

That is, $p_{1,m}$ is computed by using $p_{1,m}, p_{2,m} \cdots p_{i-1,m}$ and $p_{i,m-1}, p_{i+1,m-1}, \ldots, p_{n,m-1}$. This modification of Algorithm 3-1 is commonly called **Seidel's method of iteration.**

EXAMPLE 3-3. Apply the Seidel method of iteration to Example 3-2. Take $P_0 = (0, 0)$.

Solution. Here

$$x_1 = \frac{x_1}{4} + \frac{x_2}{8} + \frac{11}{8}$$

$$x_2 = \frac{-x_1}{8} + \frac{x_2}{4} + 1$$

Thus

$$p_{1,m} = \frac{p_{1,m-1}}{4} + \frac{p_{2,m-1}}{8} + \frac{11}{8}$$

$$p_{2,m} = -\frac{p_{1,m}}{8} + \frac{p_{2,m-1}}{4} + 1$$

The Seidel iteration gives

$m = 1$:

$$p_{1,1} = 11/8$$

$$p_{2,1} = \frac{-(11/8)}{8} + 0 + 1 = \frac{53}{64}$$

or $P_1 = (11/8,\ 53/64)$

$m = 2$:

$$p_{1,2} = \frac{11/8}{4} + \frac{53/64}{8} + \frac{11}{8} = \frac{933}{512}$$

$$p_{2,2} = -\frac{933}{512 \times 8} + \frac{53}{64 \times 4} + 1 = \frac{4,011}{4,096}$$

Using the approximation $P_2 = (1.8223, .97925)$ we continue.

$m = 3$:

$$p_{1,3} = \frac{1.8223}{4} + \frac{.97925}{8} + \frac{11}{8} = 1.9530$$

$$p_{2,3} = -\frac{1.9530}{8} + \frac{.97925}{4} + 1 = 1.00068$$

or $P_3 = (1.9530, 1.0007)$. A similar calculation gives $P_4 = (1.9883, 1.0016)$. An error table (Table 3-2) similar to that of Example 3-2 is given.

TABLE 3-2

i	$|p_{1i} - 2|$	$|p_{2i} - 1|$
0	2.000	1.000
1	0.875	.172
2	.178	.021
3	.147	.001
4	.012	.002

The **rate of convergence** in the above example seems to be about the same as for the ordinary fixed-point iteration. In many problems of general interest it turns out that the Seidel iteration converges somewhat more

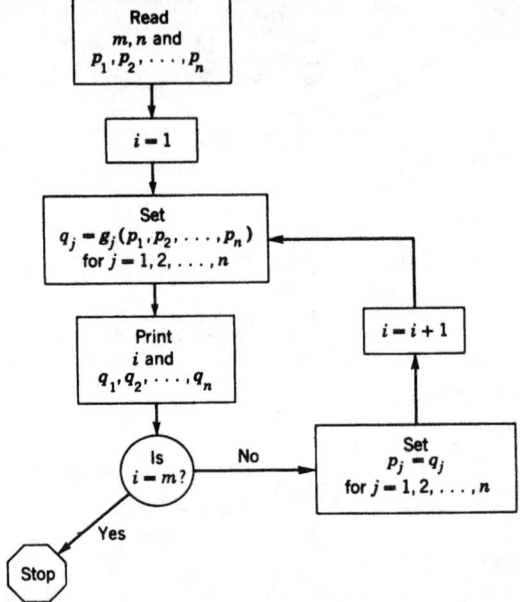

FIG. 3-2   Flow chart for Algorithm 3-1.

rapidly than the ordinary iteration. However, there are several additional reasons for using the Seidel iteration.

The Seidel iteration is somewhat easier to implement on a computer than Algorithm 3-1. This can be seen by studying the flow charts (Figs. 3-2 and 3-3). In addition, the Seidel iteration requires less storage space.

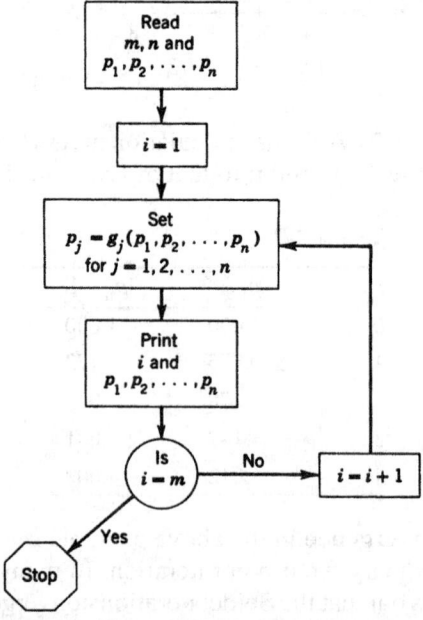

FIG. 3-3   Flow chart for Seidel iteration.

In many problems if Algorithm 3-1 produces a convergent sequence then so does the Seidel iteration, and vice versa. However, there are cases where one of the methods produces a convergent sequence and the other does not. Finally, under appropriate conditions the Seidel iteration can be modified to give a more rapidly convergent process. Extensive use is made of the Seidel iteration with relaxation in the numerical solution of partial differential equations (see Varga).

EXERCISE 8. Calculate P_1, P_2 in Example 3-2, starting with $P_0 = (1, 1)$. Use the Seidel method of iteration.

EXERCISE 9. Find P_1, P_2, P_3, using $P_0 = (1/4, -1/4)$ and the Seidel iteration in the problem of Exercise 7. Compare your results with those obtained in Exercise 7.

3-2   FUNCTIONS OF n VARIABLES

In this section, we shall give a brief discussion of some basic results concerning functions of n variables where $n > 1$. Such topics as limit of a sequence, limit of a function at a point, and continuity are included. The discussion of limit of a sequence is needed for a proper understanding of Sec. 3-3. The material on continuity is needed in Sec. 3-4. In addition, much of Chapter 7 assumes a knowledge of this material.

We shall restrict our attention to systems $F(X) = 0$, which are written in the

$$f_1(x_1, x_2, \ldots, x_n) = 0$$
$$f_2(x_1, x_2, \ldots, x_n) = 0$$
$$\ldots\ldots\ldots\ldots\ldots\ldots\ldots\ldots \tag{3-4}$$
$$f_n(x_1, x_2, \ldots, x_n) = 0$$

A shorthand notation for Eqs. (3-4) is given by

$$f_1(X) = 0$$
$$f_2(X) = 0$$
$$\ldots\ldots\ldots\ldots$$
$$f_n(X) = 0$$

The function $F(X)$ has domain and range in E^n. Each of the functions $f_i(X)$ has domain in E^n and range in E^1. It is common to discuss functions $F(X)$ in terms of their component functions $f_i(X)$. We shall define continuity and limit for such functions.

Recall from Chapter 1 that the continuity of a function at a point was defined by using the concept of limit of the function at a point. We must therefore define

$$\lim_{X \to P} f(X).$$

for a real-valued function $f(X)$ defined on a domain in E^n. To do this, we must know what it means to have $X \rightarrow P$. Thus, we must first define distance between points in E^n. Let

$$X = (x_1, x_2, \ldots, x_n)$$
$$Y = (y_1, y_2, \ldots, y_n)$$

DEFINITION 3-1. The distance between X and Y is denoted by $\|X - Y\|$ and is given by

$$\|X - Y\| = [(x_1 - y_1)^2 + (x_2 - y_2)^2 + \cdots + (x_n - y_n)^2]^{1/2}$$

The notation $\|X - Y\|$ is read as "The norm of X minus Y."

The norm just defined is the Euclidean norm, or Euclidean distance function. It should be familiar to the reader for the cases $n = 2, 3$.

EXAMPLE 3-4. If $X = (1, -2, 3, 6)$ and $Y = (-2, 1, 0, 1)$, find $\|X - Y\|$ and $\|Y - X\|$.

Solution.

$$\|X - Y\| = \left[(1+2)^2 + (-2+1)^2 + (3)^2 + (6-1)^2\right]^{1/2}$$
$$= (52)^{1/2}$$
$$\|Y - X\| = \left[(-2-1)^2 + (1+2)^2 + (3)^2 + (1-6)^2\right]^{1/2}$$
$$= (52)^{1/2}$$

EXERCISE 10.

(a) If X, Y are arbitrary vectors in E^n, prove that

$$\|X - Y\| = \|Y - X\|$$

(b) Prove that $\|X - Y\| = 0$ if and only if $X = Y$.

The notation $\|X - Y\|$ is similar to the absolute-value notation. In case $n = 1$, the norm reduces to just the absolute value. In writing $X - Y$ we are thinking of X and Y as vectors. Thus $X - Y$ is the vector with components $(x_1 - y_1, x_2 - y_2, \ldots, x_n - y_n)$. The notation $\|X\|$ is shorthand for

$$\|X - 0\| = \left(x_1^2 + x_2^2 + \cdots + x_n^2\right)^{1/2} = \|X\|$$

Once distance has been defined, we can define what is meant by a **neighborhood of a point**.

DEFINITION 3-2. A neighborhood of the point P, with radius δ, is the set

$$\|X - P\| < \delta$$

DEFINITION 3-3. A **deleted neighborhood** of the point P, with radius δ, is the set of points X such that

$$0 < \|X - P\| < \delta$$

EXAMPLE 3-5. Draw the graph of the neighborhood of $P = (1, 2)$ of radius $\delta = \frac{1}{2}$ in E^2.

Solution. This neighborhood is the set of points $X = (x_1, x_2)$ such that

$$\left[(x_1 - 1)^2 + (x_2 - 2)^2 \right]^{1/2} < 1/2$$

Thus

$$(x_1 - 1)^2 + (x_2 - 2)^2 < (1/2)^2$$

The relation $(x_1 - 1)^2 + (x_2 - 2)^2 = (1/2)^2$ has as solution a set of points forming a circle in E^2, with center at $(1, 2)$ and radius $1/2$. The desired neighborhood is the interior of this circle. (See Fig. 3-4.)

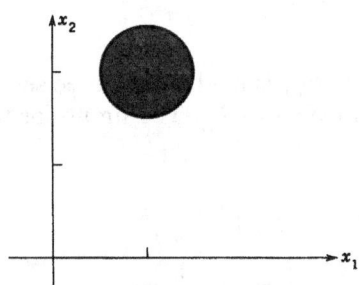

FIG. 3-4

EXERCISE 11. Find $\|X\|$, $\|Y\|$, $\|X + Y\|$, $\|X - Y\|$, where

$$X = (-1, -1, 2, 0) \quad \text{and} \quad Y = (0, -2, 4, -7)$$

EXERCISE 12. Draw the graph of each of the following sets in E^2:

(a) $\left\{ X : \|X\| = 2 \right\}$

(b) $\left\{ X : \|X - P\| = 1 \right\}$, where $P = (-2, 1)$

(c) $\left\{ X : 0 < \|X\| < 1/2 \right\}$

EXERCISE 13. In E^n let S be a neighborhood of P, with radius r_1, and let T be a neighborhood of P, with radius r_2. Suppose that $r_1 < r_2$. Prove that each point of S is a point of T.

EXERCISE 14. Describe the following sets:

(a) $\left\{X \in E^3 : \|X\| \leq 1\right\}$

(b) $\left\{X \in E^2 : 1 \leq \|X\| \leq 2\right\}$

(c) $\left\{X \in E^4 : \|X\| \leq 1\right\}$

We can now define limit of a sequence, limit of a function at a point, and continuity. The definitions are highly similar to those given in Chapter 1.

DEFINITION 3-4. We write

$$\lim_{m \to \infty} P_m = P$$

to mean that for each $\varepsilon > 0$ there exists an integer M (which often depends on ε) such that

$$\|P_m - P\| < \varepsilon \text{ whenever } m > M.$$

EXAMPLE 3-6. Suppose that $n = 3$ and

$$P_m = \left[1 - \frac{1}{m}, \ 2 + \frac{1}{m}, \ \cos\frac{1}{m}\right]$$

Find $\lim_{m \to \infty} P_m$.

Solution. The sequence $\{P_m\}$ is made up of three sequences of real numbers, one for each component of P_m. The limits for these sequences are easily determined.

$$\lim_{m \to \infty}\left(1 - \frac{1}{m}\right) = 1$$

$$\lim_{m \to \infty}\left(2 + \frac{1}{m}\right) = 2 \qquad (3\text{-}5)$$

$$\lim_{m \to \infty}\cos\frac{1}{m} = 1$$

Thus, we expect that the limit for the sequence will exist and will be $P = (1, 2, 1)$. It remains to prove this.

Let $\varepsilon > 0$ be arbitrary. Then for the three limits in Eqs. (3-5) there exist integers M_1, M_2, M_3, respectively, such that

$$\left|\left(1 - \frac{1}{m}\right) - 1\right| < \frac{\varepsilon}{3^{1/2}} \qquad \text{if } m > M_1$$

$$\left|\left(2 + \frac{1}{m}\right) - 2\right| < \frac{\varepsilon}{3^{1/2}} \qquad \text{if } m > M_2 \qquad (3\text{-}6)$$

$$\left|\cos\frac{1}{m} - 1\right| < \frac{\varepsilon}{3^{1/2}} \qquad \text{if } m > M_3$$

Set M equal to the largest of the three integers M_1, M_2, M_3. Then, if $m > M$, all the inequalities (3-6) are satisfied. Hence

$$\|P_m - P\| < \left[\left(\frac{\varepsilon}{3^{1/2}}\right)^2 + \left(\frac{\varepsilon}{3^{1/2}}\right)^2 + \left(\frac{\varepsilon}{3^{1/2}}\right)^2\right]^{1/2} = \varepsilon$$

Thus, Definition 3-4 is satisfied.

EXERCISE 15. Let $P_m = [(m-1)/m, \sin(1/m)]$. Find $\lim\limits_{m\to\infty} p_m$, and prove that your result is correct.

In the treatment of Example 3-6 it was seen that the sequence $\{P_m\}$ could be considered as n sequences of real numbers. We shall establish this as a general result. Let

$$p_m = (p_{1m}, p_{2m}, \ldots, p_{nm})$$
$$p = (p_1, p_2, \ldots, p_n)$$

THEOREM 3-2-1. $\lim\limits_{m\to\infty} P_m = P$ if and only if, for each i, $1 \le i \le n$,

$$\lim\limits_{m\to\infty} p_{im} = p_i$$

Proof. The proof consists of two distinct part. Suppose first that

$$\lim\limits_{m\to\infty} p_{im} = p_i \qquad i = 1, \ldots, n$$

Let $\varepsilon > 0$ be arbitrary. Then there exist integers $M_1, M_2, \ldots, M_n$ such that, if $m > M_i$, then

$$|p_{im} - p_i| < \frac{\varepsilon}{n^{1/2}} \qquad i = 1, \ldots, n$$

Hence, if M is the largest of $M_1, M_2, \ldots, M_n$ and $m > M$,

$$\|P_m - P\| = [(p_{1m} - p_1)^2 + \cdots + (p_{nm} - p_n)^2]^{1/2}$$

$$< \left[\left(\frac{\varepsilon}{n^{1/2}}\right)^2 + \left(\frac{\varepsilon}{n^{1/2}}\right)^2 + \cdots + \left(\frac{\varepsilon}{n^{1/2}}\right)^2\right]^{1/2} = \varepsilon$$

The second half of the proof consists in showing that if

$$\lim\limits_{m\to\infty} P_m = P$$

then

$$\lim\limits_{m\to\infty} p_{im} = p_i \qquad i = 1, \ldots, n$$

Let $\varepsilon > 0$ be arbitrary. Then, by definition, there exists an M such that if $m > M$ then

$$\|P_m - P\| = [(p_{1m} - p_1)^2 + \cdots + (p_{nm} - p_n)^2]^{1/2} < \varepsilon$$

It follows immediately that if $m > M$ them

$$\left[(p_{im} - p_i)^2\right]^{1/2} < \varepsilon \qquad i = 1, \ldots, n$$

That is, $m > M$ implies that

$$|p_{im} - p_i| < \varepsilon \qquad i = 1, \ldots, n$$

Because that is the definition of $\lim\limits_{m \to \infty} p_{im} = p_i$, the proof is complete.

EXAMPLE 3-7. Let

$$p_{10} = 1/4, p_{20} = -1/4, \text{ and}$$

$$p_{1,i+1} = p_{1i}^{\,2} - p_{2i}^{\,2}$$

$$p_{2,i+1} = \frac{p_{1i} - p_{2i}}{3}$$

(See Exercise 7.) Prove that

$$\lim_{i \to \infty} p_{1i} = 0$$

$$\lim_{i \to \infty} p_{2i} = 0$$

Solution. The following inequalities hold:

$$|p_{1,i+1}| \le |p_{1i}|^2 + |p_{2i}|^2$$

$$|p_{2,i+1}| \le \frac{|p_{1i}| + |p_{2i}|}{3}$$

In particular, using the given values of p_{10}, p_{20}, it follows that

$$|p_{11}| \le 1/8$$
$$|p_{21}| \le 1/6$$
$$|p_{12}| \le (1/8)^2 + (1/6)^2 = 25/576$$
$$|p_{22}| \le \frac{1/8 + 1/6}{3} = \frac{7}{72}$$

While it may now be fairly clear that the sequences converge to zero, we still need a formal proof.

To give a formal proof, note that by using a weaker inequality on $|p_{11}|$ we obtain

$$|p_{11}| \le 1/6$$
$$|p_{21}| \le 1/6$$

Then, using these inequalities,

$$|p_{12}| \le (2)(1/6)^2 = 1/18$$
$$|p_{22}| \le (2/3)(1/6) = 1/9$$

As before, we use a weaker inequality on $|p_{12}|$ to write

$$|p_{12}| \le 1/9$$
$$|p_{22}| \le 1/9$$

A pattern now emerges. By induction one can prove that

$$|p_{1,i+1}| < |p_{2i}|$$

and

$$|p_{2,i+1}| \le \frac{2|p_{2i}|}{3}$$

From this last inequality it follows that

$$|p_{2i} - 0| \le (2/3)^i |p_{20}|$$

and hence both sequences converge to zero. That details are left as Exercise 17.

EXERCISE 16. Using Theorem 3-2-1, find $\lim\limits_{m \to \infty} P_m$, where

$$P_m = \left(\frac{1 + \cos m}{m}, \frac{m^2 + 1}{m^2 - 2}, \frac{\sin m^2}{m} \right)$$

EXERCISE 17. Complete the details of Example 3-7.

EXERCISE 18. Prove that, if $p_{10} = \frac{1}{2}$, $p_{20} = \frac{1}{2}$ in Example 3-7, convergence to $(0,0)$ occurs.

EXERCISE 19. Let $P_0 = (p_{10}, p_{20}) = (1, 1)$. Prove that the fixed-point algorithm for the system

$$x_1 = \frac{x_1 - x_2}{4}$$

$$x_2 = \frac{x_1 + x_2}{6}$$

converges to $P = (0, 0)$. *Hint:* Show that

$$|p_{1,i+1}| + |p_{2,i+1}| \le 5/12 (|p_{1i}| + |p_{2i}|)$$

Next we define what is meant by the limit of a function $f(X)$ at a point $P \in E^n$.

DEFINITION 3-5. Suppose that $f(x)$ is a real-valued function defined on a domain D in E^n and that every **deleted neighborhood** of P contains points of D. Then

$$\lim_{X \to P} f(X) = v$$

if for each $\varepsilon > 0$ there exists a $\delta > 0$ such that

$$|f(X) - v| < \varepsilon$$

whenever $0 < \| X - P \| < \delta$ and $X \in D$.

EXAMPLE 3-8. Let $f(X) = x_1^2 - x_2^2 + 2$ and $P = (0, 0)$. Find $\lim\limits_{X \to P} f(X)$.

Solution. In this case, it is relatively easy to guess that 2 is the value of the limit. We shall show that Definition 3-5 is satisfied. Let $\varepsilon > 0$ be arbitrary. Then we must examine

$$|f(X)-2| = |x_1^2 - x_2^2|$$

for small values of $\|X - P\|$. In this case

$$\|X - P\| = (x_1^2 + x_2^2)^{1/2}$$

Now

$$|x_1^2 - x_2^2| \leq |x_1^2 + x_2^2|$$

Hence, let $\delta = \varepsilon^{1/2}$. Then, if $\|X - P\| = (x_1^2 + x_2^2)^{1/2} < \delta$, it follows that

$$|f(X)-2| = |x_1^2 + x_2^2| \leq |x_1^2 + x_2^2| < \delta^2 = \varepsilon$$

Finally we define continuity.

DEFINITION 3-6. Let $f(X)$ be a real-valued function defined on a domain D contained in E^n. Then $f(X)$ is continuous at a point $P \in D$ if

$$\lim_{X \to P} f(X) = f(P)$$

If $f(X)$ is continuous at each point $P \in D$, then $f(X)$ is said to be continuous on D.

Example 3-8 illustrates a continuous function. The proposed limit was just $f(0, 0) = 2$, and we proved that

$$\lim_{X \to (0,0)} f(X) = f(0, \ 0)$$

EXAMPLE 3-9. Prove that $f(X) = 2x_1 - 3x_2 + 6$ is continuous at $P = (0, -4)$.

Solution. The function $f(X)$ is defined for all $X \in E^2$.

$$|f(X) - f(P)| = |2(x_1 - 0) - 3(x_2 + 4)|$$

We must show that given $\varepsilon > 0$ arbitrary there exists a $\delta > 0$ such that if

$$\left[(x_1 - 0)^2 + (x_2 + 4)^2 \right]^{1/2} < \delta$$

then

$$|2(x_1 - 0) - 3(x_2 + 4)| < \varepsilon$$

Note that

$$|2(x_1 - 0) - 3(x_2 + 4)| \leq 2|x_1 - 0| + 3|x_2 + 4|$$

Let $\delta = \varepsilon/5$. Then $\|X - P\| < \delta$ implies that

$$\left[(x_1 - 0)^2 \right]^{1/2} < \frac{\varepsilon}{5} \qquad \left[(x_2 + 4)^2 \right]^{1/4} < \frac{\varepsilon}{5}$$

so that

$$2|x_1 - 0| + 3|x_2 + 4| < \varepsilon$$

This completes the proof.

EXERCISE 20. Let $f(x_1, x_2, x_3) = x_1 - 2x_2 - x_3 + 5$. Prove that $f(X)$ is continuous at $P = (0, -1, 2)$.

EXERCISE 21. Let $f(x_1, x_2, x_3) = ax_1 + bx_2 + cx_3 + d$. Let $P = (p_1, p_2, p_3)$ be arbitrary. Prove that $f(X)$ is continuous at P for all choices of a, b, c, d.

Usually a considerable amount of work is required to prove that a function $f(X)$ is continuous at a point P by use of the definition of continuity. In working with functions of one real variable, appeal is made to the fact that a differentiable function is continuous. For functions of more than one variable, no such simple result exists. However, various theorems can be given which aid in proving a function continuous. As for functions of one variable, it is not difficult to prove that the sum, difference, and product of continuous functions is continuous. The quotient of continuous functions is continuous at points where the denominator is non-zero. These results are left as exercises. The following theorem is one giving a sufficient condition for continuity in terms of partial derivatives. It can often be used to prove that a function is continuous on a particular domain.

THEOREM 3-2-2. Suppose that $f(X)$ is defined in some neighborhood of a point P. If there exists a constant $c < \infty$ such that

$$\left| \frac{\partial f}{\partial x_i} \right| \le c$$

for $i = 1, \ldots, n$ and for all X in this neighborhood, then $f(X)$ is continuous at P.

Proof. For illustrative purposes, we shall give a proof for the case $n = 2$. The methods used can be extended to the more general case. The proof is based upon the **mean-value theorem** for functions of one variable.

Let $Y \ne P$ denote an arbitrary point in the given neighborhood of P, and let

$$P = (p_1, p_2)$$
$$Y = (y_1, y_2)$$

(See Fig. 3-5). Observe that the point $Q = (y_1, p_2)$ is also in the given neighborhood, as are the lines joining P to Q and Q to Y.

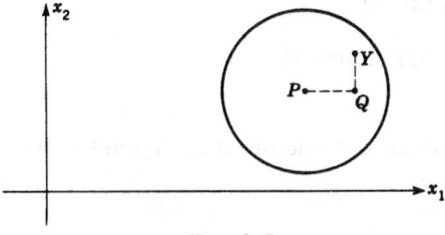

FIG. 3-5

To prove continuity, we must show that $|f(P) - f(Y)|$ is arbitrarily small provided that $\|Y - P\|$ is small enough.

$$|f(P)-f(Y)|=|f(p_1,p_2)-f(y_1,y_2)|$$
$$=|f(p_1,p_2)-f(y_1,p_2)+f(y_1,p_2)-f(y_1,y_2)|$$
$$\leq|f(p_1,p_2)-f(y_1,p_2)|+|f(y_1,p_2)-f(y_1,y_2)|$$

Now consider the two functions $f(x_1,p_2)$ and $f(y_1,x_2)$. If p_2 and y_1 are held fixed, each is a function of one variable. Consider the first function, for example. The function $f(x_1,p_2)$ is defined on the interval with end points (p_1,p_2) and (y_1,p_2). Since

$$\frac{\partial f}{\partial x_1}(x_1,p_2)$$

exists at each point of this interval, the mean-value theorem is applicable.

Hence there exists a point t between p_1 and y_1 such that

$$|f(p_1,p_2)-f(y_1,p_2)|=|p_1-y_1|\left|\frac{\partial f(t,p_2)}{\partial x_1}\right|\leq|p_1-y_1|c$$

By using a similar argument on the function $f(y_1,x_2)$ and the given bound on the partial derivatives, it follows that

$$|f(p_1,p_2)-f(y_1,p_2)|+|f(y_1,p_2)-f(y_1,y_2)|$$
$$\leq|p_1-y_1|c+|p_2-y_2|c=c(|p_1-y_1|+|p_2-y_2|)$$

Finally, let $\varepsilon>0$ be arbitrary and δ be the smaller of $\varepsilon/2c$ and the radius of the given neighborhood. (If $c=0$, then δ is the radius of the given neighborhood.) Observe that

$$|p_1-y_1|\leq\left[(p_1-y_1)^2+(p_2-y_2)^2\right]^{1/4}=\|P-Y\|$$
$$|p_2-y_2|\leq\left[(p_1-y_1)^2+(p_2-y_2)^2\right]^{1/2}=\|P-Y\|$$

Hence

$$c(|p_1-y_1|+|p_2-y_2|)\leq 2c\|P-Y\|$$

Thus, if $\|P-Y\|<\delta$, then

$$|f(P)-f(Y)|<\varepsilon$$

and the proof is complete.

EXAMPLE 3-10. Prove that the function

$$f(x_1,x_2,x_3)=ax_1+bx_2+cx_3+d$$

for a, b, c, d arbitrary, is continuous at each point E^3 (see Exercise 21).

Solution.

$$\frac{\partial f}{\partial x_1}=a \qquad \frac{\partial f}{\partial x_2}=b \qquad \frac{\partial f}{\partial x_3}=c$$

Let t be the largest of $|a|$, $|b|$, $|c|$. Then, for *any* neighborhood of any point $P \in E^3$, it follows that

$$\left|\frac{\partial f}{\partial x_i}\right| \le t \quad \text{for } i = 1, 2, 3$$

for all X in the neighborhood. Thus, by Theorem 3-2-2, the function is continuous at each point.

Example 3-11. Prove that

$$f(x_1, x_2) = x_1^2 - 2x_1 x_2 + 4x_2^2 - 7$$

is continuous at each point of E^2.

Solution. Let P be an arbitrary point in E^2.

$$\frac{\partial f}{\partial x_1} = 2x_1 - 2x_2$$

$$\frac{\partial f}{\partial x_2} = -2x_1 - 8x_2$$

Consider a neighborhood of P with radius 1. If $P = (p_1, p_2)$, then it can be seen that in this neighborhood

$$\left|\frac{\partial f}{\partial x_1}\right| \le 2(|p_1| + 1) + 2(|p_2| + 1)$$

$$\left|\frac{\partial f}{\partial x_2}\right| \le 2(|p_1| + 1) + 8(|p_2| + 1)$$

Thus, with $c = 10 + 2|p_1| + 8|p_2|$, the hypotheses of Theorem 3-2-2 are satisfied.

Exercise 22. Prove that the function

$$f(x_1, x_2, x_3) = x_1^2 + x_2^2 + x_3^2$$

is continuous at each point of E^3.

Exercise 23. Prove that the function

$$f(x_1, x_2, \ldots, x_n) = b_1 x_1 + b_2 x_2 + \cdots + b_n x_n + b_{n+1}$$

is continuous for arbitrary $b_1, b_2, \ldots, b_{n+1}$ at each point of E^n.

We conclude this section with a definition of continuity for a function $F(X)$ with domain and range contained in E^n. Our definition is based upon the component functions $f_1(X), f_2(X), \ldots, f_n(X)$.

Definition 3-7. Let $F(X)$ have domain D, and suppose that $P \in D$. Then $F(X)$ is continuous at P if each of the functions $f_1(X), f_2(X), \ldots, f_n(X)$ is continuous at P.

EXERCISE 24. Prove that the function given below is continuous for all $X \in E^3$:

$$f_1(x_1, x_2, x_3) = 2x_1 x_2 \sin x_3$$
$$f_2(x_1, x_2, x_3) = x_1 - x_2$$
$$f_3(x_1, x_2, x_3) = 3x_1 x_2 x_3$$

EXERCISE 25. Make up an example of a continuous function whose domain is E^2 and whose range is contained in the first quadrant of E^2.

3-3   ITERATION FOR LINEAR SYSTEMS

The theory underlying the fixed-point Algorithm 3-1 is considerably more involved than for the corresponding algorithm in E^1. Even for linear systems, proofs of some of the basic results are beyond the scope of this text. These results, however, are easily applied and are valuable tools of the numerical analyst. In this section we shall state, discuss, and illustrate various results concerning the iterative solution of a linear system. Section 3-4 contains a corresponding treatment for **nonlinear systems**. Proofs of the basic results given in this section may be found in Faddeev and Faddeeva.

The reader may wonder why the study of iterative procedures for solving linear systems is important. The methods of Chapter 2 are often used to solve systems of 100 or more simultaneous linear equations. However, with iterative procedures, it is not uncommon to solve certain special systems of 1,000 to 10,000 and even more equations. It is not difficult to show that such problems are beyond the scope of a general elimination procedure such as was discussed in Chapter 2. The number of computational operations involved in the elimination procedure varies as n^3, where n is the order of the system. Thus a system of 10,000 equations would require 1,000,000 times as many computations as a system of 100 equations. Even the fastest computers in use today can require as much as a second to solve 100 simultaneous linear equations by elimination. Because 1,000,000 seconds is approximately 11 1/2 days, the times involved would be unreasonably large; not to mention the truncation errors introduced throughout the operations.

One standard source of large systems of simultaneous linear equations is in the numerical solution of partial differential equations. Such systems of linear equations have the property that most of the coefficients are zero. There are two standard approaches to solving such systems. One approach is by means of special-purpose elimination algorithms which are designed to take into account the structure of the equations and the fact that most coefficients are zero. The second approach is by fixed-point iteration. The time to perform one iteration is greatly shortened if the presence of the zero coefficients is properly taken into account. In addition, usually most of the equations are generated by some very simple rule. This allows the

equations to be generated as needed, rather than to be stored in the computer memory throughout the computation. Thus large systems can be solved using relatively little memory space. An even greater savings in storage occurs if the Seidel iteration is used.

Iterative procedures have one additional advantage. In physical problems one is often content to obtain an answer correct to two or three significant figures. The accuracy obtained in a convergent iterative procedure is governed by the number of iterations performed. If less accuracy is desired, the result is a direct savings in the number of iterations needed, and hence in the computational time involved. This is not true for an elimination procedure.

To begin with, suppose that one has a linear system in the form $X = G(X)$. In matrix notation, such a system can be written as

$$X = CX + D$$

where $C = (c_{ij})_{n \times n}$, $D = (d_i)_{n \times 1}$.

DEFINITION 3-8. The **row norm** of a square matrix C is the number

$$\max_{1 \leq i \leq n} \sum_{j=1}^{n} |c_{ij}|$$

In is denoted by $\|C\|_r$.

DEFINITION 3-9. The **column norm** of a square matrix C is the number

$$\max_{1 \leq j \leq n} \sum_{i=1}^{n} |c_{ij}|$$

It is denoted by $\|C\|_c$.

EXAMPLE 3-12. Find the row and column norms of the matrix

$$A = \begin{bmatrix} 1 & -1 & 3 \\ -5 & 2 & 1 \\ 1 & 3 & -1 \end{bmatrix}$$

Solution. To find the row norm, we examine the sums of the absolute values of the row elements of A. These are 5, 8, 5 respectively. The maximum of these sums is 8; so $\|A\|_r = 8$, The sums of the absolute values of the column elements are 7, 6, 5, respectively. Thus $\|A\|_c = 7$.

EXERCISE 26. Find the row norm and column norm of each of the matrices below:

$$A = \begin{bmatrix} 1 & 0 & 0 \\ 0 & 1 & 0 \\ 0 & 0 & 1 \end{bmatrix} \qquad B = \begin{bmatrix} .5 & .3 & .1 \\ -.2 & -.2 & -.2 \\ -.1 & -.3 & -.5 \end{bmatrix}$$

Using row and column norms, one can give sufficient conditions for the sequence $\{P_m\}$ generated by Algorithm 3-1 to converge to a solution

of $X = CX + D$. It is important to keep in mind that these are sufficient conditions rather than necessary conditions. The sequence generated by the algorithm might converge even though the conditions of the following theorem are not satisfied.

THEOREM 3-3-1. Let k be the minimum of $\|C\|_r$ and $\|C\|_c$. If $k < 1$, then:

(a) $X = CX + D$ has a unique solution, P.
(b) For any P_0, the sequence generated by Algorithm 3-1 and the sequence generated by the Seidel iteration converge to P.

Proof. Certain aspects of the proof are beyond the scope of this text. However, we shall be able to understand some of the main ideas involved.

We begin by considering the sequence $\{P_1\}$ for the ordinary iteration

$$
\begin{aligned}
P_1 &= CP_0 + D \\
P_2 &= CP_1 + D = C(CP_0 + D) + D = C^2 P_0 + (I + C)D \\
P_3 &= CP_2 + D = C\left[C^2 P_0 + (I + C)D\right] + D \\
&= C^3 P_0 + (I + C + C^2)D
\end{aligned}
\tag{3-7}
$$

$$\cdots\cdots\cdots\cdots\cdots\cdots$$

$$P_m = C^m P_0 + (I + C + C^2 + \cdots + C^{m-1})D$$

Next let us solve the equation $X = CX + D$ using matrix notation. The result is

$$(I - C)X = D$$

$$X = (I - C)^{-1} D$$

provided that $I - C$ is nonsingular.

Finally, in studying the expressions for P_m and X we may be reminded of the result

$$\frac{1}{1-a} = (1-a)^{-1} = 1 + a + a^2 + a^3 + \cdots \qquad \text{if } |a| < 1$$

It turn out that a similar result holds for matrix. In paticular, if C is such that

$$\lim_{m \to \infty} C^m = 0 \tag{3-8}$$

then

$$(I - C)^{-1} = I + C + C^2 + C^3 + \cdots \tag{3-9}$$

A sufficient condition for Eq. (3.8) to hold is that $\|C\|_r < 1$ or $\|C\|_c < 1$ (see Faddeev and Faddeeva). Finally, upon examining the expression for P_m in the light of Eqs. (3-8) and (3-9), it follows that

$$\lim_{m \to \infty} C^m P_0 = 0$$

$$\lim_{m \to \infty} (I + C + C^2 + \cdots + C^{m-1}) D = (I - C)^{-1} D$$

so that

$$\lim_{m \to \infty} P_m = (I - C)^{-1} D$$

which is the solution to the linear system.

EXAMPLE 3-13. Discuss the solution of

$$x_1 = 0x_1 - 0.2x_2 + 0.8$$
$$x_2 = 0.25x_1 + 0x_2 - 1.25$$

by fixed-point iteration.

Solution. The system can be written as

$$\begin{bmatrix} x_1 \\ x_2 \end{bmatrix} = \begin{bmatrix} 0 & -0.2 \\ 0.25 & 0 \end{bmatrix} \begin{bmatrix} x_1 \\ x_2 \end{bmatrix} + \begin{bmatrix} 0.8 \\ -1.25 \end{bmatrix}$$

The matrix of coefficients

$$C = \begin{bmatrix} 0 & -0.2 \\ 0.25 & 0 \end{bmatrix}$$

has $\|C\|_r = 0.25$ and $\|C\|_c = 0.25$. From Theorem 3-3-1 it follows that the system has a unique solution, and this solution can be found by use of Algorithm 3-1 for any starting value P_0.

Starting with $P_0 = (0, 0)$, values of $P_1, P_2, \ldots$ for this problem are given in Table 3-3.

TABLE 3-3

i	p_{1i}	p_{2i}
1	.8000000000	−1.2500000000
2	1.0500000000	−1.0500000000
3	1.0100000000	−.9875000000
4	.9975000000	−.9975000000
5	.9995000000	−1.0006250000
6	1.0001250000	−1.0001250000
7	1.0000250000	−.9999687500
8	.9999937500	−.9999937500
9	.9999987500	−1.0000015625
10	1.0000003125	−1.0000003125
11	1.0000000625	−.9999999219
12	.9999999844	−.9999999844
13	.9999999969	−1.0000000039
14	1.0000000008	−1.0000000008
15	1.0000000002	−.9999999998
16	1.0000000000	−.9999999999
17	1.0000000000	−1.0000000000

EXAMPLE 3-14. Discuss the iterative solution of

$$x_1 = .3x_1 - .2x_2 + .1x_3 - .5$$
$$x_2 = .5x_1 + .1x_2 + .1x_3 + 1.7$$
$$x_3 = .4x_1 - .2x_2 - .2x_3 - .8$$

Solution. The matrix of coefficients has row norm equal to .8 and column norm equal to 1.2. Because one of these, namely, the row norm, is less than 1.0, the system can be solved by the fixed-point iterative procedure.

Starting with $P_0 = (0,0,0)$, values of $P_1, P_2, \ldots$ are given in Table 3-4.

TABLE 3-4

i	p_{1i}	p_{2i}	p_{3i}
1	$-.5000000000E + 00$	$.1700000000E + 01$	$-.8000000000E + 00$
2	$-.1070000000E + 01$	$.1540000000E + 01$	$-.1180000000E + 01$
3	$-.1247000000E + 01$	$.1201000000E + 01$	$-.1300000000E + 01$
4	$-.1244300000E + 01$	$.1066600000E + 01$	$-.1279000000E + 01$
5	$-.1214510000E + 01$	$.1056610000E + 01$	$-.1255240000E + 01$
6	$-.1201199000E + 01$	$.1072882000E + 01$	$-.1246078000E + 01$
7	$-.1199543900E + 01$	$.1082080900E + 01$	$-.1245840400E + 01$
8	$-.1200863390E + 01$	$.1083852100E + 01$	$-.1247065660E + 01$
9	$-.1201736003E + 01$	$.1083246949E + 01$	$-.1247702644E + 01$
10	$-.1201940455E + 01$	$.1082686429E + 01$	$-.1247803262E + 01$
11	$-.1201899749E + 01$	$.1082518089E + 01$	$-.1247752815E + 01$
12	$-.1201848824E + 01$	$.1082526653E + 01$	$-.1247712954E + 01$
13	$-.1201831273E + 01$	$.1082556958E + 01$	$-.1247702269E + 01$
14	$-.1201831000E + 01$	$.1082559832E + 01$	$-.1247703447E + 01$
15	$-.1201833611E + 01$	$.1082571138E + 01$	$-.1247705677E + 01$
16	$-.1201834879E + 01$	$.1082569740E + 01$	$-.1247706537E + 01$
17	$-.1201835065E + 01$	$.1082568881E + 01$	$-.1247706592E + 01$
18	$-.1201834955E + 01$	$.1082568696E + 01$	$-.1247706484E + 01$
19	$-.1201834874E + 01$	$.1082568744E + 01$	$-.1247706424E + 01$
20	$-.1201834853E + 01$	$.1082568795E + 01$	$-.1247706414E + 01$
21	$-.1201834856E + 01$	$.1082568811E + 01$	$-.1247706418E + 01$
22	$-.1201834861E + 01$	$.1082568811E + 01$	$-.1247706421E + 01$
23	$-.1201834863E + 01$	$.1082568809E + 01$	$-.1247706422E + 01$
24	$-.1201834863E + 01$	$.1082568807E + 01$	$-.1247706422E + 01$
25	$-.1201834862E + 01$	$.1082568807E + 01$	$-.1247706422E + 01$
26	$-.1201834862E + 01$	$.1082568807E + 01$	$-.1247706422E + 01$

The discussion of a proof of Theorem 3-3-1 contains information which allows us to find an expression for the error in stopping the iteration after m steps. We have

$$P = (I + C + C^2 + C^3 + \cdots)D$$
$$P_m = C^m P_0 + (I + C + C^2 + \cdots + C^{m-1})D$$

Thus

$$P - P_m = (C^m + C^{m+1} + C^{m+2} + \cdots)D - C^m P_0$$
$$= C^m(I + C + C^2 + \cdots)D - C^m P_0$$
$$= C^m(I - C)^{-1}D - C^m P_0$$
$$= C^m P - C^m P_0$$
$$= C^m(P - P_0)$$

From the result $P - P^m = C^m(P - P_0)$ it can be seen that the more rapidly C^m tends to zero the more rapidly P_m tends to P. In particular it can be shown that $\|P - P_m\|$ is bounded by a constant times k^m, where

$$k = \max(\|C\|_r, \|C\|_c)$$

If k is too close to 1, then convergence may be quite slow and Algorithm 3-1 is probably a poor way to attempt to solve the linear system. In such cases the previously mentioned Seidel iteration with relaxation is often used (see Varga).

EXERCISE 27. Using Theorem 3-3-1, show that the iteration in Example 3-2 converges.

EXERCISE 28. Give a careful proof of Theorem 3-3-1 for the case $n = 1$.

EXERCISE 29. If $n = 1$ and $k < 1$ in Theorem 3-3-1, find a bound on the error after m iterations in terms of k, m, and the initial error. Hint: Use results from Chapter 1.

EXERCISE 30. Suppose that Algorithm 3-1 is applied to a system $X = CX + D$, where C is a diagonal matrix. If C is $n \times n$, where $n > 1$, show that the results are similar to doing n individual problems each involving only one variable.

We shall conclude this section with a discussion of the problem of transforming a system $AX + B = 0$ into the form $X = CX + D$. This can be accomplished in a number of ways. However, unless one of the norms of the resulting matrix C is less than 1, little is gained by making such a transformation.

In some systems $AX + B = 0$, it turns out that the diagonal elements of the matrix A are much larger in absolute value than the other elements in each row. The following definition is used to describe this more precisely.

DEFINITION 3-10. A matrix A is **diagonally dominant** if

$$|a_{ii}| > \sum_{\substack{j=1 \\ j \neq i}}^{n} |a_{ij}| \qquad \text{for } i = 1, \ldots, n$$

EXAMPLE 3-15. Show that the matrix given below is diagonally dominant:

$$A = \begin{bmatrix} 5 & 1 & -2 \\ 3 & -7 & 1 \\ 2 & -1 & 4 \end{bmatrix}$$

Solution. In each row, we must show that the absolute value of the diagonal element is greater than the sum of the absolute values of the remaining elements.

Thus

$$5 > 1 + 2 = 3$$
$$7 > 3 + 1 = 4$$
$$4 > 2 + 1 = 3$$

If a matrix A is diagonally dominant, then the system $AX + B = 0$ can easily be written in the form $X = CX + D$ in a way which makes $|C|_r < 1$. The procedure is to solve the first equation for x_1, the second for x_2, the third for x_3, etc., in terms of the other variables in their respective equations.

EXAMPLE 3-16. Apply the procedure described above to the system $AX + B = 0$, where A is given in Example 3-15 and $B = (-8, 0, 3)$.

Solution. The system is given by

$$5x_1 + x_2 - 2x_3 - 8 = 0$$
$$3x_1 - 7x_2 + x_3 \quad = 0$$
$$2x_1 - x_2 + 4x_3 + 3 = 0$$

The first equation, when solved for x_1, yields

$$x_1 = \frac{-x_2}{5} + \frac{2x_3}{5} + \frac{8}{5}$$

The second equation, when solved for x_2, yields

$$x_2 = \frac{3x_1}{7} + \frac{x_3}{7}$$

The third equation, when solved for x_3, yields

$$x_3 = \frac{-x_1}{2} + \frac{x_2}{4} - \frac{3}{4}$$

Thus, we have

$$\begin{bmatrix} x_1 \\ x_2 \\ x_3 \end{bmatrix} = \begin{bmatrix} 0 & -1/5 & 2/5 \\ 3/7 & 0 & 1/7 \\ -1/2 & 1/4 & 0 \end{bmatrix} \begin{bmatrix} x_1 \\ x_2 \\ x_3 \end{bmatrix} + \begin{bmatrix} 8/5 \\ 0 \\ -3/4 \end{bmatrix}$$

It is easy to see that the row norm of the above matrix is less than 1. The above example illustrates the following important theorem.

THEOREM 3-3-2. If the matrix A of the system $AX + B = 0$ is diagonally dominant, then:

(a) The system has a unique solution.
(b) If the system is written in the form $X = CX + D$ by solving the ith equation for x_i $(i = 1, \ldots, n)$, then $\|C\|_r < 1$ (see Theorem 3-3-1).

Proof. Consider the ith equation of the system.

$$a_{i1}x_1 + a_{i2}x_2 + \cdots + a_{in}x_n + b_i = 0$$

From the definition of diagonal dominance, it follows that

$$|a_{ii}| > |a_{i1}| + \cdots + |a_{i,i-1}| + |a_{i,i+1}| + \cdots + |a_{in}| \qquad (3\text{-}10)$$

In particular, $|a_{ii}| > 0$. Thus

$$x_i = \frac{-a_{i1}}{a_{ii}} x_1 + \cdots + \frac{-a_{i,i-1}}{a_{ii}} x_{i-1} + \frac{-a_{i,i+1}}{a_{ii}} x_{i+1} + \cdots + \frac{-a_{in}}{a_{ii}} x_n - \frac{b_i}{a_{ii}}$$

This now gives the ith equation in the system $X = CX + D$. We must show that the sum of the absolute values of the elements in the ith row of C is less than 1. It is given by

$$\left| \frac{a_{i1}}{a_{ii}} \right| + \cdots + \left| \frac{a_{i,i-1}}{a_{ii}} \right| + \left| \frac{a_{i,i+1}}{a_{ii}} \right| + \cdots + \left| \frac{a_{in}}{a_{ii}} \right|$$

$$= \left| \frac{1}{a_{ii}} \right| \left(|a_{i1}| + \cdots + |a_{i,i-1}| + |a_{i,i+1}| + \cdots + |a_{in}| \right)$$

From Eq. (3-10) it is easily seen that this expression is less than 1. Thus, we have proved assertion (b). Assertion (a) follows by use of Theorem 3-3-1.

It should be evident to the reader that not all systems $AX + B = 0$ can be written in the form $X = CX + D$ with a norm of C less than 1. This is because not all systems $AX + B = 0$ have a solution. On the other hand, if $AX + B = 0$ has a unique solution, then the inverse of the matrix A exists. If we use this inverse, we can write

$$A^{-1}AX + A^{-1}B = 0$$

This simplifies to

$$X = -A^{-1}B$$

which may also be written as

$$X = 0X + (-A^{-1}B)$$

where 0 is a matrix of all zeros. Clearly $\|0\|_r = 0$, and regardless of the initial guess P_0, the exact solution is found in one step of the iterative procedure.

From the previous discussion, it follows that any system $AX + B = 0$ having a unique solution can be written in a form so that the solution can be found by fixed-point iteration. Thus, what is desired is a simple procedure (simpler than solving the system) for converting $AX + B = 0$ into the form $X = CX + D$ with a norm of C less than 1. Except in certain special cases as illustrated by diagonal dominance, no general procedure for doing this is known.

EXAMPLE 3-17. Convert the following system into the form

$$X = CX + D$$

with $\|C\|_r < 1$:

$$5x_1 - x_2 - x_3 + 7 = 0$$
$$x_1 - 2x_2 + 8x_3 + 2 = 0$$
$$-x_1 + 8x_2 - x_3 + 5 = 0$$

Solution. The matrix of coefficients is not diagonally dominant. However, if the order of the second and third equations is interchanged, the resulting system has a diagonally dominant matrix. The remainder, of the example is left as an exercise for the student.

EXERCISE 31. Complete Example 3-17.

EXERCISE 32. Rewrite the following system so that it can be solved by fixed-point iteration, and find P_1, P_2 for $P_0 = (0, 0)$:

$$x_1 + 20x_2 = 19$$
$$40x_1 - 2x_2 = -42$$

Compare the results with the exact solution $x_1 = -1$, $x_2 = 1$.

EXERCISE 33. Write a program which reads in n and the elements a_{ij} of an $n \times n$ matrix A, and determines whether or not A is diagonally dominant.

EXERCISE 34. Write a segment of a program which writes the diagonally dominant system $AX + B = 0$, where A, B are stored in the computer, into the form $X = CX + D$, using the procedure described in Theorem 3-3-2.

When the **Seidel** iteration is applied to a system of linear equations, it is called the **Gauss-Seidel** method. Suppose that a linear system is given in the form

$$\sum_{j=1}^{n} a_{ij} x_j + b_i = 0 \qquad i = 1, \ldots, n \qquad (3-11)$$

Further, suppose that the matrix $A = (a_{ij})$ is diagonally dominant. Then we know that if the ith equation is solved for x_i the Gauss-Seidel iteration will converge. The resulting algorithm is given by:

1. Make initial guess $(p_{10}, p_{20}, \ldots, p_{n0})$.
2. Compute $(p_{1m}, p_{2m}, \ldots, p_{nm})$ by

$$p_{im} = \frac{\sum_{j=1}^{i-1} a_{ij} p_{jm} + \sum_{j=i+1}^{n} a_{ij} p_{j,m-1} + b_i}{-a_{ii}} \qquad \text{for } i = 1, \ldots, n$$

In programming this procedure, it is more convenient to make a slight change in (2) above. The new p_{im} is computed by the equivalent formula

$$p_{im} = p_{i,m-1} + \frac{\sum_{j=1}^{i-1} a_{ij} p_{jm} + \sum_{j=i}^{n} a_{ij} p_{j,m-1} + b_i}{-a_{ii}}$$

As will be seen in the flow chart below (Fig. 3-6), this allows convenient usage of a loop in the summation. In addition, the quantity

$$\left| p_{im} - p_{i,m-1} \right|$$

is easily computed. The magnitude of the sum

$$\sum_{i=1}^{n} |p_{im} - p_{i,m-1}| \tag{3-12}$$

can be used in a convergence criterion. That is, the iteration is stopped when the quantity (3-12) becomes less than some preassigned $\epsilon > 0$. The choice of a proper ϵ depends upon the problem being solved, the desired accuracy, and the number of significant digits carried in the calculation.

In the flow chart for the **Gauss-Seidel** iteration the matrix of coefficients is stored in A, and the b_i's are stored in B. The notation of Eq. (3-11) is used for the system. The test for convergence is that

$$TR = \sum_{i=1}^{n} |p_{im} - p_{i,m-1}|$$

be less than a specified ϵ. In implementing this algorithm it would be wise to make provisions for the case when the iteration fails to converge or converges very slowly.

EXERCISE 35. Using the Gauss-Seidel method, find P_1, P_2, starting with $P_0 = (0, 0, 0)$ for the system

$$5x_1 + x_2 - 2x_3 + 12 = 0$$
$$3x_1 - 7x_2 + x_3 - 23 = 0$$
$$2x_1 - x_2 + 4x_3 - 15 = 0$$

EXERCISE 36. Write a program for the Gauss-Seidel iteration. Use Exercise 35 as a test problem.

3-4   NONLINEAR SYSTEMS OF EQUATIONS

In this section we shall discuss the problem of solving nonlinear systems of the form

$$X = G(X)$$

and $F(X) = 0$

by fixed-point iteration. We shall state several important theorems without proof. Unless otherwise indicated, a discussion of the general underlying theory may be found in Henrici (1964) and/or Todd.

To begin, we need to know what is meant by a **Jacobian matrix** of a function with domain and range in E^n.

DEFINITION 3-11. Suppose that a function $G(X)$ has components given by

$$g_1(X)$$
$$g_2(X)$$
$$\cdots\cdots\cdots$$
$$g_n(X)$$

and for some points P,

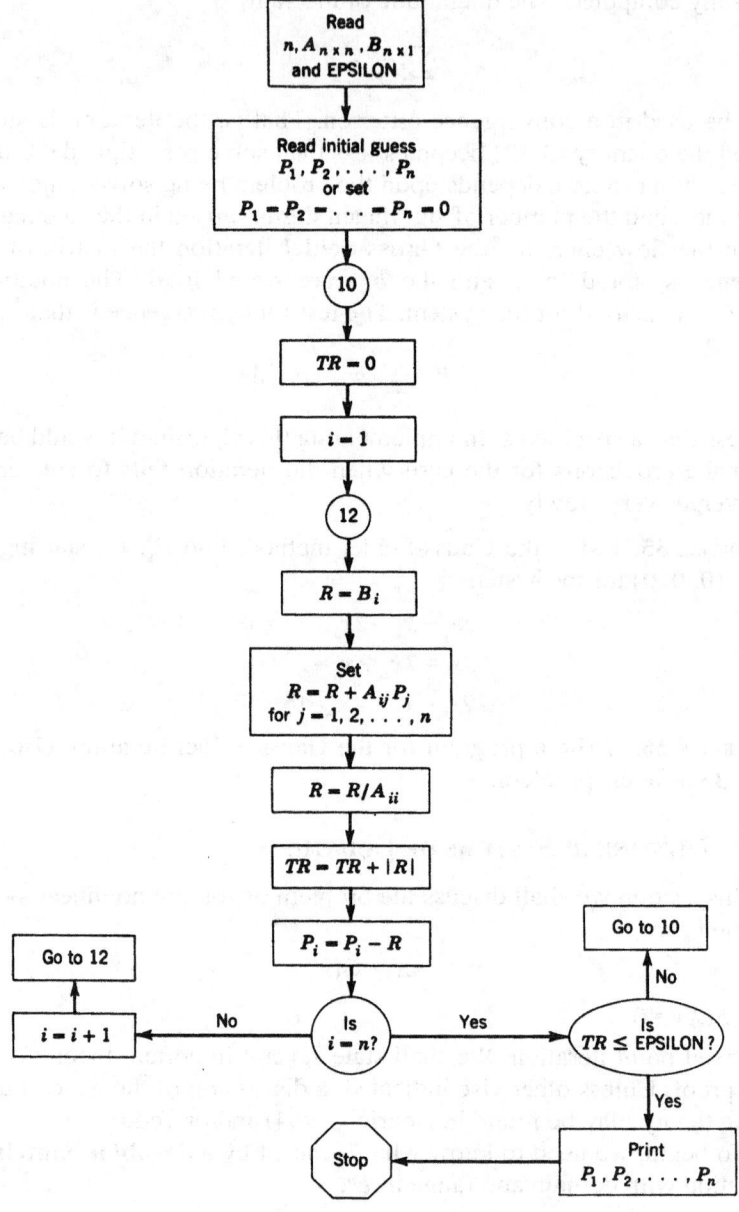

FIG. 3.6   Flow chart for Gouss-Seidel iteration

$$\left.\frac{\partial g_i}{\partial x_j}\right|_P \equiv \frac{\partial g_i(P)}{\partial x_j}$$

exists for $1 \le i, j \le n$. Then the Jacobian matrix of $G(X)$ at the point P is given by

$$J(P) = \left[\frac{\partial g_i(P)}{\partial x_j} \right]_{n \times n}$$

EXAMPLE 3-18. Find the Jacobian matrix for the funtion $G(X)$ giving below at an arbitrary point X and at the point $P = (-2, 1, 0)$.

$$g_1(x_1, x_2, x_3) = x_1^2 + x_2^2 + x_3^2 + 1$$
$$g_2(x_1, x_2, x_3) = x_1 - 4x_2 + x_3 - 2$$
$$g_3(x_1, x_2, x_3) = -x_1^2 + x_1 x_2$$

Solution.

$$\frac{\partial g_1}{\partial x_1} = 2x_1 \qquad \frac{\partial g_1}{\partial x_2} = 2x_2 \qquad \frac{\partial g_1}{\partial x_3} = 2x_3$$

$$\frac{\partial g_2}{\partial x_1} = 1 \qquad \frac{\partial g_2}{\partial x_2} = -4 \qquad \frac{\partial g_2}{\partial x_3} = 1$$

$$\frac{\partial g_3}{\partial x_1} = -2x_1 + x_2 \qquad \frac{\partial g_3}{\partial x_2} = x_1 \qquad \frac{\partial g_3}{\partial x_3} = 0$$

$$J(X) = \begin{bmatrix} 2x_1 & 2x_2 & 2x_3 \\ 1 & -4 & 1 \\ -2x_1 + x_2 & x_1 & 0 \end{bmatrix}$$

$$J(P) = \begin{bmatrix} -4 & 2 & 0 \\ 1 & -4 & 1 \\ 5 & -2 & 0 \end{bmatrix}$$

EXERCISE 37. Find the Jacobian matrix for the function $F(X)$, where

$$f_1(x_1, x_2) = \sin x_1 x_2$$
$$f_2(x_1, x_2) = x_1^2 + 2\cos x_2$$

Find $J(P)$, where $P = (0, \pi/2)$.

EXERCISE 38. Find the Jacobain matrix for the function $G(X)$, where

$$g_1(x_1, x_2, x_3) = a_{11}x_1 + a_{12}x_2 + a_{13}x_3 + b_1$$
$$g_2(x_1, x_2, x_3) = a_{21}x_1 + a_{22}x_2 + a_{23}x_3 + b_2$$
$$g_3(x_1, x_2, x_3) = a_{31}x_1 + a_{32}x_2 + a_{33}x_3 + b_3$$

Find $J(P)$, where $P = (p_1, p_2, p_3)$ is arbitrary.

Using the Jacobian matrix for a system $X = G(X)$, we can give a theorem giving sufficient conditions for convergence of the fixed-point iteration on the system.

THEOREM 3-4-1. Suppose that $G(X)$ is such that the system $X = G(X)$ has a solution P. Further, suppose that for some $\delta > 0$ there exists a neighborhood of radius δ about P in which

$$\frac{\partial g_i}{\partial x_j} \qquad i = 1, \ldots, n; j = 1, \ldots, n$$

exist for each point of the neighborhood. Finally, suppose that there exists a constant $k < 1$ such that the row norm or the column norm of $J(X)$ does not exceed k, for all X in the above neighborhood. Then, if $\| P_0 - P \| < \delta$ and

$$P_{m+1} = G(P_m) \qquad m = 0, 1, \ldots$$

it follows that

$$\lim_{m \to \infty} P_m = P$$

Theorem 3-4-1 implies that there is a unique solution in the neighborhood under consideration. When applied to the case $n = 1$, the theorem is similar to some of the results of Chapter 1. A proof of this theorem for the general case is found in Todd.

EXERCISE 39. Prove Theorem 3-4-1 for the case $n = 1$.

EXAMPLE 3-19. Apply Theorem 3-4-1 to the general linear system $X = CX + D$.

Solution Here $G(X) = CX + D$; so for each i,

$$g_i(x_1, x_2, \ldots, x_n) = c_{i1}x_1 + c_{i2}x_2 + \cdots + c_{in}x_n + d_i$$

Hence

$$\frac{\partial g_i}{\partial x_j} = c_{ij}$$

Thus the Jacobian matrix J is just exactly the matrix of coefficients C. The conclusion of Theorem 3-4-1 then is similar to conclusion (b) of Theorem 3-3-1.

EXAMPLE 3-20. Apply Theorem 3-4-1 to the following equation, for the root $P = 0$:

$$x_1 = x_1^2 - x_2^2$$
$$x_2 = x_1^2 + x_2^2$$

Solution. Here

$$g_1(x_1, x_2) = x_1^2 - x_2^2$$
$$g_2(x_1, x_2) = x_1^2 + x_2^2$$

$$\frac{\partial g_1}{\partial x_1} = 2x_1 \qquad \frac{\partial g_1}{\partial x_2} = -2x_2$$

$$\frac{\partial g_2}{\partial x_1} = 2x_1 \qquad \frac{\partial g_2}{\partial x_2} = 2x_2$$

$$J = \begin{bmatrix} 2x_1 & -2x_2 \\ 2x_1 & 2x_2 \end{bmatrix}$$

It can be seen that, at $P = (0, 0)$, $J(P)$ is a matrix of all zeros. Hence the row (and column) norm of $J(P)$ is zero. By continuity, one would expect that for some neighborhood of P the row norm of J would be less than 1. Let us investigate a neighborhood of radius 1/3. Consider

$$S = \left\{ X : \|X - P\| \le 1/3 \right\} = \left\{ X : \|X\| \le 1/3 \right\}$$

This set contains the neighborhood under consideration and is chosen to simplify the mathematical concepts needed in the problem. The largest possible row norm of J on S is given by the larger of

$$\max_{x \in S} \left(|2x_1| + |-2x_2| \right) \quad \text{and} \quad \max_{x \in S} \left(|2x_1| + |2x_2| \right)$$

This simplifies to

$$\max_{x \in S} \left(|2x_1| + |2x_2| \right) = \frac{2(2)^{1/2}}{3}$$

(see Exercise 40). The last result is by application of maximization techniques from elementary calculus.

Because $2(2)^{1/2}/3 < 1$, the iterative procedure will converge to the solution (0, 0) for any starting value satisfying $\|X\| < 1/3$.

For example, suppose that $P_0 = (-1/4, 1/8)$. Then

$$P_1 = \left[(-1/4)^2 - (1/8)^2, \; (-1/4)^2 + (1/8)^2 \right] = (3/64, 5/64)$$

$$P_2 = \left[(3/64)^2 - (5/64)^2, \; (3/64)^2 + (5/64)^2 \right] = \left(\frac{-1}{256}, \frac{12}{2{,}048} \right)$$

EXERCISE 40. Verify that $2(2)^{1/2}/3$ is indeed

$$\max_{x \in S} \left(|2x_1| + |2x_2| \right)$$

in Example 3-20.

EXERCISE 41. In Example 3-20 set $P_0 = (0, 1/4)$, and compute P_1, P_2.

Theorem 3-4-1 is not, as it stands, a readily applicable result. The reason is obvious. In order to check the hypotheses of the theorem one needs to know the solution to the system $X = G(X)$ under consideration. There are a number of generalizations of Theorem 3-4-1 which exclude the undesirable hypothesis. We shall discuss one such result to illustrate the more desirable type of theorem.

The reader may recall that the **Cartesian product** of two intervals is a rectangle and the Cartesian product of three intervals is a rectangular solid. This idea extends to E^n.

DEFINITION 3-12. Let $I_1, I_2, \ldots, I_n$ be intervals in E^1. The set

$$R = I_1 \times I_2 \times \cdots \times I_n$$

in E^n (the Cartesian product of the given intervals) is the set in E^n defined by

$$R = \left\{ (x_1, x_2, \ldots, x_n) : x_1 \in I_1, x_2 \in I_2, \ldots, x_n \in I_n \right\}$$

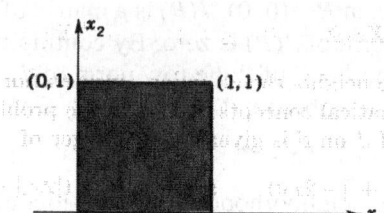

FIG. 3-7

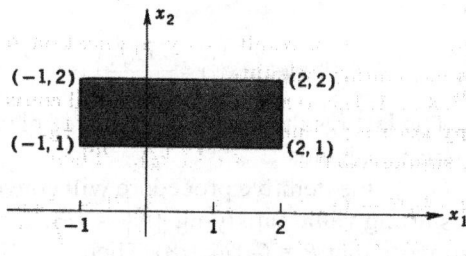

FIG.   3-8

EXAMPLE 3-21. If $I_1 = [0, 1]$ and $I_2 = [0, 1]$, then $R = I_1 \times I_2$ is a unit square in E^2 (see Fig. 3-7). If $I_1 = [-1, 2]$ and $I_2 = [1,2]$, then $I_1 \times I_2$ is the rectangular region in Fig. 3-8.

EXAMPLE 3-22. If $I_1 = [1, 2]$ and $I_2 = [-\infty, \infty]$, then $I_1 \times I_2$ is an infinite strip, (Fig. 3-9).

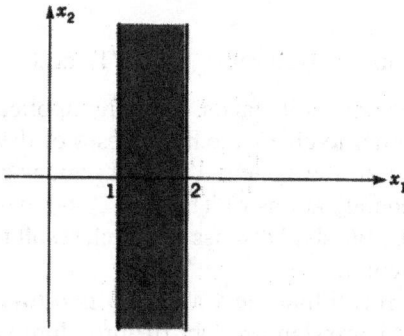

FIG.   3-9

EXERCISE 42. Draw graphs of the regions $I_1 \times I_2, I_1 \times I_3, I_2 \times I_3$, where

$$I_1 = [-3, -2]$$
$$I_2 = [1, 2]$$
$$I_3 = [0, \infty)$$

EXERCISE 43. Describe the regions $I_1 \times I_2$, $I_1 \times I_3$, $I_2 \times I_3$, where

$$I_1 = [0, \infty)$$
$$I_2 = (-\infty, 0]$$
$$I_3 = (-\infty, \infty)$$

THEOREM 3-4-2. Let $R = I_1 \times I_2 \times \cdots \times I_n$ be a **cartesian product** of closed intervals of finite length. Let $G(X)$ be a function with domain R and range contained in R.

(a) If $G(X)$ is continuous, then $X = G(X)$ has at least one solution in R.
(b) Suppose that there exists a constant $k < 1$ such that the row norm or column norm of the **Jacobian matrix** of $G(X)$ does not exceed k, for all $X \in R$. Then $X = G(X)$ has a unique solution $P \in R$, and if

$$P_0 \in R$$
$$P_{m+1} = G(P_m) \qquad m = 0, 1, \ldots$$

then

$$\lim_{m \to \infty} P_m = P$$

EXAMPLE 3-23. Apply Theorem 3-4-2 to the function $G(X)$ below, defined on the unit square $I_1 \times I_2$, where $I_1 = I_2 = [0, 1]$.

$$g_1(x_1, x_2) = \frac{1}{2} + \frac{x_1}{4} + \frac{x_1^2 + x_2^2}{16}$$

$$g_2(x_1, x_2) = \frac{1}{2} + \frac{x_2}{4} - \frac{x_1^2 + x_2^2}{16}$$

Solution. We shall show that $G(X)$ has range contained in the same unit square. If

$$0 \le x_1 \le 1$$
$$0 \le x_2 \le 1$$

then it is easily seen that

$$1/2 \le g_1(x_1, x_2) \le 7/8$$
$$7/16 \le g_2(x_1, x_2) \le 11/16$$

The function $G(X)$ can be seen to be continuous in E^2 by use of the methods of Sec. 3-2. Hence the equation $X = G(X)$ has a solution in the unit square. The Jacobian matrix is given by

$$J(X) = \begin{bmatrix} \dfrac{1}{4} + \dfrac{x_1}{8} & \dfrac{x_2}{8} \\[2mm] -\dfrac{x_1}{8} & \dfrac{1}{4} - \dfrac{x_2}{8} \end{bmatrix}$$

For $0 \le x_1, x_2 \le 1$, the maximum row norm of this matrix is 1/2. Hence, the solution in this region is unique and may be found by fixed-point iteration.

A C++ program segment to perform the iteration is given below. Sample output for a starting value $P_0 = (0, 0)$ is given in Table 3-5.

Program 3–1. Fixed-point iteration for a system

```
# include iostream using namespae std; int main (void)
// PROGRAM ITER
  {
  cin >> x >>y;
  cout <<x<< ' ' << y;
  for(j = 1; j <= 30; ++j)
    {
    a = .5 + .25 * x + ( x*x + y*y) / 16.;
    b = .5 + .25 * y – (x*x + y*y) / 16.;
    cout<<j<<' '<<a<<' '<<b<<endl;
    x = a;
    y = b;
    }
  }
```

TABLE 3-5

i	p_{1i}	p_{2i}
1	.5000000000E + 00	.5000000000E + 00
2	.6562500000E + 00	.5937500000E + 00
3	.7130126953E + 00	.5994873047E + 00
4	.7324889321E + 00	.5956360679E + 00
5	.7388298806E + 00	.5932013694E + 00
6	.7408173112E + 00	.5921905013E + 00
7	.7414230702E + 00	.5918288829E + 00
8	.7416038673E + 00	.5917091210E + 00
9	.7416569656E + 00	.5916712815E + 00
10	.7416723639E + 00	.5916596979E + 00
11	.7416767843E + 00	.5916562312E + 00
12	.7416780429E + 00	.5916552110E + 00
13	.7416783987E + 00	.5916549148E + 00
14	.7416784988E + 00	.5916548296E + 00
15	.7416785268E + 00	.5916548054E + 00
16	.7416785346E + 00	.5916547985E + 00
17	.7416785367E + 00	.5916547966E + 00
18	.7416785373E + 00	.5916547960E + 00
19	.7416785375E + 00	.5916547959E + 00
20	.7416785375E + 00	.5916547958E + 00
21	.7416785376E + 00	.5916547958E + 00
22	.7416785376E + 00	.5916547958E + 00

The values of P_{23} to P_{30} are identical to P_{22}.

EXERCISE 44. Discuss the existence and uniqueness of a solution to $X = G(X)$ in $R = I_1 \times I_2$, where $I_1 = [-1,1]$, $I_2 = [0,1]$,

$$g_1(x_1,x_2) = -\frac{1}{4} + \frac{x_1^2 - x_2}{12}$$

$$g_2(x_1,x_2) = \frac{1}{3} + \frac{x_1 - x_2^3}{8}$$

EXERCISE 45. Write a program to solve the problem of Exercise 44. Run the program for various starting values.

EXERCISE 46. Apply Theorem 3-4-2 to the system given below:

$$I_1 = [-1,1], I_2 = [-1,1], I_3 = [-1,1]$$
$$x_1 = 1/2 - 1/8 \sin x_1 x_2 x_3$$
$$x_2 = -1/3 + 1/4 \cos x_1 x_2 x_3$$
$$x_3 = 1/6(x_1^2 + x_2^2 - x_3^2)$$

In actual practice, it may happen that the hypotheses of Theorem 3-4-2 are satisfied for some region, but the region is not readily discernible. Thus, even though $X = G(X)$ has a solution and Algorithm 3-1 converges for some starting values, such starting values may be difficult to determine. By using a computer, a number of different starting values might be tried. If a convergent sequence $\{P_m\}$ is found, recourse may be had to the following theorem. The proof is left as Exercise 48.

THEOREM 3-4-3. Suppose that Algorithm 3-1 is applied to a system $X = G(X)$ and the sequence $\{P_m\}$ converges to a point T. Then, if $G(X)$ is continuous at T, T is a solution to $X = G(X)$.

EXERCISE 47. Using the program written for Exercise 45, make an experimental investigation of the boundary of a region of starting values for which the iteration will converge in that problem.

EXERCISE 48. The proof of Theorem 3-4-3 is highly similar to the proof of the corresponding theorem for the case $n = 1$ given in Chapter 1. The main differences involve the use of $\| \ \|$ in place of $| \ |$. Write out a careful proof of this theorem.

We conclude this chapter with a discussion of the **Newton-Raphson** procedure for a system $F(X) = 0$. This procedure is considerably more complicated than it is for solving a single equation $f(x) = 0$. To find the $(m + 1)$st approximation P_{m+1} to the solution, from the mth approximation P_m, requires the solution of a system of n simultaneous linear equations. The matrix of coefficients of the system will be the Jacobian matrix of $F(X)$ evaluated at the point P_m.

ALGORITHM 3-2. Let P_0 be an approximation to a solution to $F(X) = 0$. Compute P_{m+1} from P_m by the procedure below.

(a) Let $X = D_m$ be the solution to the linear system

$$J(P_m)X = F(P_m)$$

(b) $P_{m+1} = P_m - D_m$.

As can be seen, the computation of P_{m+1} is a two-step procedure. One step involves solving a linear system of equations. If this system fails to have a solution, then the algorithm breaks down. The following theorem gives sufficient conditions for the sequence $\{P_m\}$ generated by Algorithm 3-2 to converge to a solution to $F(X) = 0$.

THEOREM 3-4-4. Suppose that $F(X) = 0$ has a solution P. Further, suppose that the functions

$$\frac{\partial f_i}{\partial x_j} \qquad i = 1, \ldots, n; j = 1, \ldots, n$$

are all continuous at the point P. Finally, suppose that $J(P)$ is non-singular. Then there exists a $\delta > 0$ such that, if $\| P_0 - P \| < \delta$, the sequence generated by Algorithm 3-2 converges to P.

For the case $n = 1$, the theorem and algorithm just given should agree with what we learned in Chapter 1. In this case, we have an equation $f(x) = 0$ and just one partial derivative

$$\frac{\partial f}{dx} = \frac{df}{dx} = f'(x)$$

Part (a) of the algorithm asks for the solution to

$$f'(p_m)x = f(p_m)$$

Assuming that $f'(p_m) \neq 0$, this is $x = f(p_m)/f'(p_m)$. Substitution of this solution into part (b) gives

$$P_{m+1} = P_m - \frac{f(p_m)}{f'(p_m)}$$

which is indeed the Newton-Raphson procedure for a function of one variable.

In Theorem 3-4-4, the hypothesis that $J(P)$ be nonsingular is an extension of the idea that $f'(p) \neq 0$. If $f'(p) \neq 0$ and $f(x)$ is continuous at $x = p$, then there is a neighborhood about p in which $f'(x) \neq 0$. The same idea holds for the non-singularity of $J(X)$ in a neighborhood of P. As with the problem of a single equation, it will usually be that we will not be able to check the hypothesis that $J(P)$ be nonsingular. This is because the point P is not known in advance. Thus attention should be paid to the possible singularity of $J(P_m)$ in the course of the computation. If the sequence $\{P_m\}$ generated in a particular problem at first seems to be converging but then fails to converge, a possible explanation may be that $J(P)$ is singular.

We shall not attempt to give a proof of Theorem 3-4-4. However, we shall give a derivation of Algorithm 3-2 for the case $n = 2$. The ideas of this derivation easily extend to an arbitrary n. Our derivation is based upon Taylor's series for a function of two variables.

THEOREM 3-4-5. (Taylor's theorem in two dimensions.) Let $f(x,y)$ have all the partial derivatives of order $n + 1$ continuous in a domain

$$D = [a,b] \times [c,d]$$

containing the point (x_0, y_0). Then for $(x, y) \in D$

$$f(x,y) = f(x_0,y_0) + \left[(x-x_0)\frac{\partial}{\partial x} + (y-y_0)\frac{\partial}{\partial y} \right] f(x_0,y_0)$$

$$+ \frac{1}{2!}\left[(x-x_0)\frac{\partial}{\partial x} + (y-y_0)\frac{\partial}{\partial y} \right]^2 f(x_0,y_0) + \cdots$$

$$+ \frac{1}{n!}\left[(x-x_0)\frac{\partial}{\partial x} + (y-y_0)\frac{\partial}{\partial y} \right]^n f(x_0,y_0)$$

$$+ \frac{1}{(n+1)!}\left[(x-x_0)\frac{\partial}{\partial x} + (y-y_0)\frac{\partial}{\partial y} \right]^{n+1} f(\xi,\eta)$$

where (ξ, η) lies on the line joining (x_0, y_0) and (x, y).

In this theorem we understand

$$\left[(x-x_0)\frac{\partial}{\partial x} + (y-y_0)\frac{\partial}{\partial y} \right] f(x_0,y_0)$$

to mean

$$(x-x_0)\frac{\partial}{\partial x} f(x,y)\bigg|_{(x_0,y_0)} + (y-y_0)\frac{\partial}{\partial y} f(x,y)\bigg|_{(x_0,y_0)}$$

To continue,

$$\left[(x-x_0)\frac{\partial}{\partial x} + (y-y_0)\frac{\partial}{\partial y} \right]^2 f(x_0,y_0)$$

means

$$(x-x_0)^2 \frac{\partial^2}{\partial x^2} f(x,y)\bigg|_{(x_0,y_0)} + 2(x-x_0)(y-y_0)\frac{\partial^2}{\partial x\,\partial y} f(x,y)\bigg|_{(x_0,y_0)}$$

$$+ (y-y_0)^2 \frac{\partial^2}{\partial y^2} f(x,y)\bigg|_{(x_0,y_0)}$$

A proof is given in Fulks.

To simplify the notation, we shall derive the Newton-Raphson iteration for a system

$$f(x,y) = 0$$
$$g(x,y) = 0$$

$$(3\text{-}13)$$

If each of the functions $f(x, y)$, $g(x, y)$ is expanded in a Taylor's series about (x_0, y_0) through the first two terms, the result is

$$f(x,y) \approx f(x_0,y_0)+(x-x_0)f_x(x_0,y_0)+(y-y_0)f_y(x_0,y_0)$$
$$g(x,y) \approx g(x_0,y_0)+(x-x_0)g_x(x_0,y_0)+(y-y_0)g_y(x_0,y_0)$$

(3-14)

Now in Eqs. (3-14) we wish to find (x, y) such that the left side becomes zero. We are led to

$$0 \approx f(x_0,y_0)+(x-x_0)f_x(x_0,y_0)+(y-y_0)f_y(x_0,y_0)$$
$$0 \approx g(x_0,y_0)+(x-x_0)g_x(x_0,y_0)+(y-y_0)g_y(x_0,y_0)$$

(3-15)

Let

$$J(x,y) = \begin{bmatrix} f_x(x,y) & f_y(x,y) \\ g_x(x,y) & g_y(x,y) \end{bmatrix}$$

If the approximations (3-15) are treated as equations, the result is a linear system in the unknown x, y. Using matrix notation,

$$\begin{bmatrix} 0 \\ 0 \end{bmatrix} = \begin{bmatrix} f(x_0,y_0) \\ g(x_0,y_0) \end{bmatrix} + J(x_0,y_0)\begin{bmatrix} x-x_0 \\ y-y_0 \end{bmatrix}$$

If $J(x_0, y_0)$ is nonsingular, then

$$\begin{bmatrix} x \\ y \end{bmatrix} = \begin{bmatrix} x_0 \\ y_0 \end{bmatrix} - J^{-1}(x_0,y_0)\begin{bmatrix} f(x_0,y_0) \\ g(x_0,y_0) \end{bmatrix}$$

Used iteratively, this is just Algorithm 3-2 for the case $n = 2$.

In applying Algorithm 3-2 for the case $n = 2$ it is standard to solve the resulting system of linear equations and thus derive an explicit formula for P_{m+1} in terms of P_m. The linear system to be solved is

$$\begin{bmatrix} \dfrac{\partial f_1(P_m)}{\partial x_1} & \dfrac{\partial f_1(P_m)}{\partial x_2} \\ \dfrac{\partial f_2(P_m)}{\partial x_1} & \dfrac{\partial f_2(P_m)}{\partial x_2} \end{bmatrix}\begin{bmatrix} x_1 \\ x_2 \end{bmatrix} = \begin{bmatrix} f_1(P_m) \\ f_2(P_m) \end{bmatrix}$$

On the assumption that the Jacobian matrix $J(P_m)$ is nonsingular the result is

$$D_m = \frac{1}{\det J(P_m)}\begin{bmatrix} f_1(P_m)\dfrac{\partial f_2(P_m)}{\partial x_2}-f_2(P_m)\dfrac{\partial f_1(P_m)}{\partial x_2} \\ -f_1(P_m)\dfrac{\partial f_2(P_m)}{\partial x_2}+f_2(P_m)\dfrac{\partial f_1(P_m)}{\partial x_1} \end{bmatrix}$$

where

$$\det J(P_m) = \frac{\partial f_1(P_m)}{\partial x_1}\frac{\partial f_2(P_m)}{\partial x_2} - \frac{\partial f_1(P_m)}{\partial x_2}\frac{\partial f_2(P_m)}{\partial x_1}$$

This may be combined with part (b) of the algorithm to give explicit formulas for the coordinates of P_{m+1}.

$$p_{1,m+1} = \frac{f_2(P_m)\left[\partial f_1(P_m)/\partial x_2\right] - f_1(P_m)\left[\partial f_2(P_m)/\partial x_2\right]}{\det J(P_m)} + p_{1m}$$

$$p_{2,m+1} = \frac{f_1(P_m)\left[\partial f_2(P_m)/\partial x_1\right] - f_2(P_m)\left[\partial f_1(P_m)/\partial x_1\right]}{\det J(P_m)} + p_{2m}$$

(3-16)

EXAMPLE 3-24. We have previously considered the system

$$x_1 = x_1^2 + x_2^2$$
$$x_2 = x_1^2 - x_2^2$$

This system has a second root near (3/4,5/12). Apply the Newton-Raphson procedure to this problem.

Solution. The problem in the form $F(X) = 0$ is

$$f_1(x_1,x_2) \equiv x_1^2 + x_2^2 - x_1 = 0$$
$$f_2(x_1,x_2) \equiv x_1^2 - x_2^2 - x_2 = 0$$

The Jacobian matrix is

$$J(X) = \begin{bmatrix} 2x_1 - 1 & 2x_2 \\ 2x_1 & -2x_2 - 1 \end{bmatrix}$$

A program was written to carry out the iteration. A copy of the program, and output for an initial approximation (.7, .4) is given below.

```
#include <iostream>
#include <cmath>
using namespace std;
// PROGRAM 3.2 NEWTON-RAPHSON ITERATION
// PROGRAM ITER
int main(void)
{
float x,y,f1,f2,d,f1x,f1y,f2y,f2x,t,a,b;
int i;
cout<<"Input ititial values for x and y: ";
cin>>x>>y;
cout<<"The input x is "<<x<<" The input y is "<<y<<endl;
```

```
for(i = 1; i<= 10; ++i)
{
f1 = x*x + y*y -x;
f2 = x*x - y*y -y;
d =(2*x -1)*(-2*y -1) - 4 *x *y;
f1x = 2*x -1;
f1y = 2*y;
f2y = -2*y -1;
f2x = 2*x;
t = f2*f1y - f1*f2y;
a = x + t/d;
t = f1*f2x - f2*f1x;
b = y + t/d;
x = a;
y = b;
cout<<i<<' '<<x<<' '<<y<<endl;
}
return(0);
}
/*
```

Input ititial values for x and y: .7 .4
The input x is 0.7 The input y is 0.4

```
1  0.779348  0.422826
2  0.771913  0.419676
3  0.771845  0.419643
4  0.771845  0.419643
5  0.771845  0.419643
6  0.771845  0.419643
7  0.771845  0.419643
8  0.771845  0.419643
9  0.771845  0.419643
10 0.771845  0.419643
```

As can be seen, the sequence generated by the fixed-point algorithm appears to converge. An approximation to the solution is the point

$$x_1 = 0.771845 \quad x_2 = 0.419643$$

In solving a system of three simultaneous equations by the Newton-Raphson procedure, one may still find it convenient to derive explicit formulas for the $p_{i,\,m+1}$ as in Eqs. (3-16). For larger values of n, it is usually more convenient to follow the two-step procedure of Algorithm 3-2 and solve the requisite linear system at each step. The program becomes quite long if n is large, owing to the need for computing all the partial derivatives. In Sec. 6-2, on Numerical Differentiation, we show how the problem can be greatly simplified by use of numerical approximations to the partial derivatives.

EXERCISE 49. The system

$$x_1^2 + x_2^2 - 1 = 0$$
$$x_1 - x_2 \quad = 0$$

is, of course, easily solved by straightforward methods. Let $P_0 = (.7, .7)$, and find P_1, using the Newton-Raphson procedure.

EXERCISE 50. Suppose that $P_0 = (.1, -.1)$ in Example 3-24. Calculate P_1. To what solution will the iteration converge in this case?

EXERCISE 51. Suppose that in the system $F(X) = 0$ only the first equation is nonlinear. That is, $f_1(x_1, \ldots, x_n) = 0$ is nonlinear, while $f_2(x_1, \ldots, x_n) = 0$, $\ldots, f_n(x_1, \ldots, x_n) = 0$ are all linear equations. Discuss how this affects the ease or difficulty of writing a program to apply the Newton-Raphson procedure to the system.

EXERCISE 52. Write a program to solve the following system:

$$x_1^2 + 2x_2^2 + x_3^2 - 4 = 0$$
$$2x_1 - x_2 + x_3 - 1 \quad = 0$$
$$x_1 + 2x_2 - x_3 + 2 \quad = 0$$

Discuss the difficulties inherent in finding a suitable initial approximation P_0 to a solution. Run the program on a computer for a number of initial approximations P_0.

Polynomials, Taylor's Series and Interpolation Theory

4–1 INTRODUCTION

In elementary mathematics the word *polynomial* refers to a function of the form $P(x) = a_0 + a_1x + a_2x^2 + \cdots + a_nx^n$, where n is an integer greater than or equal to zero. In higher mathematics, the word is used to describe a more general class of functions. Let $\phi_1(x), \phi_2(x), \ldots, \phi_m(x)$ be a given set of functions. A polynomial in the base functions $\{\phi_i(x)\}$ is any function of the form

$$P(x) = a_1\phi_1(x) + a_2\phi_2(x) + \cdots + a_m\phi_m(x) \tag{4-1}$$

where the a_i are constants. When the base functions are $1, x, x^2, \ldots, x^n$, then a polynomial

$$P(x) = a_0 + a_1x + a_2x^2 + \cdots + a_nx^n = \sum_{i=0}^{n} a_ix^i \tag{4-2}$$

is called an **algebraic polynomial** in one variable. If $a_n \neq 0$, the polynomial is said to be of degree n.

There are other commonly used types of polynomials. For example, if the **base functions** are taken to be $1, \cos(x), \sin(x), \cos(2x), \sin(2x), \ldots, \cos(nx), \sin(nx)$, then the polynomials can be written in the form

$$P(x) = a_0 + \sum_{i=1}^{n} (a_i \cos ix + b_i \sin ix) \tag{4-3}$$

Such polynomials are called **trigonometric polynomials**. If $|a_n| + |b_n| \neq 0$, then Eq. (4-3) is a trigonometric polynomial of degree n.

If the base functions are taken to be $1, x, y, x^2, xy, y^2, \ldots, x^n, x^{n-1}y, x^{n-2}y^2, \ldots, y^n$, then the polynomials can be written in the form

$$P(x) = \sum_{i=0}^{n} \sum_{j=0}^{i} a_{ij} x^{i-j} y^j \tag{4-4}$$

If

$$\sum_{j=0}^{n} |a_{nj}| \neq 0$$

then Eq. (4-4) is an algebraic polynomial of degree n in the two variables x, y.

In this chapter, we shall restrict ourselves to the study of algebraic polynomials in one variable. We shall use the word polynomial to mean algebraic polynomial.

The standard notation for a polynomial in one variable is inconvenient for use in some computer languages, because it makes use of zero as a subscript. All translating computer languages allow for a subscript of one, therefore, it is convenient to use one of the following two notations:

$$P(x) = a_1 x^n + a_2 x^{n-1} + \cdots + a_n x + a_{n+1} \qquad (4-5)$$

$$P(x) = b_1 + b_2 x + \cdots + b_n x^{n-1} + b_{n+1} x^n \qquad (4-6)$$

The particular notation chosen will depend on the nature of the problem being solved and the preference of the programmer. In general, if a problem requires addition, subtraction, or multiplication of polynomials, form (4-6) will be more convenient. Form (4-5) is more convenient for the problem of finding the value of $P(x)$ at a point.

In a typical problem, we shall be given a function $f(x)$ defined on an interval $[a,b]$. Our goal will be to determine a polynomial $P(x)$ which can be used to approximate $f(x)$ on $[a,b]$. The difference between $f(x)$ and $P(x)$ will be denoted by $R(x) = f(x) - P(x)$. Intuitively, we say that $P(x)$ is a good approximation to $f(x)$ if the function $R(x)$ is close to zero on the interval $[a, b]$.

We shall state without proof a fundamental theorem on **polynomial approximation** to a continuous function on an interval. A proof of this theorem may be found in almost any graduate-level numerical-analysis text (see Ralston).

THEOREM 4-1-1. (Weierstrass). Let $f(x)$ be continuous on $[a,b]$, and let $\varepsilon > 0$ be given. Then there exists some integer n and a polynomial $P(x)$ of degree n such that

$$|f(x) - P(x)| < \varepsilon \qquad \text{for all } x \in [a,b]$$

Basically, this theorem says that it is reasonable to seek a polynomial approximation to a continuous function on a finite interval. One can approximate such a function to any desired accuracy provided that a sufficiently high degree of polynomial is allowed. Of course in many problems the size of n needed to give a desired ε will be unreasonably large. In such a case approximations using other **base functions**, or rational approximations, are often employed. These topics are treated in more advanced courses in numerical analysis.

4-2 EVALUATION OF A POLYNOMIAL

Polynomials are relatively easy functions to deal with in any high level computer language. It is necessary to store the coefficients and to keep track of the degree of the polynomial under consideration. A standard problem is that of finding the value of a polynomial $P(x)$ at a given point. We shall study this problem in detail to show how polynomials are handled in C++.

EXAMPLE 4-1. Suppose that the degree n and the the coefficients $a_1, a_2, \ldots,$ a_{n+1} of a polynomial

$$P(x) = \sum_{i=1}^{n+1} a_i x^{n+1-i}$$

are entered and then the value of x at which $P(x)$ is to be evaluated is entered. Assume that $1 < n < 9$. The following program demonstrates this and the output is listed for a given input.

Program:

```cpp
#include <iostream>
#include <cmath> // provides pow
using namespace std;
// poly 1
int main(void)
  {
  float a[11],value,t;
  int n,m,i;
  cout<<"Input The Polynomial Degree: ";
  cin>>n;
  m = n + 1;
  cout<<"Input The Polynomial Coefficients: "<<endl;
  for(i = 1; i<= m; ++i)
   cin>>a[i];
  cout<<"Input The Point To Evaluate the Polynomial: ";
  cin>>t;
  value = a[m];
  for(i = 1; i <= n; ++i)
   value = value + a[m-i] * pow(t,i);

  cout<<"The poly evaluated at "<< t <<" is "<<value<<endl;
  return(0);
  }
```

Output:
```
/*
Input The Polynomial Degree: 2
Input The Polynomial Coefficients:
3 2 1
Input The Point To Evaluate the Polynomial: 2
The poly evaluated at 2 is 17
*/
```

The efficiency of various schemes for evaluating a polynomial at a point may be compared by determining the number of arithmetic operations involved in each evaluation. In evaluating

$$P(x) = a_1 x^n + a_2 x^{n-1} + \cdots + a_{n+1}$$

by the process used in the previous example it is necessary to calculate

$a_n x$	1 multiplication	
$a_{n-1} x\, x$	2 multiplications	
$a_{n-2} x\, x\, x$	3 multiplications	(4-7)
....	....	
$a_1 xx \ldots x$	n multiplications	

and then add the above terms to a_{n+1}. A total of

$$1+2+3+ \cdots +n = \frac{n(n+1)}{2}$$

multiplications and n additions is required.

It should be evident from looking at (4-7) that the scheme just given is highly inefficient. For example, instead of computing x^n by multiplying x times itself $n-1$ times, it would be shorter to store the result x^{n-1} previously computed and then multiply this by x.

EXAMPLE 4-2. Write a C++ program to solve the problem of Example 4-1 which incorporates the feature discussed above.

Solution.
```
#include <iostream>
using namespace std;
// poly 2
int main(void)
 {
 float a[11],value,t,tt;
 int n,m,i;
 cout<<"Input The Polynomial Degree: ";
 cin>>n;
 m = n + 1;
 cout<<"Input The Polynomial Coefficients: "<<endl;
 for(i = 1; i<= m; ++i)
  cin>>a[i];
 cout<<"Input The Point To Evaluate the Polynomial: ";
 cin>>t;
 value = a[m];
 tt = t;
 for(i = 1; i <= n; ++i)
  {
```

```
value = value + a[m-i] * tt;
tt = tt * t;
}
```

cout<<"The poly evaluated at "<< t <<" is "<<value<<endl;
return(0);
}

Now the calculation of $P(t)$, in the above program, requires $n - 1$ multiplications of tt by t, the n multiplications by the coefficients a_i ($i = 1$, ..., n), and n additions. Thus a total of $2n - 1$ multiplications and n additions is required. By preparing a short table of values of $n(n - 1)/2$ and $2n - 1$ the reader can convince himself that the method of Example 4-2 is to be preferred over the method of Example 4-1 if $n > 3$.

There exists a still more efficient method for evaluating a polynomial at a point. The process is known as **synthetic division**, or **Horner's method**. It is based upon the idea of nested multiplication. To illustrate, notice that

$$a_1x^3 + a_2x^2 + a_3x + a_4 = a_4 + x(a_3 + x(a_2 + a_1x)) \tag{4-8}$$

The evaluation of the right side of Eq. (4-8) takes just three multiplications and three additions. This can be seen to be better than the processes of Examples 4-1 and 4-2.

We shall state the general synthetic-division process as an algorithm. The proof of the accompanying theorem is given immediately after Exercises 1 to 3 below.

ALGORITHM 4-1. (synthetic division). Given $a_1, a_2, \ldots, a_{n+1}$ and t, set

$$b_1 = a_1$$
$$b_i = tb_{i-1} + a_i \quad i = 2, \ldots, n+1$$

THEOREM 4-2-1. If $P(x) = a_1x^n + a_2x^{n-1} + \cdots + a_{n+1}$ and $\{b_i\}$ are defined by Algorithm 4-1 for a given t, then

$$P(x) = (x-t)\left(b_1x^{n-1} + b_2x^{n-2} + \cdots + b_{n-1}x + b_n\right) + b_{n+1}$$

and $P(t) = b_{n+1}$.

EXAMPLE 4-3. Use Algorithm 4-1 to find the value of

$$P(x) = 3x^4 - 5x^2 + 3x - 15$$

at the point $t = 2$.

Solution.

$$b_1 = 3$$
$$b_2 = (2)(3) + 0 = 6$$
$$b_3 = (2)(6) - 5 = 7$$
$$b_4 = (2)(7) + 3 = 17$$
$$b_5 = (2)(17) - 15 = 19$$

Thus $P(x) = (x - 2)(3x^3 + 6x^2 + 7x + 17) + 19$, and $P(2) = 19$.

If the computation is to be performed by hand, it is often convenient to arrange the work horizontally (Fig. 4-1). The first row contains the coefficients of $P(x)$. The order of the calculations is indicated by the arrows, starting from the left. Each vertical arrow corresponds to an addition, and each diagonal to a multiplication by the value of t.

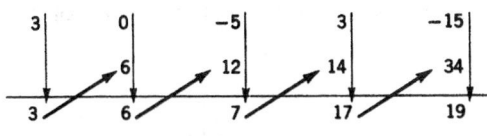

FIG. 4-1

EXERCISE 1. In the above example, verify that $P(2) = 19$ by direct substitution of the number $t = 2$ into the polynomial.

EXERCISE 2. Divide $x - 2$ into $3x^4 - 5x^2 + 3x - 15$, and verify that the quotient and remainder are $(3x^3 + 6x^2 + 7x + 17)$ and 19, respectively.

EXERCISE 3. Using Algorithm 4-1, find the value of

$$P(x) = 5x^5 - 7x^4 - 3x^2 + 4x - 7$$

at the points $t = 2$ and $t = -3$.

The synthetic-division algorithm requires just n multiplications and n additions. Thus it is more efficient than the other two procedures considered. In addition, this algorithm generates the coefficients of the quotient $P(x)/(x - t)$. Finally, it is easily programmed. A program using **synthetic division** is given in Example 4-4.

We shall now give a proof of Theorem 4-2-1.

The main problem is to verify that

$$a_1 x^n + a_2 x^{n-1} + \cdots + a_{n+1} \tag{4-9}$$
$$= (x - t)(b_1 x^{n-1} + b_2 x^{n-2} + \cdots + b_n) + b_{n+1}$$

where the b_i's are generated by Algorithm 4-1. If (4-9) is true one sees immediately that $P(t) = b_n + 1$. Two polynomials are the same if and only if the corresponding coefficients are identical. Thus we shall determine the coefficients of the various powers of x on the right side of Eq. (4-9) and compare these with those on the left side.

$$(x - t)\left(b_1 x^{n-1} + b_2 x^{n-2} + \cdots + b_{n-1} x + b_n\right) + b_{n+1}$$
$$= x\left(b_1 x^{n-1} + \cdots + b_n\right) - t\left(b_1 x^{n-1} + \cdots + b_n\right) + b_{n+1}$$
$$= \left(b_1 x^n + \cdots + b_n x\right) - \left(t b_1 x^{n-1} + \cdots + t b_n\right) + b_{n+1}$$
$$= b_1 x^n + \left(b_2 - t b_1\right) x^{n-1} + \left(b_3 - t b_2\right) x^{n-2} + \cdots + \left(b_{n+1} - t b_n\right)$$

The two sides of Eq. (4-9) are identical only if

$$a_1 = b_1$$
$$a_2 = b_2 - tb_1$$
$$a_3 = b_3 - tb_2$$
$$\cdots\cdots\cdots\cdots\cdots$$
$$a_{n+1} = b_{n+1} - tb_n$$

$$(4\text{-}10)$$

Now solve the first of Eqs. (4-10) for b_1, the second for b_2, etc. The result is

$$b_1 = a_1$$
$$b_2 = a_2 + tb_1$$
$$b_3 = a_3 + tb_2$$
$$\cdots\cdots\cdots\cdots\cdots$$
$$b_{n+1} = a_{n+1} + tb_n$$

Because these are exactly the values generated by Algorithm 4-1, the proof is complete.

In performing a **Newton-Raphson** iteration on a polynomial equation $P(x) = 0$ at a point t, it is necessary to compute both $P(t)$ and $P'(t)$. One method of computing $P'(t)$ would be by synthetic division on the polynomial $P'(t)$, where

$$P'(t) = \sum_{i=1}^{n} (n+1-i)a_i x^{n-i}$$

However, a more efficient scheme is given below.

Using synthetic division, one can write $P(x)$ in the form

$$P(x) = (x-t)Q(x) + b_{n+1}$$

where $Q(x) = b_1 x^{n-2} + b_2 x^{n-2} + \cdots + b_n$. Using the rule for differentiating a product, we obtain

$$P'(x) = Q(x) + (x-t)Q'(x)$$

It can be seen that $P'(t) = Q(t) + (t-t)Q'(t) = Q(t)$. Thus $P'(t)$ can be found by synthetic division on the polynomial $Q(x)$ of degree $n-1$.

EXAMPLE 4-4. Write a program to find both $P(x)$ and $P'(x)$ in the problem of Example 4-1. Assume that $2 < n < 9$.

Solution

```
#include <iostream>
using namespace std;
// PROGRAM POLY 3
int main(void)
{
float a[11],b[11],c[10],t;
int m,n,i;
cout<<"Input the poly degree ";
 cin>>n;
m − n + 1;
```

```
cout<<"Input the coefficients of poly "<<endl;
for(i = 1; i<=m; ++i)
  cin>>a[i];
b[1] = a[1];
c[1] = a[1];
cout<<"Input the value of t: ";
  cin>>t;
for(i = 2; i<=m; ++i)
  b[i] = b[i-1] * t + a[i];
for(i = 2; i<=n; ++i)
  c[i] = c[i-1]*t + b[i];
cout<<"The value of poly and derivative at "<<t<< " is "<<endl;
cout<<"          "<<b[m]<<"  " <<c[n]<<endl;
  return 0;
}
/*
Input the poly degree 1
Input the coefficients of poly
3 2
Input the value of t: 1
The value of poly and derivative at 1 is
        5 3
*/
```

EXERCISE 4. Using the idea illustrated in Example 4-4, compute $P(3)$, $P'(3)$; where $P(x) = 2x^5 - 7x^3 + 4x - 5$.

EXERCISE 5.
  (a) Suppose that for the real number α

$$P_n(x) = (x - \alpha)P_{n-1}(x) + P_n(\alpha)$$

and

$$P_{n-1}(x) = (x - \alpha)P_{n-2}(x) + P_{n-1}(\alpha)$$

Show that

$$P''_n(x)\big|_{x=\alpha} = 2P_{n-2}(\alpha)$$

[Does $P_n''(x) = 2P_{n-2}(x)$ for all x?]
  (b) Using the idea of (a), find $P(2)$, $P'(2)$, $P''(2)$, $P(-1)$, $P'(-1)$, and $P''(-1)$, where

$$P(x) = 7x^4 + 3x^2 - 5x^2 - 4x + 5$$

EXERCISE 6. Suppose that $P(x) = a_1 x^n + a_2 x^{n-1} + \cdots + a_{n+1}$, where $n \geq 3$ and $D(x) = x^2 - rx - s$. Define $b_1, \ldots, b_{n+1}$ by

$$b_1 = a_1$$
$$b_2 = a_2 + rb_1$$
$$b_3 = a_3 + rb_2 + sb_1$$

In general $b_i = a_i + rb_{i-1} + sb_{i-2}$ for $i = 4, \ldots, n+1$

Prove that

$$P(x) = (x^2 - rx - s)(b_1 x^{n-2} + b_2 x^{n-3} + \cdots + b_{n-1}) + b_n(x - r) + b_{n+1}$$

Thus prove that $D(x)$ evenly divides into $P(x)$ if and only if

$$b_n = b_{n+1} = 0.$$

EXERCISE 7. Use the algorithm derived in Exercise 6 to find the quotient and remainder when

$$P(x) = 2x^4 - 9x^2 + 9x - 2$$

is divided by (a) $x^2 - 3x + 1$; (b) $x^2 + 2x - 4$.

EXERCISE 8. Write a program which multiplies a polynomial

$$P(x) = \sum_{i=1}^{m+1} a_i x^{m+1-i}$$

by a linear factor $x - t$ and prints out the coefficients of the resulting polynomial.

EXERCISE 9. Write a program which reads in the degree n and the real roots $x_1, x_2, x_3, \ldots x_n$ of a polynomial

$$P(x) = x^n + a_2 x^{n-1} + a_3 x^{n-2} + \cdots + a_{n+1}$$

and calculates $a_2, a_3, \ldots, a_{n+1}$. Hint: Use the method of Exercise 8 repeatedly.

The ideas developed so far are quite useful in the problem of finding the roots of a polynomial equation. In such a problem it is helpful to have a bound on the roots. Using the synthetic-division algorithm, we shall prove the following result.

THEOREM 4-2-2. Let

$$P(x) = x^n + a_2 x^{n-1} + a_3 x^{n-2} + \cdots + a_{n+1}$$

Let $T = 1 + \max_{2 \le i \le n+1} |a_i|$. Then, if $|x| \ge T$, it follows that $|P(x)| \ge 1$.

Proof. The theorem implies that all roots (both real and complex) of $P(x) = 0$ lie inside a circle of radius T. To prove the theorem, we shall show that when the synthetic-division algorithm 4-1 is applied to $P(x)$ for $|x| \ge T$ each of the resulting coefficients b_i satisfies $|b_i| \ge 1$.

Recall that from the triangular inequality one can prove (see Exercise 10) that

$$|c + d| \ge ||c| - |d||$$

To begin, suppose the synthetic-division algorithm is applied to $P(x)$ at a point x. Then $b_1 = 1$ and

$$b_2 = b_1 x + a_2 = x + a_2$$

Now, assuming that $|x| \ge T$, we have

$$|b_2| = |x + a_2| \geq ||x| - |a_2|| \geq 1$$

$$|b_3| = |b_2 x + a_3| \geq ||b_2 x| - |a_3|| \geq 1$$

$$\cdots\cdots\cdots\cdots\cdots\cdots\cdots\cdots\cdots\cdots$$

$$|b_{n+1}| = |b_n x + a_{n+1}| \geq ||b_n x| - |a_{n+1}|| \geq 1$$

EXAMPLE 4-5. Find a bound on the roots of

$$P(x) = 6x^3 - 14x^2 - 24x + 56$$

Solution. To apply Theorem 4-2-2, the leading coefficient of the polynomial must be 1. But the solutions to $P(x) = 0$ are not changed if both sides of the equation are divided by 6. Thus we consider

$$x^3 \; 7/3 \, x^2 - 4x + 9 \, \tfrac{1}{3} = 0$$

The conclusion is that no root exceeds 10 1/3 in absolute value. In this example the roots are ±2 and 7/3.

EXERCISE 10. Using the triangular inequality $|a + b| \leq |a| + |b|$, prove that $|c + d| \geq |c| - |d|$.

EXERCISE 11. Write a C++ function which performs the Newton-Raphson iteration on a polynomial. The function should have as arguments the degree of the polynomial, N, the coefficient array, A, the number of iterations to be performed, M, the initial point, X, and the final result XX.

EXERCISE 12. It is proved in Stiefel that if $P(x)$ is a polynomial with all real distinct roots and the Newton-Raphson iteration is applied to $P(x) = 0$ with an x_0 greater than any root used as an initial guess, then convergence will occur. Write a C++ program which attempts to find all the roots of a polynomial having all real distinct roots.

4-3   THE TAYLOR EXPANSION

One of the most fundamental and useful tools in numerical analysis is the Taylor expansion of a function. In this section we shall state the basic theorem of Taylor. A proof may be found in any standard text on elementary calculus. In Exercises 38 and 39 an outline is given of a proof based upon the ideas developed in Secs. 4-4 and 4-5. A number of examples illustrating applications to numerical analysis will be discussed.

THEOREM 4-3-1. (Taylor). Suppose that $f(x) \in C^{n+1}[a, b]$ and $x_0 \in [a, b]$. Define the plynomial $P(x)$ to be

$$P(x) = f(x_0) + f'(x_0)(x - x_0) + \frac{f''(x_0)(x - x_0)^2}{2!}$$

$$+ \; \cdots \; + \frac{f^n(x_0)(x - x_0)^n}{n!}$$

(4-11)

Then for cach $x \in [a, b]$ there exists a number ξ (which usually depends on x) between x and x_0 such that

$$f(x) = P(x) + \frac{f^{(n+1)}(\xi)(x-x_0)^{n+1}}{(n+1)!} \tag{4-12}$$

The notation $\xi(x)$ is often used in place of ξ to denote that ξ is a function of x. Expression (4-12) is often written out in detail, giving

$$f(x) = f(x_0) + f'(x_0)(x-x_0) + \frac{f''(x_0)(x-x_0)^2}{2!}$$
$$+ \cdots + \frac{f^{(n)}(x_0)(x-x_0)^n}{n!} + \frac{f^{(n+1)}(\xi)(x-x_0)^{n+1}}{(n+1)!} \tag{4-13}$$

It is called the *Taylor expansion with remainder* of degree n about the point x_0 for the function $f(x)$. The term

$$R(x) = \frac{f^{(n+1)}(\xi)(x-x_0)^{n+1}}{(n+1)!} \tag{4-14}$$

is called the *remainder*, or *truncation error*. The function $R(x)$ is by definition the function $f(x) - P(x)$, where $P(x)$ is given by Eq. (4-11). Expression (4-14) for $R(x)$ is useful in obtaining a bound on $|R(x)|$.

$$f(x) - P(x) = \frac{f^{(n+1)}(\xi)(x-x_0)^{n+1}}{(n+1)!}$$
$$|f(x) - P(x)| = \frac{|f^{(n+1)}(\xi)| \cdot |x-x_0|^{n+1}}{(n+1)!} \tag{4-15}$$

Let $M = \max\limits_{x\in[a,b]} |f^{n+1}(x)|$. Then

$$|f(x) - P(x)| \le \frac{M|x-x_0|^{n+1}}{(n+1)!} \qquad \text{for } x \in |a, b| \tag{4-16}$$

Several points should be emphasized. A Taylor expansion consists of a polynomial part and a remainder. The polynomial has as base functions the functions $1, x - x_0, (x - x_0)^2, \ldots, (x - x_0)^n$. The coefficients of the polynomial are the constants $f(x_0), f'(x_0), f''(x_0)/2!, \ldots, f^{(n)}(x_0)/n!$. Thus, by suitable algebraic simplification, the polynomial can be written in the form (4-2).

The quantity ξ in the remainder (4-14) depends on $x, x_0, n,$ and $f(x)$. In all but the simplest cases, this dependence is exceedingly complicated. The value of ξ always lies between x_0 and x and thus is in $[a, b]$.

The polynomial $P(x)$ given in Eq. (4-11) is called the **Taylor polynomial**. It is characterized by the following corollary.

COROLLARY 4-3-1. If $P(x)$ is the polynomial defined in Eq. (4-11), then

$$P(x_0) = f(x_0)$$
$$P'(x_0) = f'(x_0)$$
$$P''(x_0) = f''(x_0)$$
$$\cdots\cdots\cdots\cdots\cdots\cdots$$
$$P^{(n)}(x_0) = f^{(n)}(x_0)$$

Proof. The result $P(x_0) = f(x_0)$ follows immediately from substituting $x = x_0$ into both sides of Eq. (4-11). In general, suppose that both sides of Eq. (4-11) are differentiated k times, where $1 \le k \le n$. The result for $k = 2$, for example, is

$$P''(x) = f''(x_0) + \frac{f'''(x_0)(x - x_0)}{1!} + \cdots + \frac{f^{(n)}(x_0)(x - x_0)^{n-2}}{(n-2)!}$$

The general result is

$$P^{(k)}(x) = f^{(k)}(x_0) + \frac{f^{(k+1)}(x_0)(x - x_0)}{1!} + \cdots + \frac{f^{(n)}(x_0)(x - x_0)^{n-k}}{(n-k)!}$$

Substitution of $x = x_0$ into the general result gives $P^{(k)}(x_0) = f^{(k)}(x_0)$ and completes the proof.

EXAMPLE 4-6. Obtain a first degree polynomial approximation to

$$f(x) = (1 + x)^{1/2}$$

by means of the Taylor expansion about $x_0 = 0$.

Solution.

$$f'(x) = \frac{1}{2(1+x)^{1/2}} \qquad f''(x) = \frac{-1}{4\left[(1+x)^{1/2}\right]^3} \qquad f(0) = 1 \qquad f'(0) = \frac{1}{2}$$

Thus the Taylor expansion with remainder term is

$$(1 + x)^{1/2} = 1 + \frac{x}{2} - \frac{x^2}{8\left[(1+\xi)^{1/2}\right]^3}$$

(Here ξ lies between 0 and x.) Thus the first-degree polynomial approximation to $(1 + x)^{1/2}$ is $1 + x/2$.

The accuracy of this approximation depends on what set of values x is allowed to assume. If $x \in [0, .1]$, for example, we have

$$\left| (1+x)^{1/2} - \left(1 + \frac{x}{2}\right) \right| \le \frac{.1^2}{8\left[(1+\xi)^{1/2}\right]^3} \le \frac{.1^2}{8} = .00125$$

EXAMPLE 4-7. Obtain a second-degree polynomial approximation to $f(x) = e^{-(x*x)}$ over $[0, .1]$ by means of the Taylor expansion about $x_0 = 0$. Use the expansion to approximate $f(.05)$, and bound the error.

Solution.

$$f(x) = e^{-x^2} \qquad f'(x) = -2xe^{-x^2}$$
$$f''(x) = (-2 + 4x^2)e^{-x^2} \qquad f'''(x) = (12x - 8x^3)e^{-x^2}$$
$$f(0) = 1 \qquad f'(0) = 0 \qquad f''(0) = -2$$

Thus

$$f(x) = 1 - x^2 + \frac{f'''(t)x^3}{6} \qquad \text{where } 0 < t < .1$$

Using $f(x) \approx 1 - x^2$, we have $f(.05) \approx .9975$. The truncation error is bounded by

$$|f(.05) - .9975| \leq \frac{(.05)^3}{6} \max_{t \in I} |f'''(t)|$$

where $I = [0, .05]$. To obtain a bound on $|f'''(t)|$, we use the result that, for $t \in I$,

$$|f'''(t)| = \left|12 - 8t^2\right| \cdot |t| \cdot \left|e^{-t^2}\right|$$
$$< \max_{t \in I}\left|12 - 8t^2\right| \cdot \max_{t \in I}|t| \cdot \max_{t \in I}\left|e^{-t^2}\right|$$

Hence

$$|f(0.5) - .9975| < \frac{(.05)^3}{6}(12)(.05)(1.0) = 1.25 \times 10^{-5}$$

EXAMPLE 4-8. Obtain an approximate value of

$$\frac{1}{2}\int_{-1}^{1} \frac{\sin x}{x}\, dx$$

by means of the Taylor expansion.

Solution. The expansion of $\sin(x)$ in a Taylor series about $x_0 = 0$ through terms of degree 6 is

$$\sin x = x - \frac{x^3}{3!} + \frac{x^5}{5!} - \frac{x^7}{7!}\cos \xi$$

Thus

$$\frac{\sin x}{x} = 1 - \frac{x^2}{6} + \frac{x^4}{120} - \frac{x^6}{7!}\cos \xi$$

$$\frac{1}{2}\int_{-1}^{1} \frac{\sin x}{x}\, dx = \frac{1}{2}\left(x - \frac{x^3}{18} + \frac{x^5}{600}\right)\Big|_{-1}^{1} - \frac{1}{2}\int_{-1}^{1} \frac{x^6}{7!}\cos \xi\, dx$$

This simplifies to

$$\frac{1}{2}\int_{-1}^{1} \frac{\sin x}{x}\, dx = \frac{1,703}{1,800} - \frac{1}{2}\int_{-1}^{1} \frac{x^6}{7!}\cos \xi\, dx$$

To bound the error, we compute

$$\left| \frac{1}{2} \int_{-1}^{1} \frac{x^6}{7!} \cos \xi \, dx \right| \leq \frac{1}{2} \int_{-1}^{1} \frac{x^6}{7!} |\cos \xi| \, dx \leq \frac{1}{2} \int_{-1}^{1} \frac{x^6}{7!} \, dx = \frac{1}{(7)(7!)}$$

Hence

$$\left| \frac{1}{2} \int_{-1}^{1} \frac{\sin x}{x} \, dx - \frac{1,703}{1,800} \right| \leq \frac{1}{(7)(7!)} < 3.0 \times 10^{-5}$$

EXAMPLE 4-9. An approximate value for the solution of the differential equation $y' = y^2 + x^2$, $y(0) = 1$, is desired at .1. Obtain the first three terms of the Taylor expansion about $x_0 = 0$ for $y(x)$, and calculate an approximate value for $y(.1)$.

Solution.

$$y' = x^2 + y^2 \quad y''(x) = 2x + 2yy'$$
$$y'''(x) = 2 + 2(y')^2 + 2yy''$$
$$y(0) = 1 \quad y'(0) = 1 \quad y''(0) = 2$$

Thus

$$y(x) = 1 + x + x^2 + \frac{x^3}{6} y'''(\xi)$$

$$y(.1) = 1 + .1 + .01 + \frac{.1^3}{6} y'''(\xi)$$

$$= 1.11 + \frac{.1^3}{6} y'''(\xi)$$

Thus the approximate value for $y(.1)$ is 1.11. Since y is an unknown function, it is difficult to make a worthwhile statement on a bound for $|y'''(\xi)|$.

EXAMPLE 4-10. Find an approximation to the solution to $x = \cos(x)$ by use of Taylor's series.

Solution. The general idea here is that the equation $x = \cos(x)$ is to be replaced by an equation of the form

$$x = 1 - \frac{x^2}{2!} + \frac{x^4}{4!} - \frac{x^6}{6!} \cdots + (-1)^n \frac{x^{2n}}{(2n)!} \qquad (4\text{-}17)$$

Taking $n = 1$, for example, we solve

$$x = 1 - \frac{x^2}{2}$$

to obtain $x = -1 \pm 3^{1/2}$. Since the only solution to $x - \cos(x)$ lies in the interval $[0, 1]$, we obtain $-1 + 3^{1/2}$ as an approximate solution. This value can be used as an initial guess in a fixed-point iteration to solve the equation. The iteration might be performed on the original equation $x = \cos(x)$ or on the approximate equation (4-17) for some $n > 1$.

EXAMPLE 4-11. Expand $\ln(x)$ in a Taylor expansion about $x_0 = 1$, through terms of degree 4. Obtain a bound on the **truncation error** when approximating. $1n(1.2)$ using this expansion.

Solution.

$$f(x) = \ln x \quad f'(x) = \frac{1}{x} \quad f''(x) = -\frac{1}{x^2}$$

$$f'''(x) = \frac{2}{x^3} \quad f^{(4)}(x) = -\frac{6}{x^4} \quad f^{(5)}(x) = \frac{24}{x^5}$$

Thus

$$\ln x = (x-1) - \frac{(x-1)^2}{2} + \frac{(x-1)^3}{3} - \frac{(x-1)^4}{4} + \frac{(x-1)^5}{5\xi^5}$$

$$(\xi \text{ between } 1 \text{ and } x)$$

$$\text{In } 1.2 = .2 - \frac{.2^2}{2} + \frac{.2^3}{3} - \frac{.2^4}{4} + \frac{.2^5}{5\xi^5}$$

or In $1.2 \approx .18227$. A bound on the truncation error for $x \geq 1$ is

$$\left| \frac{.2^5}{5\xi^5} \right| < \frac{.2^5}{5} = 6.4 \times 10^{-5}$$

EXERCISE 13. Obtain a second-degree polynomial approximation to $\ln(1 + x)$ by expanding the function in a Taylor series about $x_0 = 0$. Calculate an approximate value for $\ln(1.2)$, and obtain a bound on the truncation error.

EXERCISE 14. Obtain a third-degree polynomial approximation to $1/(1 + x^2)$ by expanding $f(x) = 1/(1 + x^2)$ in a Taylor expansion about $x_0 = 0$. Find the remainder term for the above expansion.

EXERCISE 15. Use the results of Exercise 14 to calculate an approximate

$$\int_0^1 \frac{dx}{1 + x^2}$$

and obtain a bound on the error. Compare the result with the exact value $\pi/4$.

EXERCISE 16. Using the idea of Example 4-10, obtain a first approximation for the solution to the equation $x = e^{-(x*x)}$

EXERCISE 17. Given $y' = x + y$, $y(1) = 2$. Obtain an approximate value for $y(1.05)$ by using first two, then three, then four terms of the Taylor expansion about $x_0 = 1$.

EXERCISE 18. Use the Taylor expansion for e^{-t} about $t_0 = 0$ through terms of order 5 to obtain an approximate value of

$$\int_0^1 \frac{1-e^{-t}}{t}\, dt$$

Find a bound on the error.

Exercise 19. The approximation $\sin x \approx x$ is often used for small x. For what values of x does the approximation have an error not exceeding .001 in absolute value?

Exercise 20. The function $f(x) = [1/(2\pi)^{1/2}]e^{-x^2/2}$ is called the *normal curve of error*. Obtain the Taylor expansion to

$$\varphi(x) = \frac{1}{(2\pi)^{1/2}} \int_0^x e^{-t^2/2}\, dt$$

through the first three nonzero terms, and calculate an approximate value for $\varphi(.2)$. Compare your result with the tabulated value .0793.

4-4   Linear Interpolation

In addition to Taylor polynomials there are other readily **constructable polynomial** approximations of a function. In this section we introduce the ideas of interpolation theory by studying **linear interpolation**. More general aspects of this subject are given in the next section.

Let x_0, x_1 be two distinct real numbers, and suppose that $f(x)$ is defined at these points. Then the linear function $P(x)$ which satisfies

$$P(x_0) = f(x_0)$$
$$P(x_1) = f(x_1)$$

is given by

$$P(x) = \frac{x - x_1}{x_0 - x_1} f(x_0) + \frac{x - x_0}{x_1 - x_0} f(x_1) \qquad (4\text{-}18)$$

This is readily verified by the substitution of x_0, x_1 into the right side of Eq. (4-18). The use of the approximation

$$P(x) \approx f(x)$$

is called **linear interpolation**. The following theorem gives an expression for the remainder, or **truncation error**, in linear interpolation. In this theorem and in other places to follow we use the notation min $(x_0, x_1, \ldots, x_n)$ to mean the smallest of the real numbers $x_0, x_1, \ldots, x_n$. The notation max $(x_0, x_1, \ldots, x_n)$ means the largest of the real numbers $x_0, x_1, \ldots x_n$.

THEOREM 4-4-1. Let $f(x) \in C^2[a, b]$, where $a \le x_0 < x_1 \le b$. Then for each $x \in [a, b]$ there exists a number ξ, depending upon x and satisfying

$$\min (x_0, x_1, x) < \xi < \max (x_0, x_1, x) \qquad (4\text{-}19)$$

such that

$$f(x) = \frac{x - x_1}{x_0 - x_1} f(x_0) + \frac{x - x_0}{x_1 - x_0} f(x_1) + \frac{(x - x_0)(x - x_1)}{2} f''(\xi) \quad (4\text{-}20)$$

Proof. To simplify the argument, we shall first prove a lemma. This lemma is a simple extension of Rolle's theorem.

Lemma 4-4-1. Let $g(t) \in C^2[p_1, p_3]$, where $p_1 < p_2 < p_3$. If $g(p_i) = 0$ for $i = 1, 2, 3$, then there exists a $\xi \in (p_1, p_2)$ such that $g''(\xi) = 0$.

Proof. The hypotheses of Rolle's theorem are satisfied on $[p_1, p_2]$ and on $[p_2, p_3]$. Hence there exist $\xi_1 \in (p_1, p_2)$ and $\xi_2 \in (p_2, p_3)$ such that $g'(\xi_1) = 0$ and $g'(\xi_2) = 0$.

To continue, on the interval $[\xi_1, \xi_2]$ we have $g'(\xi_1) = 0$, $g'(\xi_2) = 0$ and $g''(x)$ exist. Thus by Rolle's theorem there exists a $\xi \in (\xi_1, \xi_2)$ such that $g''(\xi) = 0$. This prove Lemma 4-4-1.

To prove Theorem 4-4-1, we shall construct a special function satisfying the hypotheses of the above lemma. First notice that if $x = x_0$ or $x = x_1$ then Eq. (4-20) holds independently of the choice of $\xi \in [a, b]$. Let $x \in [a, b]$, $x \neq x_0$, $x \neq x_1$, be fixed. Then for this x define a function $g(t)$ by

$$g(t) = f(t) - P(t) - [f(x) - P(x)] \frac{(t - x_0)(t - x_1)}{(x - x_0)(x - x_1)}$$

Here $P(t)$ is the linear interpolation polynomial to $f(t)$ at x_0, x_1. By direct substigution it is easy to verify that $g(t) = 0$ at the three distinct points $t = x_0$, $t = x_1$, $t = x$. Moreover, $g''(t)$ can be seen to exist on $[a, b]$ since

$$g''(t) = f''(t) - \frac{f(x) - P(x)}{(x - x_0)(x - x_1)} (2) \quad (4\text{-}21)$$

Now by Lemma 4-4-1 there exists a ξ satisfying Eq. (4-19) such that $g''(\xi) = 0$. Using this in Eq. (4-21), we get

$$0 = f''(\xi) - 2 \frac{f(x) - P(x)}{(x - x_0)(x - x_1)}$$

Hence, solving for $f(x)$, we conclude

$$f(x) = P(x) + 2 \frac{(x - x_0)(x - x_1)}{2} f''(\xi)$$

The truncation error in linear interpolation is

$$\frac{(x - x_0)(x - x_1) f''(\xi)}{2} \quad (4\text{-}22)$$

As in Taylor's remainder formula the ξ usually depends on x and is often written as $\xi(x)$. The following examples illustrate typical uses of **linear interpolation**.

EXAMPLE 4-12. Given $f(x) = \sin x$, $f(.1) = 0.9983$, $f(.2) = .19867$, use the method of linear interpolation to find an approximate value for $f(.16)$, and obtain a bound on the truncation error.

Solution. We have

$$P(.16) = \frac{.16-.2}{.1-.2}(.09983) + \frac{.16-.1}{.2-.1}(.19867)$$

$$= (.4)(.9983) + (.6)(.19867) = .159134$$

$$\approx .15913$$

Because $f''(x) = -\sin x$, the truncation error is

$$\frac{(x-.1)(x-.2)(-\sin\xi)}{2}$$

The maximum value of $|-\sin x|$ on $[.1,.2]$ is given by $\sin .2 = .19867$.
Thus

$$\left|f(.16) - P(.16)\right| < (.19867)\left|\frac{(.16-.1)(.16-.2)}{2}\right|$$

$$= (.19867)(.0012) \approx .00024$$

The exact result given to five decimal places is .15932. For convenience we often denote the number of decimal places by the notation "to n D." This means that the number is given correctly to n decimal places. For example, .159324 to 5 D is .15932.

EXAMPLE 4-13. For $f(x) = 5^x$ one easily obtains $f(0) = 1$ and $f(.5) = 2.23608$. Find an approximation to $f(.3)$ using **linear interpolation**, and find a bound on the truncation error.

Solution.

$$P(.3) = 2/5(1.00000) + 3/5(2.23608) = 1.741648$$

To bound the truncation error, we need an expression for $f''(x)$. Using

$$f(x) = 5^x = e^{x \ln 5}$$

we obtain

$$f'(x) = (\ln 5)e^{x \ln 5}$$
$$f''(x) = (\ln 5)^2 e^{x \ln 5}$$

Thus

$$\max_{0 \le x \le .5} |f''(x)| = (\ln 5)^2 e^{.5 \ln 5} = 5^{1/2}(\ln 5)^2$$

Using the result $\ln 5 < 1.61$ obtained from a table we conclude

$$\left|f(.3) - P(.3)\right| < (.03)5^{1/2}(\ln 5)^2 \approx .17$$

EXAMPLE 4-14. Suppose that $x_0 < x < x_1$ and $|f''(x)| < M$ on $[x_0, x_1]$. Find a bound on the worst possible truncation error for linear interpolation on $[x_0, x_1]$.

Solution. The truncation error is bounded by

$$\max_{x_0 \le x \le x_1} \left| \frac{(x - x_0)(x - x_1)}{2} \right| \max_{x_0 \le \xi \le x_1} \left| f''(\xi) \right|$$

By elementary calculus it can be verified that the maximum value of $|(x - x_0)(x - x_1)|$ occurs at the point $(x_0 + x_1)/2$. Hence we obtain the result

$$|f(x) - P(x)| \le \frac{M(x_1 - x_0)^2}{8}$$

EXAMPLE 4-15. A table of a certain function $f(x)$ is to be constructed for equally spaced x on $[a,b]$ so that the maximum truncation error using linear interpolation in the table is less than a given $\xi > 0$. Indicate how to find a suitable spacing for the table.

Solution. The function $f(x)$ will be tabulated at points $a, a + h, a + 2h, \ldots,$ $a + nh = b$, where $h = (b - a)/n$. Here h is the spacing, or **step size**, of the table. By using the result of Example 4-14 it follows that the truncation error for interpolation between successive tabular values is bounded by

$$\frac{h^2}{8} \max_{a \le \xi \le b} \left| f''(\xi) \right|$$

If a table for $f(x) = \sin x$ is desired over $[1, 2]$ with an interpolation truncation error not exceeding 5×10^{-4}, a suitable h can be determined by noting that

$$\frac{h^2}{8} \max_{\xi, [1,2]} \left| f''(\xi) \right| \le \frac{h^2}{8}$$

This requires that $h^2/8 \le 5 \times 10^{-4}$ or $h \le 0.063$ Probably an $h = .05$ would be used in order to make the table more convenient.

EXERCISE 21. Verify that Eq. (4-18) is the equation of the straight line joining the points $(x_0, f(x_0))$, $(x_1, f(x_1))$.

EXERCISE 22. Let $f(x) = \cos x, x_0 = 0, x_1 = .1$. Use linear interpolation to calculate an approximate value for $f(.04)$, and obtain a bound on the truncation error.

EXERCISE 23. Do the same for $f(x) = \ln (1 + x)$, where $x_0 = 1, x_1 = 1.1,$ $x = 1.04$.

EXERCISE 24. Do the same for $f(x) = (1 + x)^6$, where $x_0 = .01, x_{1,} = .03,$ $x = .02$.

EXERCISE 25. Do the same for $f(x) = e^x, x_0 = 0, x_1 = .3, x = .25$.

EXERCISE 26. Using $\sin (1.0) = .84147$ and $\sin(1.2) = .93204$, find an approximate value for $\sin(1.4)$, by linear interpolation. Obtain a bound on the truncation error.

EXERCISE 27. Determine an appropriate step size to use in the construction of a table of $f(x) = e^x$ on $[0,1]$. The truncation error for linear interpolation is to be bounded by .00005.

EXERCISE 28. Suppose that $g(t) \in C^n[t_1, t_{n+1}]$, where $t_1 < t_2 < \ldots < t_{n+1}$, and $g(t_i) = 0$ ($i = 1, \ldots, n + 1$). Prove that there exists a $\xi \in (t_1, t_{n+1})$ such that $g^{(n)}(\xi) = 0$.

4-5 LAGRANGE INTERPOLATION

In this section, we shall generalize the results of Sec. 4.4. Suppose that x_0, $x_1, \ldots, x_n$ are $n + 1$ distinct points and that the values of a function $f(x)$ are known on these points. The problem is to find a polynomial

$$P(x) = a_1 + a_2 x + a_3 x^2 + \cdots + a_{n+1} x^n$$

with the property that

$$P(x_i) = f(x_i) \quad \text{for } i = 0, 1, 2, \ldots, n \tag{4-23}$$

The resulting polynomial is called the **Lagrange interpolating polynomial** to $f(x)$ on the points $x_0, x_1, x_2, \ldots, x_n$.

First we shall formulate the problem as a system of simultaneous linear equations. The coefficients $a_1, a_2, \ldots, a_{n+1}$ are to be determined, and we want

$$\sum_{j=1}^{n+1} a_j x_i^{j-1} = f(x_i) \quad \text{for } i = 0, 1, 2, \ldots, n \tag{4-24}$$

In matrix notation this becomes

$$\begin{bmatrix} 1 & x_0 & x_0^2 & \cdots & x_0^n \\ 1 & x_1 & x_1^2 & \cdots & x_1^2 \\ \multicolumn{5}{c}{\cdots\cdots\cdots\cdots\cdots\cdots} \\ 1 & x_n & x_n^2 & \cdots & x_n^n \end{bmatrix} \begin{bmatrix} a_1 \\ a_2 \\ \cdots \\ a_{n+1} \end{bmatrix} = \begin{bmatrix} f(x_0) \\ f(x_1) \\ \cdots \\ f(x_n) \end{bmatrix} \tag{4-25}$$

The matrix of coefficients is of a special sort and is called the **Vandermonde matrix** on $x_0, x_1, \ldots, x_n$. It can be proved, using standard techniques of matrix algebra, that this matrix has an inverse, and consequently that Eq. (4-25) has a unique solution. However, we shall take a different approach.

We shall show that the system (4-23) has a solution by actually constructing a polynomial with the required properties. Then we shall show that this is the only polynomial of degree $< n$ with these properties. From this, it follows that the system (4-23) has a unique solution. Then, by the results of Chapter 2 it follows that the Vandermonde matrix has an inverse.

Let us begin by considering a simpler problem. Let us suppose that x_0, $x_1, \ldots, x_n$ are distinct given points and that one wants to find a polynomial $L_i(x)$ of degree n with the property that

$$L_i(x_j) = 0 \quad j \neq i; j = 0, 1, \ldots, n$$

$$L_i(x_i) = 1 \tag{4-26}$$

The problem of constructing a polynomial with a certain set of roots is simple. If the roots are to be $r_1, r_2, \ldots, r_n$ then such a polynomial is given by

$$Q(x) = k(x - r_1)(x - r_2) \cdots (x - r_n) \tag{4-27}$$

If $k \neq 0$, then $r_1, r_2, \ldots, r_n$ are the only roots of $Q(x)$.

We can find a polynomial satisfying Eqs. (4-26) by following the idea of Eq. (4-27) and choosing k in the proper manner. The roots are to be

$$x_0, x_1, \ldots, x_{i-1}, x_{i+1}, \ldots, x_n$$

Hence the solution must be of the form

$$L_i(x) = k(x - x_0)(x - x_1) \cdots (x - x_{i-1})(x - x_{i+1}) \cdots (x - x_n) \tag{4-28}$$

The constant k is to be selected so that $L_i(x_i) = 1$. It is easy to see that

$$L_i(x_i) = k(x_i - x_0)(x_i - x_1) \cdots (x_i - x_{i-1})(x_i - x_{i+1}) \cdots (x_i - x_n) \tag{4-29}$$

If we set Eq. (4-29) equal to 1 and solve for k, the final result is

$$L_i(x) = \frac{(x - x_0)(x - x_1) \cdots (x - x_{i-1})(x - x_{i+1}) \cdots (x - x_n)}{(x_i - x_0)(x_i - x_1) \cdots (x_i - x_{i-1})(x_i - x_{i+1}) \cdots (x_i - x_n)} \tag{4-30}$$

This is clearly a solution to the problem (4-26).

It is convenient to use a more compact notation. two commonly used notations for Eq. (4-30) are

$$L_i(x) = \frac{\prod\limits_{\substack{j=0 \\ j \neq i}}^{n} (x - x_j)}{\prod\limits_{\substack{j=0 \\ j \neq i}}^{n} (x_i - x_j)}$$

$$L_i(x) = \prod\limits_{\substack{j=0 \\ j \neq i}}^{n} \frac{x - x_j}{x_i - x_j}$$

EXAMPLE 4-16. Suppose that $n = 2$, $x_0 = 1$, $x_1 = 1.5$, $x_2 = 3$. Find $L_0(x)$, $L_1(x)$, $L_2(x)$.

Solution. By substitution into Eq. (4-30) we get

$$L_0(x) = \frac{(x - x_1)(x - x_2)}{(x_0 - x_1)(x_0 - x_2)} = \frac{(x - 1.5)(x - 3)}{1}$$

$$L_1(x) = \frac{(x - x_0)(x - x_2)}{(x_1 - x_0)(x_1 - x_2)} = \frac{(x - 1)(x - 3)}{-.75}$$

$$L_2(x) = \frac{(x - x_0)(x - x_1)}{(x_2 - x_0)(x_2 - x_1)} = \frac{(x - 1)(x - 1.5)}{3}$$

DEFINITION 4-1. The functions $L_i(x)$ defined by Eq. (4-30) are called the Lagrange coefficient polynomials.

Using the Lagrange coefficient polynomials, we can easily solve problem (4-21).

THEOREM 4-5-1. Let $x_0, x_1, \ldots x_n$ be distinct points, and suppose that $f(x)$ is given at these points. The unique polynomial $P(x)$ of degree $\leq n$ with the property that

$$P(x_i) = f(x_i) \qquad i = 0, 1, 2, \ldots, n$$

is given by

$$P(x) = L_0(x)f(x_0) + L_1(x)f(x_1) + \cdots + L_n(x)f(x_n) \qquad (4\text{-}31)$$

Proof. The $L_i(x)$ are polynomials, each of degree n. Hence Eq. (4.31) is a polynomial of degree $\leq n$. [Keep in mind that the $f(x_i)$ are known values.] We must show that this polynomial has the required properties.

First we shall show that $P(x_0) = f(x_0)$. We know from Eqs. (4-26) that

$$L_0(x_0) = 1$$
$$L_1(x_0) = 0$$
$$L_2(x_0) = 0$$
$$\ldots\ldots\ldots\ldots$$
$$L_n(x_0) = 0$$

Thus

$$P(x_0) = \sum_{j=0}^{n} L_j(x_0)f(x_j)$$
$$= 1f(x_0) + 0f(x_1) + \cdots + 0f(x_n)$$
$$= f(x_0)$$

More generally now let $0 \leq i \leq n$ be arbitrary. Then

$$P(x_i) = \sum_{j=0}^{n} L_j(x_i)f(x_j)$$
$$= 0 \in f(x_0) + \cdots + 0 \in f(x_{i-1}) + 1 \in f(x_i)$$
$$\quad + 0 \in f(x_{i+1}) + \cdots + 0 \in f(x_n)$$
$$= f(x_i)$$

Next we shall prove that $P(x)$ is the unique solution. Suppose that there are two polynomials $P(x)$, $Q(x)$, each of degree $< n$ with the properties

$$\left.\begin{array}{l} P(x_1) = f(x_2) \\ Q(x_2) = f(x_2) \end{array}\right\} \qquad i = 0, 1, \ldots, n$$

Then the difference of these two polynomials $D(x) = P(x) - Q(x)$ is itself a polynomial of degree $< n$. It has the properties

$$D(x_i) = P(x_i) - Q(x_i) = f(x_i) - f(x_i) = 0 \qquad i = 0, 1, \ldots, n$$

That is, $D(x)$ is a polynomial of degree $< n$ which is zero at $x_0, x_1, \ldots, x_n$. We know that a polynomial of degree n can have at most n zeros unless all its coefficients are zero. But, if all the coefficients of $D(x)$ are zero, then $P(x)$ and $Q(x)$ must be exactly the same polynomial. Hence there is only one polynomial which satisfies the desired conditions.

COROLLARY 4-5-1. The Vandermonde matrix on the distinct points x_0, $x_1, \ldots, x_n$ has an inverse.

EXAMPLE 4-17.   Find a polynomial of degree < 2 with the properties $P(1) = 5$, $P(1.5) = -3$, $P(3) = 0$.

Solution. The Lagrange coefficient polynomials for this problem were computed in Example 4-16. Thus, using Eq. (4-31), we get

$$P(x) = \frac{(x-1.5)(x-3)(5)}{1} + \frac{(x-1)(x-3)(-3)}{-.75} + \frac{(x-1)(x-1.5)(0)}{3}$$

This, of course, can be simplified; the result is

$$P(x) = 9x^2 - 38.5x + 34.5$$

The remainder formula for **Lagrange interpolation** on $n + 1$ points is similar in nature to the remainder for the Taylor expansion of degree n.

THEOREM 4-5-2.   Let n be fixed, and suppose that $f(x)$ is given on the distinct points $x_0, x_1, \ldots, x_n$. Further suppose that x is a point at which $f(x)$ is to be approximated by the interpolating polynomial $P(x)$. Let I be the smallest interval containing $x, x_0, x_1, \ldots, x_n$. Finally, suppose that $f(x) \in C^{n+1}[I]$. Then there exists a ξ in I which depends on $x, x_0, x_1, \ldots, x_n$ and $f(x)$ such that

$$f(x) = \sum_{i=0}^{n} L_2(x) f(x_i) + R(x)$$

where

$$R(x) = (x - x_0)(x - x_1) \cdots (x - x_n) \frac{f^{(n+1)}(\xi)}{(n+1)!} \qquad (4\text{-}32)$$

Proof. The proof parallels the proof given for linear interpolation. First one obtains an extension of Rolle's theorem to $n + 2$ points (see Exercise 28). Thus, if $g(t) \in C^{n+1}[I]$ and has $n + 2$ distinct zeros in I, then there exists a point $\xi \in I$ such that $g^{n+1}(\xi) = 0$.

Remainder (4-32) is zero for $x = x_0, x_1, x_2, \ldots, x_n$ regardless of the value of $\xi \in I$. Thus in these cases the theorem is proved. Now suppose that x is distinct from the points $\{x_i\}$. We define a function $g(t)$ by

$$g(t) = f(t) - P(t) - [f(x) - P(x)] \frac{(t - x_0)(t - x_1) \ldots (t - x_n)}{(x - x_0)(x - x_1) \ldots (x - x_n)} \qquad (4\text{-}33)$$

Then it is a simple matter to verify that $g(t) = 0$ for $t = x_0, x_1, \ldots, x_n, x$. Hence there exists an $\xi \in I$ such that $g^{n+1}(\xi) = 0$. This gives (see Exercise 37)

$$0 = f^{(n+1)}(\xi) - \frac{[f(x) - P(x)](n+1)!}{(x - x_0)(x - x_1) \cdots (x - x_n)} \tag{4-34}$$

Upon solving this equation for $f(x)$, we obtain

$$f(x) = P(x) + \frac{(x - x_0)(x - x_1) \cdots (x - x_n) f^{(n+1)}(\xi)}{(n+1)!}$$

EXAMPLE 4-18. Calculate an approximate value for ln 1.2 given

$$\ln 1.0 = 0$$
$$\ln 1.1 = .09531$$
$$\ln 1.3 = .26236$$

Find a bound on the truncation error.

Solution Here $x_0 = 1.0$, $x_1 = 1.1$, $x_2 = 1.3$. If we let $P(1.2)$ denote the value given by interpolation, then

$$P(1.2) = \frac{(1.2 - 1.1)(1.2 - 1.3)(0)}{(1 - 1.1)(1 - 1.3)} + \frac{(1.2 - 1)(1.2 - 1.3)(.09531)}{(1.1 - 1)(1.1 - 1.3)}$$
$$+ \frac{(1.2 - 1)(1.2 - 1.1)(.26236)}{(1.3 - 1)(1.3 - 1.1)}$$
$$= .09531 + \frac{.26236}{3} \approx .18276$$

To bound the truncation error, we first compute the third derivative of $f(x) = \ln x$ (see Example 4-11). Then the absolute value of the remainder is

$$\frac{|(1.2 - 1)(1.2 - 1.1)(1.2 - 1.3)|}{6} \frac{2}{\xi^3}, \qquad \text{where } 1 \leq \xi \leq 1.3$$

Selecting ξ to give the maximum of the above expression (that is, $\xi = 1$), we have

$$|\ln 1.2 - P(1.2)| \leq \frac{(.2)(.1)(.1)(2)}{6} = \frac{2}{3} \times 10^{-3}$$

EXAMPLE 4-19. Tables of values of the function

$$f(x) = \int_0^x e^{t^2} dt$$

are often available. Suppose that $f(1.15)$ is to be computed from a table giving $f(1, 0)$, $f(1.1)$, $f(1.2)$. Find a bound on the truncation error which results from using an interpolating polynomial to $f(x)$ at the three points indicated, to approximate $f(1.15)$.

Solution.

$$f'(x) = e^{z^2} \qquad f''(x) = 2xe^{z^2} \qquad f'''(x) = (4x^2 + 2)e^{z^2}$$

Thus, $\max\limits_{1\le x\le 1.2}|f'''(x)| = [4(1.2)^2 + 2]e^{1.44} < 33$, and the truncation error is bounded by

$$|(1.15-1)(1.15-1.1)(1.15-1.2)|33/6 < .0021$$

EXAMPLE 4-20. Suppose that one wishes to construct a table of values of sin(x) on $[0, \pi/4]$ at the points $0, h, 2h, 3h, \ldots, mh = \pi/4$. The size of h (and hence m) is to be determined so that interpolation on any three successive points (a second-degree polynomial being used) to approximate a value sin(x) in the interval of the three points, will give a **truncation error** of less than .000001.

Solution. Let $nh, (n + 1)h, (n + 2)h$ denote three successive points. Since $|f''(x)| \le 1$ on $[0, \pi/4]$, the truncation error is bounded by

$$\frac{|(x - nh)(x - h - nh)(x - 2h - nh)|}{6}$$

where $nh < x < (n + 2)h$.

To simplify the work in finding the maximum of this expression, we make the change of variable $t = x - h - nh$. The problem then is to find the maximum of $|(t + h)(t)(t - A)|/6$, where $-h < t < h$. This calculus problem is left as an exercise for the reader. The result is that

$$\max_{-h\le t\le h} \frac{|(t+h)(t)(t-h)|}{6} = \frac{h^3}{(9)(3)^{1/2}}$$

Thus, we want $h^3/(9)(3)^{1/2} < .000001$; so $h^3 < (9)(3)^{1/2} \times 10^{-6}$, and $h < 3^{1/2} 10^{-2} \approx .0173$.

EXAMPLE 4-21. Suppose that we wish to compute sin(1.2) from a table of values which gives $\sin(x_0)$ only at $x = .1, .2, .3, \ldots, .9, 1.0$. The computation is to be based on interpolation on the points

$$x_0 = 1.0$$
$$x_1 = .9$$
$$\ldots\ldots\ldots$$
$$x_n = 1 - .1n.$$

How big should n be to guarantee a truncation error of less than .0001?

Solution. For the function sin (x) we know that $|(d^{n+1} \sin x)/dx^{n+1}| < 1$ for all $n > 0$. Hence, the truncation error is bounded by

$$T_n = \frac{|(1.2-1.0)(1.2-.9) \cdots [1.2-1+.1n]|}{(n+1)!}$$

A short table of values of the above expression is easily computed.

n	T_n
2	.004
3	.0005
4	.0006

Hence, $n = 4$ gives the desired accuracy.

EXERCISE 29. The Bessel function $J_0(x)$ has the values $J_0(2.0) = .2239$, $J_0(2.1) = .1666$, $J_0(2.2) = .1104$. Find $J_0(2.03)$ by interpolation, and find a bound on the truncation error. Use the fact that $|J_0'''(x)| < 1$ for all x.

EXERCISE 30. The values of $(1 + x)^{50}$ for $x_0 = .01$, $x_1 = .0125$, $x_2 = .015$ can be found in a compound interest table. Find an approximate value for this function at $x = .013$, and find a bound on the truncation error.

EXERCISE 31. Suppose that $f(x) = \sin(x)$ is to be approximated on $[0,1]$ by an interpolating polynomial on $n + 1$ equally spaced points $0 = x_0 < x_1 < x_2$, $\ldots < x_n = 1$. Determine n so that the truncation error will be less than .01 for x in this interval.

EXERCISE 32. Suppose that $x_0, x_1, \ldots, x_n$ are distinct and that $L_i(x)$ for $i = 0$, $1, \ldots, n$ are the Lagrange coefficient polynomials. Prove that

$$\sum_{i=0}^{n} L_i(x) \equiv 1$$

Hint: Consider the interpolating polynomial which approximates the function $f(x) = 1$ [that is, $f(x)$ is just the constant 1].

EXERCISE 33. Suppose that $P(x)$ is a polynomial of degree $< n$. Prove that, if $x_0, x_1, \ldots, x_n$ are distinct, then

$$\sum_{i=0}^{n} L_i(x) P(x_i) \equiv P(x)$$

EXERCISE 34. The function

$$f(x) = \int_0^x e^{t^2} dt$$

has values $f(.9) = 1.2155$, $f(.91) = 1.2382$, $f(.92) = 1.2613$. Find an approximate value for $f(.912)$, and bound the truncation error.

EXERCISE 35. Values of $\Phi(x) = [1/(2\pi)^{1/2}]e^{-(x^2)/2}$ can be found in mathematics tables. By interpolation on the points $x_0 = 0$, $x_1 = .05$, $x_2 = .10$, find an approximate value of $\Phi(.04)$. Compare your result with the tabulated value, and find a bound on the truncation error.

EXERCISE 36. The equation $x - 9^{-x} = 0$ has a solution in $[0,1]$. Find the interpolating polynomial on $x_0 = 0$, $x_1 = .5$, $x_2 = 1$. By setting the interpolating polynomial equal to zero and solving, find an approximate solution to $x - 9^{-x} = 0$.

Exercise 37. Prove that, if the function $g(t)$ in Eq. (4-33) is differentiated $n + 1$ times with respect to t, the result is

$$g^{(n+1)}(t) = f^{(n+1)}(t) - \frac{[f(x) - P(x)](n+1)!}{(x - x_0)(x - x_1) \, \cdots \, (x - x_n)}$$

Exercise 38. Let $g(t) \in C^{n+1}[I]$, and suppose, for two distinct points x_0 and $x \in I$, that $g(x) = 0$, $g(x_0) = 0$, $g'(x_0) = 0$, $g''(x_0) = 0, \ldots, g^{(n)}(x_0) = 0$. Prove that there exists a $\xi \in I$ such that $g^{(n+1)}(\xi) = 0$. Hint: Use Rolle's theorem on the interval with end points x_0, x to get an ξ_1 between x_0 and x such that $g'(\xi_1) = 0$. Then repeat the process on the interval with end points x_0 and ξ_1.

Exercise 39. Using the result of Exercise 38 and the special function

$$g(t) = f(t) - P(t) - [f(x) - P(x)] \frac{(t - x_0)^{n+1}}{(x - x_0)^{n+1}}$$

where $x \neq x_0$ and $P(t)$ is a Taylor polynomial, derive remainder (4-14) for Taylor's series.

4-6   Iterated Interpolation

In this section we shall give a convenient means for implementing **Lagrange interpolation** on a computer. The ideas to be discussed are due to Aitken and Neville. It should be evident that a computer program could be written to evaluate Eq. (4-31) at a particular point. Indeed, there is room for considerable ingenuity in developing an efficient scheme for evaluating each of the Lagrange coefficient polynomials $L_i(x)$. This is because there is a considerable savings in computation if certain results in computing any one of the $L_i(x)$ are saved and used in evaluating subsequent coefficient polynomials.

In many applications of Lagrange interpolation the proper choice of the number of points to use is not readily evident. Then it is common to perform the interpolation for a sequence of points. The results obtained are compared, and the computation is stopped when two successive interpolated values are sufficiently close together. It can be seen that form (4-31) of the Lagrange interpolation polynomial requires that a substantial amount of the calculation for the case n be repeated in order to calculate the case $n + 1$. The **iterated interpolation** scheme to be discussed below is particularly efficient in such a case.

Let $f(x)$ be given on a set of distinct points $x_1, x_2, \ldots, x_m$. The points need not have any particular arrangement. If $n_1, n_2, \ldots, n_k$ denote k distinct integers in $[1, m]$, then let

$$P_{n_1, n_2, \ldots, \, n_k}(x)$$

denote the Lagrange interpolation polynomial of degree $\leq k - 1$ to $f(x)$ on $x_{n1}, x_{n2}, \ldots, x_{nk}$. For example

$$P_{1,3,7}(x)$$

denotes a polynomial of degree ≤ 2 which interpolates $f(x)$ on $x_1 \ x_3, x_7$.

THEOREM 4-6-1. Using the above definition,

$$P_{1,2,\ldots,k,k+1}(x) = \frac{1}{x_{k+1}-x_k}\begin{vmatrix} x-x_k & P_{1,2,\ldots,k}(x) \\ x-x_{k+1} & P_{1,2,\ldots,k-1,k+1}(x) \end{vmatrix} \tag{4-35}$$

Proof. This theorem gives a formula for an interpolation polynomial on $k + 1$ points in terms of formulas for interpolation polynomials for k points. Since

$$P_{1,2,\ldots,k}(x)$$
$$P_{1,2,\ldots,k-1,k+1}(x)$$

are each polynomials of degree $\leq k - 1$, the right side of Eq. (4-35) is a polynomial of degree $\leq k$. We must show that it agrees with $f(x)$ at $x_1, x_2, \ldots, x_{k+1}$

First consider a point $x = x_i$, where $1 < i < k - 1$. Then the right side of Eq. (4-35) has the value

$$\frac{1}{x_{k+1}-x_k}\begin{vmatrix} x_i-x_k & f(x_i) \\ x_i-x_{k+1} & f(x_i) \end{vmatrix} = f(x_i)$$

Next consider the case $x = x_k$. Then the right side of Eq. (4-35) is

$$\frac{1}{x_{k+1}-x_k}\begin{vmatrix} 0 & f(x_k) \\ x_k-x_{k+1} & P_{1,2,\ldots,k-1,k+1}(x_k) \end{vmatrix} = f(x_k)$$

The verification for the case $x = x_{k+1}$ is similar to the case $x = x_k$ and is left as Exercise 40.

Table 4-1 illustrates the arrangement of the work needed to construct $P_{1,2,3,4}(x)$. This table may be constructed one column at a time or one row at a time. In any case the steps of the construction are justified by repeatedly appealing to Theorem 4-6-1. Equations (4-36) to (4-38) give the steps for constructing Table 4-1 one row at a time. Later we give a program implementing this construction in the more general case. It is left as Exercise 47 to write a program for the column-by-column implementation of the iterated interpolation algorithm.

TABLE 4-1

x_1	$x-x_1$	$P_1(x)$			
x_2	$x-x_2$	$P_2(x)$	$P_{1,2}(x)$		
x_3	$x-x_3$	$P_3(x)$	$P_{1,3}(x)$	$P_{1,2,3}(x)$	
x_4	$x-x_4$	$P_4(x)$	$P_{1,4}(x)$	$P_{1,2,4}(x)$	$P_{1,2,3,4}(x)$

Construction of second row of Table 4-1

$$P_{1,2}(x) = \frac{1}{x_2-x_1}\begin{vmatrix} x-x_1 & P_1(x) \\ x-x_2 & P_2(x) \end{vmatrix} \tag{4-36}$$

Construction of third row of Table 4-1

$$P_{1,3}(x) = \frac{1}{x_3 - x_1} \begin{vmatrix} x - x_1 & P_1(x) \\ x - x_3 & P_3(x) \end{vmatrix}$$

$$\qquad(4\text{-}37)$$

$$P_{1,2,3}(x) = \frac{1}{x_3 - x_2} \begin{vmatrix} x - x_2 & P_{1,2}(x) \\ x - x_3 & P_{1,3}(x) \end{vmatrix}$$

Construction of fourth row of Table 4-1

$$P_{1,4}(x) = \frac{1}{x_4 - x_1} \begin{vmatrix} x - x_1 & P_1(x) \\ x - x_4 & P_4(x) \end{vmatrix}$$

$$P_{1,2,4}(x) = \frac{1}{x_4 - x_2} \begin{vmatrix} x - x_2 & P_{1,2}(x) \\ x - x_4 & P_{1,4}(x) \end{vmatrix} \qquad(4\text{-}38)$$

$$P_{1,2,3,4}(x) = \frac{1}{x_4 - x_3} \begin{vmatrix} x - x_3 & P_{1,2,3}(x) \\ x - x_4 & P_{1,2,4}(x) \end{vmatrix}$$

EXAMPLE 4-22. Find some approximations to the value of log 4.5, using the values tabulated below:

x	4.0	4.2	4.4	4.6	4.8
log x	.60206	.62325	.64345	.66276	.68124

Solution. First we shall complete Table 4-1, using $x_1 = 4.0$, $x_2 = 4.2$, $x_3 = 4.4$, $x_4 = 4.6$, and $x = 4.5$. The result is

4.0	.5	.60206			
4.2	.3	.62325	.65504		
4.4	.1	.64345	.65380	.65318	
4.6	−.1	.66276	65264	.65324	.65321

Each of the entries in the last three columns is an interpolation approximation to log 4.5. Perhaps of most interest are the rightmost entries in each row. These give us

$$P_{1,2}(4.5) = .65504$$
$$P_{1,2,3}(4.5) = .65318$$
$$P_{1,2,3,4}(4.5) = .65321$$

Suppose, now, that we wish to compute $P_{1,2,3,4,5}(4.5)$ using the additional point $x_5 = 4.8$. All that is necessary is the computation of an additional row in the table. The additional row is

4.8	−.3	.68124	.65155	.65330	.65321	.65321

Hence $P_{1,2,3,4,5}(4.5) = .65321$. Because $P_{1,2,3,4}(4.5) = P_{1,2,3,4,5}(4.5)$, we take this to be our answer. In this particular example the result obtained agrees exactly with the value of log(4.5) obtained from a five-place table. The Program 4-1 shows how iterated interpolation may be implemented in C++. In this program n (the number of points), xx (the point at which $f(x)$ is to be approximated), and $(x_i, f(x_i))$ $(i = 1, \ldots, n)$ are input to the program. The first two columns of a table similar to Table 4-1 are stored in $x(10)$ and $xm(10)$. Subsequent columns of the table are stored in $p(10,10)$.

```
// PROGRAM 4 - 1 INTERPOLATION
#include <iostream>
using namespace std;
// INTERP
int main(void)
  {
  float x[11],xm[11],p[11][11],xx,t;
  int n,i,j;
  cout<<"Input n and xx: ";
  cin>>n>>xx;
  cout<<"The value of n is: " <<n<< " The value of xx is: "<<xx<<endl;

  for(i = 1; i<=n; ++i)
    {
    cout<<"Input x[i] and p[i][1]:";
    cin>>x[i]>>p[i][1];
    xm[i] = xx - x[i];
    cout<<"x[i] and xm[i]: ";
    cout<<x[i]<<' '<<xm[i]<<endl;
    }
  for(i = 2; i<=n; ++i)
    for(j=2; j<=i; ++j)
      {
      t = xm[j-1]*p[i][j-1] - xm[i]*p[j-1][j-1];
      p[i][j] = t / (x[i] - x[j-1] );
      }

  for(i=1; i<=n; ++i)
    {
    for(j=1; j<= i; ++j)
      cout<<p[i][j]<<' ';
    cout<<endl;
    }

    return(0);
  }
```

We shall conclude this chapter with an application of **iterated interpolation** to the inverse of a given function. Suppose that a function $y = f(x)$ is given on some set of points and that the inverse function $x = f^{-1}(y)$

exists on this set. Ordinarily one would be given a table of values of the function $y = f(x)$. That is, a set of ordered pairs (x_i, y_i) is given. Then a table for the inverse function is given by the ordered pairs (y_i, x_i).

TABLE 4-2		TABLE 4-3	
x	$f(x)$	y	$f^{-1}(y)$
x_1	y_1	y_1	x_1
x_2	y_2	y_2	x_2
...	...	...	...
x_n	y_n	y_n	x_n

EXAMPLE 4-23. Let $f(x) = x^2 + x - 12$. Prepare a table of $f(x)$ and $f^{-1}(y)$ for $x = 2.7, 2.9, 3.2, 3.5$. Find some approximations to $f^{-1}(0)$.

Solution

x	$f(x)$	y	$f^{-1}(y)$
2.7	−3.01	−3.01	2.7
2.9	−.69	−.69	2.9
3.2	1.44	1.44	3.2
3.5	3.75	3.75	3.5

In the above example it can be seen that the equation $f(x) = 0$ has a solution in $[2.7, 3.5]$. To find the solution is equivalent to finding the value of $f^{-1}(0)$. Thus a reasonable way to approximate the solution to $f(x) = 0$ is to find the approximate value of $f^{-1}(0)$ by interpolation in the $f^{-1}(y)$ table. Using iterated interpolation, we get

−3.01	3.01	2.7			
−.69	.69	2.9	2.9595		
1.44	−1.44	3.2	3.0382	2.9850	
3.75	−3.75	3.5	3.0562	2.9745	2.9915

Thus three successive approximations to the desired solution of $f(x) = 0$ are

$$P_{1,2} = 2.9595$$
$$P_{1,2,3} = 2.9850$$
$$P_{1,2,3,4} = 2.9915$$

The exact solution is $x = 3.0000$.

EXERCISE 40. Complete the proof of Theorem 4-6-1.

EXERCISE 41. Write out a table similar to Table 4-1 for the case $n = 5$.

EXERCISE 42. Using iterated interpolation, find $P_{1,2,3}(4.25)$ where $[x_i, f(x_i)]$ are given by

x_i	4.0	4.1	4.2
$f(x_i)$	.60206	.61278	.62325

EXERCISE 43. Find $P_{1,2,3,4}(4.25)$, using the data of Exercise 42 and the additional data $x_4 = 4.3$, $f(x_4) = .63347$. The data for these two exercises are from the function $f(x) = \log(x)$. The value of $\log(4.25)$ to five places is .62839. Compare your result with this.

EXERCISE 44. Using the values $e^{-.50} = .60653$, $e^{-.60} = .54881$, find an approximation to the solution to $x - e^{-x} = 0$.

EXERCISE 45. Repeat Exercise 44, using the additional data

$$e^{-.55} = .57695$$

EXERCISE 46. The Bessel function $J_0(x)$ has values $J_0(2.3) = .0555$, $J_0(2.4) = .0025$, $J_0(2.5) = -.0484$. Using inverse interpolation, find an approximation to the root of $J_0(X) = 0$ which lies in $[2.4, 2.5]$.

EXERCISE 47. Write a program which constructs a table of the form of Table 4-1 one column at a time.

5

Errors and Floating - Point Arithmetic

5-1 INTRODUCTION

Error analysis is one of the most important aspects of numerical analysis. Basically, the problem is that of trying to determine the accuracy, correctness, or meaning of computed results. With a computer it is easy to input vast numbers of data, perform an immense number of calculations, and output reams of results. The following story illustrates how such capabilities should be used only with care, if at all: A few years ago, a student performed an extensive computation on a computer and brought the results to his major professor. The results were truly astounding, and the professor refused to accept them. The student returned to the computer lab and slightly changed the format of the output. The program was run again, giving the same results as before, along with the printed message: "The CDC 1604 says the above results are correct." The professor was duly impressed by this assertion and accepted the results without further question.

Errors may occur at almost any stage of the process of solving a problem. If the statement of the problem is incorrect or contains incorrect data or assumptions, then the solution, no matter how accurately computed, may be meaningless. Often many simplifying assumptions are made in the mathematical analysis of a physical problem. These assumptions must be carefully analyzed and justified. Even after a computer program has been written and the problem relegated to a computer, errors may still occur. Some round-off error may be made in each arithmetic operation. In the solution of some problems a single round-off error at a critical stage of the calculation will significantly affect the final result. In general, then, the cumulative effect of thousands or millions of such errors is usually very difficult to predict. Thus the computed results may not be a solution, or even approximately a solution, to the numerical problem presented to the computer.

Error analysis encompasses the consideration of all the problems mentioned above. Considerable current research is going on in this area. In this chapter we shall be concerned mainly with errors introduced in preparing a problem for solution on a computer and with errors introduced in the actual computation. To study the latter we shall study **floating point arithmetic**, which is the type of arithmetic most often used in extensive scientific computation. An excellent reference for the material discussed in this chapter is Wilkinson.

5-2 ERRORS

In numerical analysis the term **error** does not mean a blunder or mistake. A modern high-speed computer can follow a program involving many millions of calculations without making a mistake.

The concept of error is formalized by two basic definitions.

$$\text{Error} = \text{true value} - \text{approximate value}$$

$$\text{Relative error} = \frac{\text{error}}{\text{true value}}$$

(5-1)

For example, if 2/3 is approximated by .667, then

$$\text{Error} = \frac{2}{3} - \frac{667}{1,000} = -\frac{1}{3} \times 10^{-3}$$

$$\text{Relative error} = -\frac{1}{2} \times 10^{-3}$$

Usually one is interested in obtaining a bound on the absolute value of the error or relative error in an approximate solution to a problem.

In this section we shall list and discuss three sources of error. These are called **inherent error, analytic error**, and **round-off error**. The classification of sources of errors into these three categories is based upon an orderly flow of the solution process, from the statement of the problem, to the analysis of the problem, to the computation. Essentially, the inherent error is that which occurs in the statement of the problem; the analytic error is that occurring in the analysis or translation of the problem into a computational problem; round-off errors are the errors made in solving the computational problem. These terms are discussed in more detail below.

When a problem is first presented to a numerical analyst, it may contain certain data or parameters. If the data or parameters are in some way determined by physical measurement, they will probably differ from the "exact" values. Errors inherent to the statement of a problem are called **inherent errors**. Often one can obtain a bound on these errors by a careful consideration of how the data and parameters were determined.

To illustrate inherent error, suppose that a person wishes to determine the height of a tree growing on level ground (see Fig. 5-1). A distance d from the base of the tree is measured; then the angle θ is measured. The assumption is made that $a = 90°$, and the right triangle is solved for h. The

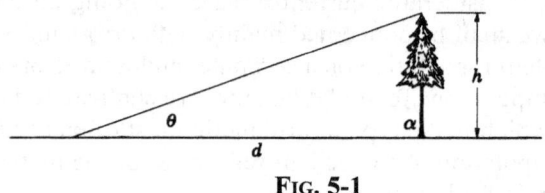

FIG. 5-1

data used are

$$\alpha = 90°(\text{assumed})$$
$$\theta = \text{measured}$$
$$d = \text{measured}$$

Each of these is probably in error. Thus, even if the triangle with these measurements is solved exactly, the resulting value of h may well differ from the exact height of the tree. That is, the final computed answer will be in error because of the errors inherent to the initial data.

Once a problem has been carefully stated, it is time to begin the analysis of the problem. This often involves making certain simplifying assumptions. Examples include neglecting the effect of friction, neglecting the curvature of the earth, assuming that air is homogeneous, neglecting the variation of gravity with the height above the earth's surface, etc. Other simplifying assumptions include approximation of complicated functions by simpler ones (cf. Chap. 4) and the replacement of infinite processes by finite ones. We shall define **analytic error** to be the error introduced by a mathematical analysis transforming a physical or mathematical problem into a computational problem.

$$\sin x \approx x - \frac{x^3}{3!} + \frac{x^5}{5!}$$

involves an analytic error. Similarly, the transformation of the equation

$$e^{-x} - x = 0$$

into the equation

$$\left(1 - x + \frac{x^2}{2!} - \frac{x^3}{3!} + \frac{x^4}{4!} - \frac{x^5}{5!}\right) - x = 0$$

The computation of $\sin(x)$ by the formula involves an **analytic error**. The solution in $[0,1]$ to this equation will be only an approximation to the solution to the equation

$$e^{-x} - x = 0.$$

As a final example, consider again the problem of finding the height of a tree. Suppose that it were known that $a = 89°$. To simplify the calculation, the numerical analyst might still assume the triangle to be a right triangle. This would be an analytic error.

EXERCISE 1. In the triangle in Fig. 5-2 suppose that $\theta = 30°$ and $d = 150$ feet. Compute h for $\alpha = 89°$, $\alpha = 90°$, $\alpha = 91°$.

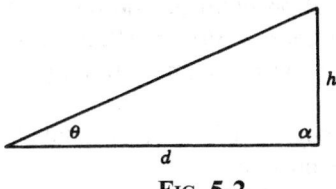

FIG. 5-2

EXERCISE 2. In the above problem, suppose that $\alpha = 90°$ and $d = 150$ feet. Compute h for $\theta = 29°, 30°, 31°$.

EXERCISE 3. In the above problem, suppose that $\theta = 30°$ and $\alpha = 90°$. Compute h for $d = 149$ feet, 150 feet, 151 feet.

EXERCISE 4. Suppose in the above problem that $89° \le \alpha \le 91°$, $29° \le \theta \le 31°$, and 149 feet $\le d \le 151$ feet. Discuss (and solve if you can) the problem of finding the smallest interval in which h must lie.

EXERCISE 5. In determining the area of a circle, it is decided to make an analytic error by using $\pi = 22/7$. Find a good bound on the absolute value of the relative error in the answer obtained.

EXERCISE 6. In Fig. 5-3 suppose that $r_1 = 100$ ohms $\pm 5\%$, while $r_2 = 200$ ohms $\pm 5\%$. Find the largest and smallest possible value of the effective resistance between A and B.

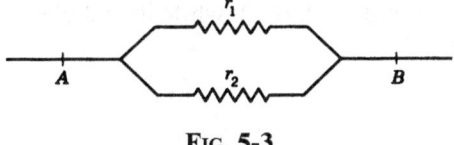

FIG. 5-3

EXERCISE 7. Suppose that 1.414 is used as an approximation to $2^{1/2}$. Find bounds on the absolute value of the error and relative error.

Each of the computer operations +, −, *, / and raising to a power is subject to possible round-off error. We shall use the term **computational error** to denote the cumulative effect of all round-off errors in the computation of a solution to a given computational problem. For any given finite computational problem the computational error can be made arbitrarily small by carrying all calculations to a sufficiently high degree of precision. In actual practice, then, the computational error will depend on the algorithms used and on the precision of the actual computations. We shall discuss this in some detail in the next section. We conclude this section with several additional examples of errors.

In general, most problems formulated by a mathematician as mathematical problems will contain no inherent error. Thus, the problems of finding $5^{1/2}$ or the inverse of

$$H = \begin{bmatrix} 1 & 1/2 \\ 1/2 & 1/3 \end{bmatrix}$$

contain no inherent error.

Given the mathematical problem of finding $5^{1/2}$, we might proceed as follows: Let $p_0 = 2$, $p_{i+1} = .5(p_i + 5/p_i)$, $i = 0, \ldots, 5$, and let p_6 be the answer. The replacement of the problem of finding $5^{1/2}$ by the problem of finding p_6 is an analytic error. To continue, $p_1, p_2, \ldots, p_6$ are computed. The calculations may be performed exactly or may be rounded to a certain number of figures after each step.

To modify the problem slightly, suppose that we wish to solve

$$x^2 - 5x + 5 = 0$$

The solutions are

$$x = \frac{5 + 5^{1/2}}{2} \qquad x = \frac{5 - 5^{1/2}}{2}$$

Considered in this manner, the problem has no inherent or analytic error. Now if a program is written to perform the computation, $5^{1/2}$ will be computed by the computer using SQRT. It is convenient to classify the error in computing $5^{1/2}$ using SQRT as a **computational error**; then it is merely part of the computational error in evaluating $(5 \pm 5^{1/2})/2$.

In Chap. 4, considerable time was spent discussing the analytic error in Lagrange interpolation. This was called the remainder, or truncation error. Often the functional values on which interpolation is performed are themselves in error. Thus, inherent error is also present. We shall illustrate for the case of **linear interpolation**. Suppose that the true values of a function f at x_0, x_1 are $f(x_0), f(x_1)$, while the tabulated values are y_0, y_1, respectively. Thus

$$f(x_0) = y_0 + \varepsilon_0$$
$$f(x_1) = y_1 + \varepsilon_1$$

where ε_0, ε_1 are the errors in the tabulated values. Now suppose that linear interpolation is used on the tabulated values to approximate $f(x_2)$, where x_2 is distinct from x_0, x_1. With the notation of Chap. 4, the exact value of $f(x_2)$ is given by

$$f(x_2) = L_0(x_2)f(x_0) + L_1(x_2)f(x_1) + \frac{(x_2 - x_0)(x_2 - x_1)f''(\xi)}{2}$$

where ξ is in the smallest interval containing x_0, x_1, x_2. The approximate value of $f(x_2)$ is

$$P(x_2) = L_0(x_2)y_0 + L_1(x_2)y_1$$

The total error is thus

$$f(x_2) - P(x_2) = L_0(x_2)\varepsilon_0 + L_1(x_2)\varepsilon_1 + \frac{(x_2 - x_0)(x_2 - x_1)f''(\xi)}{2}$$

The error consists of two parts. The second part

$$\frac{(x_2 - x_0)(x_2 - x_1)f''(\xi)}{2}$$

is the analytic error. The quantity

$$L_0(x_2)\varepsilon_0 + L_1(x_2)\varepsilon_1$$

is the error due to the inherent errors in the data. It is common to call this the inherent errror in linear interpolation.

EXAMPLE 5-1. Suppose that

$$f(1.0) = 4.1 + \varepsilon_0$$
$$f(1.6) = 4.9 + \varepsilon_1$$

where $|\xi_0| < .05$, $|\xi_1| < .05$. Find a bound on the inherent error in evaluating $f(1.4)$ by **linear interpolation.**

Solution. The inherent error is

$$L_0(1.4)\varepsilon_0 + L_1(1.4)\varepsilon_1 = \frac{1.4-1.0}{1.6-1.0}\varepsilon_0 + \frac{1.4-1.6}{1.0-1.6}\varepsilon_1$$
$$= 2/3\varepsilon_0 + 1/3\varepsilon_1$$

Hence, in absolute value, the inherent error does not exceed

$$\left|\frac{2\varepsilon_0}{3} + \frac{\varepsilon_1}{3}\right| \le \frac{2}{3}(.05) + \frac{.05}{3} = .05$$

This example illustrates another point. The approximate value of $f(1.4)$ by linear interpolation is

$$P(1.4) = \frac{2}{3}(4.1) + \frac{4.9}{3} = \frac{13.1}{3} \approx 4.3667$$

Suppose that it is known that $f''(x) = 0$; that is, $f(x)$ is linear. Then there is no analytic error. If the result is stated as

$$P(1.4) = \frac{13.1}{3}$$

then we know that

$$|f(1.4) - P(1.4)| \le .05$$

However, if the result is rounded to one decimal place, an additional error is introduced. With

$$P(1.4) = 4.4$$

the total error $|f(1.4) - P(1.4)|$ may now be as large as the sum of the inherent and rounding errors, namely, .08333 ---. Thus, in order to avoid loss of accuracy, it is often desirable to carry more significant figures in an answer than are given in the problem.

EXAMPLE 5-2. Suppose that a table of $f(x) = e^x$ is given to two decimal places for $x = 1.0$, 1.1, 1.2, . . . , 2.0. Find a bound on the total error in evaluating $f(x)$ at some point in this interval, using linear interpolation.

Solution. Assuming that the table is properly constructed, its values have errors of magnitude not exceeding .005. For a given x, the value of $f(x)$ is approximated by interpolating in an interval $[x_0, x_1]$, where $x \in [x_0, x_1]$ and $x_0 - x_1 = .1$. Thus the total error will be

$$L_0(x)\varepsilon_0 + L_1(x)\varepsilon_1 + \frac{(x-x_0)(x-x_1)e^{\xi}}{2}$$

where $|\xi_0| \le .005$, $|\xi_1| \le .005$, and $1.0 < x_0 < \xi < x_3 < 2.0$. The argument followed in Example 5-1 holds; so the inherent error does not exceed .005. The **analytic error** is bounded by

$$\left| \frac{(x-x_0)(x-x_1)}{2} \right| e^{\xi} \le \frac{.1^2}{8} e^2 < .01$$

Hence, the total error is less than .015. If the interpolated value is rounded to two decimal places, then the total error might be almost as large as .02.

EXERCISE 8. Suppose that $f(x_0), f(x_1)$ are given, with errors not exceeding some given e in magnitude. Prove that the inherent error in approximating $f(x)$ by **linear interpolation** in $[x_0, x_1]$ does not exceed e in magnitude.

EXERCISE 9. Suppose in Exercise 8 that $x \in [x_0, x_1]$. Show that the magnitude of the inherent error may exceed e. How large can the inherent error be if no restriction is placed on x?

EXERCISE 10. Suppose that it is known that, for a certain range of weights, the distance a spring is stretched varies linearly with the weight applied. The following data are obtained:

Weight, g	20	40
Distance, cm	5.0	61

The distances measured have errors which do not exceed .1 centimeter, while the weights are assumed to be exact. Find the distance corresponding to a weight of 50 grams and a bound on the inherent error. Will there be an analytic error in this problem?

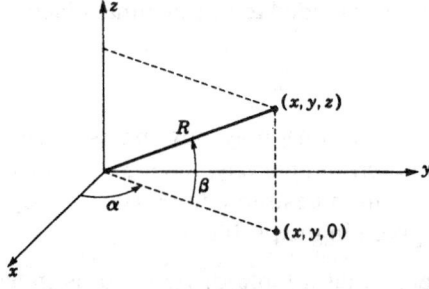

FIG 5-4

EXERCISE 11. A radar tracking system (Fig. 5-4) measures range R, azimuth angle α, and elevation angle β to an airplane or missile. These values are then used to compute ground point (x, y) and altitude z.

(a) Find (x, y, z) in terms of R, α, and β.
(b) Suppose that R is measured in meters accurate to ± 50 meters and α and β are measured in radians accurate to ± 0.001 radian. Determine an approximate bound on the inherent errors in computing x, y, and z when R measures 20,000 meters and α and β measure $\pi/4$. (Hint: Use the approximations $\cos(\delta) \approx 1 - \delta^2/2$ and $\sin(\delta) \approx \delta$ for δ small.)

EXERCISE 12. Suppose that $f(x_i) = y_i + e_i$ for $i = 0, 1, 2$ and **parabolic interpolation** is used on the values y_i to approximate $f(x)$. Find an expression for the inherent error.

EXERCISE 13. Suppose that parabolic interpolation is used in the problem of Example 5-2. Find a bound on the total error in approximating $f(1.15)$.

EXERCISE 14. The value of π is to be approximated by the following experiment: A cylindrical can is partly filled by a known volume of water. The circumference of the can is measured, along with the depth of the water. Discuss the various sources of error in this problem, and indicate how π may be computed (approximately) from the data.

EXERCISE 15. In Exercise 14 suppose that V and h are measured exactly and that the measured value of the circumference contains an error of δ. Find the error in the resulting approximation to π.

EXERCISE 16. In Exercise 14 suppose that 50 cubic inches of water is used, the inside circumference of the can is 13 inches, and the height of the water in the can is measured with an error of not more than 1/32 inch. Assuming that the 50-cubic-inch and 13-inch measurements are exact, find a bound on the error in the computed value of π.

EXERCISE 17. Suppose that the velocity of an object is increasing linearly with time (constant acceleration) and is given approximately by the following measurements:

t, sec	v, ft/sec
9.5	84
10.0	90

Suppose that the times are considered exact and the velocities have error not exceeding 1 percent of the stated values. Find the time at which a velocity of 88 feet per second was reached, including a bound on the possible error.

5-3 FLOATIONG-POINT ARITHMETIC

For the most part in this text we are concerned with the effects of analytic errors and round-off errors. We have seen examples of analytic errors in interpolation theory and in the study of **Taylor's series**. Further examples will be presented in the remaining chapters.

We turn now to the study of **computational error**. In scientific calculations one is often forced to handle numbers of widely varying magnitudes. For example, Avogadro's number is $.602252 \times 10^{24}$ molecules per mole, while Planck's constant is $.66256 \times 10^{-26}$ erg-second. Such numbers are most easily written as the product of a fractional part and a power of 10. Floating-point arithmetic is a systematic implementation of the idea of representing numbers as a product of a fractional part and an exponential part.

We shall present the ideas of floating-point arithmetic using base 10 notation. The ideas easily carry over to other bases. The rules we present for floating-point arithmetic are somewhat idealized. Minor variations of these rules will be found in various computer manufacturers' implementations of floating-point arithmetic.

DEFINITION 5-1. A **floating-point number** is a number represented in the form

$$a \times 10^b$$

Here a, the fractional (decimal) part of the floating-point number, is called the **mantissa**; b, an integer, is called the **exponent**. (We assume here that the sign of the number is included in the mantissa a.)

DEFINITION 5-2. A nonzero floating-point number is in normal form (is normalized) if the value of the mantissa lies in the interval $(-1, -.1]$ or in the interval $[.1, 1)$.

The number of digits in the mantissa of a number may vary with the computer being used. Also, it can be varied in a particular computer by appropriate programming. Most computer languages today allow double or extended precision arithmetic. This refers to the fact that a double-length mantissa (sometimes somewhat more than double length) can be used if desired.

Usually the allowable range of exponents suffices for the problems one wishes to consider. In our examples we shall assume that the exponents lie within the allowable range. The following examples are normalized three-decimal-digit-mantissa floating-point numbers:

$$.312 \times 10^{-5} \qquad .259 \times 10^{-6}$$
$$-.200 \times 10^{1} \qquad -.199 \times 10^{0}$$
$$-.180 \times 10^{-1}$$

The number 2. is exactly represented by each of the following:

$$200 \times 10^{-2} \quad .200 \times 10^{1}$$
$$20.0 \times 10^{-1} \quad .020 \times 10^{2}$$
$$2.00 \times 10^{0} \quad .002 \times 10^{3}$$

However, only $.200 \times 10^{1}$ is the normalized three-digit-mantissa representation of 2.

In four-digit-mantissa normalized floating-point notation we have

$$\pi \approx .3142 \times 10^1$$
$$e \approx .2718 \times 10^1$$
$$2^{1/2} \approx .1414 \times 10^1$$

The representation to be used for zero is somewhat arbitrary. On some computers it is written with a zero mantissa and with the largest allowable negative exponent. Such a representation is motivated by the discussion and results in Exercise 25. It is evident that zero is represented exactly by any floating-point number with a zero mantissa. However, such a representation is not a number in normal form.

It often happens that the sum, product, difference, or quotient of two t-digit-mantissa floating-point numbers will have a mantissa of more than t digits. For example, using three digits for the mantissa,

$$.863 \times 10^3$$
$$\underline{+.328 \times 10^3}$$
$$1.191 \times 10^3 = .1191 \times 10^4$$

In computers after each binary arithmetic operation $(+, -, *, /)$, the result is stored in some memory location. If memory locations hold numbers with t-digit mantissas, then we must learn how to reduce numbers not in such a form to numbers with t-digit mantissas.

DEFINITION 5-3. A nonzero floating-point number is in t-digit-mantissa standard form if it is normalized and its mantissa consists of exactly t digits.

We shall use the term standard form in place of the longer t-digit-mantissa standard form. A number is placed in standard form by first normalizing, then rounding, and finally normalizing again if necessary. The rule for rounding is quite simple. If the $(t+1)$st digit of a normalized number is a 5, 6, 7, 8, or 9, then the t'th digit is increased by 1. Then in either case the first t-digits are taken as the result. The following examples illustrate the various aspects of writing a number in standard form using $t = 3$:

$$23.45 = .2345 \times 10^2 \approx .235 \times 10^2$$
$$-23.45 = -.2345 \times 10^2 \approx -.235 \times 10^2$$
$$-20.05 = -.2005 \times 10^2 \approx -.201 \times 10^2$$
$$.00015 \times 10^3 = .150 \times 10^0$$
$$.1323 \times 10^2 \approx .132 \times 10^2$$
$$.09995 \times 10^2 = .9995 \times 10^1 \approx 1.000 \times 10^1 = .100 \times 10^2$$
$$.123499 \approx .123 \times 10^0$$

EXERCISE 18. Write the following number in standard form using three-digit mantissas:

(a) 125,000   (b) .00003506

(c) .009997   (d) 25.009

(e) −.00009999   (f) −82.44

(g) 21.349

EXERCISE 19. Reduce the following to four-digit-mantissa floating-point numbers in standard form:

(a) 3.14159   (b) 2.71828

(c) $.6023 \times 10^{23}$ (d) −25.875

(e) 63.8749

Floating-point arithmetic includes a set of rules telling how two numbers are added, subtracted, multiplied, or divided. The basic idea is that in a machine allowing a t-digit mantissa, for some fixed t, arithmetic is performed in a $2t$-digit accumulator. This means that the product of two t-digit-mantissa numbers can be formed exactly. While there are variations of this in various computers, we shall assume this feature in illustrating floating-point arithmetic.

To form the product of two floating-point numbers, one finds the exact product of the mantissas and the sum of the exponents. This double-precision product is then reduced to standard form. The following examples illustrate the basic ideas involved. In these examples, and all subsequent floating-point arithmetic examples, we use the = symbol even when only approximate equality holds.

EXAMPLE 5-3. Using three-digit-mantissa arithmetic, find the product of $.340 \times 10^2$ and $.510 \times 10^2$.

Solution.

$$(.340 \times 10^2)(.510 \times 10^2) = (.340)(.510) \times 10^4$$
$$= .173400 \times 10^4$$
$$= .173 \times 10^4 \text{ (after rounding)}$$

EXAMPLE 5-4. Find the product of $.83 \times 10^3$ and $.12 \times 10^{-2}$, using two-digit mantissas.

Solution.

$$(.83 \times 10^3)(.12 \times 10^{-2}) = .0996 \times 10^1$$
$$= .996 \times 10^0$$
$$= 1.00 \times 10^0$$
$$= .10 \times 10^1$$

EXERCISE 20. Using three-digit mantissas, write the numbers of the following in standard form, and find the indicated products:

(a) 51×34   (b) 3.14×3.14

(c) .0052×1.07   (d) 100.**5

(100.**5 is the same as 100.^{5})

EXERCISE 21. Prove that, if A and B are floating-point numbers, then $A*B = B*A$.

As in ordinary arithmetic, to add or subtract two nonzero numbers, it is first necessary to align the decimal points. The number of smaller magnitude is adjusted so that its exponent is the same as that of the number of larger magnitude. The addition or subtraction is performed, and the result is reduced to standard form.

EXAMPLE 5-5.

(a) $.123×10^3 + .456×10^2$

$$.123×10^3$$
$$+ .0456×10^3$$
$$.1686×10^3 = .169×10^3$$

(b) $.905×10^2 + .500×10^{-1}$

$$.905×10^2$$
$$+ .0005×10^2$$
$$.9055×10^2 = .906×10^2$$

(c) $.821×10^2 - .513×10^{-2}$

$$.821×10^2$$
$$- .0000513×10^2$$
$$.8209487×10^2 = .821×10^2$$

(d) $-.895×10^1 - .772×10^1$

$$-.895×10^1$$
$$-.772×10^1$$
$$-1.667×10^1 = -.1667×10^2 = -.167×10^2$$

One can formulate certain rules to simplify the implementation of addition and subtraction. One such is suggested by (c) in Example 5-5. Here the exponent of one of the numbers exceeds that of the other by 4, which is more than the number of digits being used in the mantissas. The result is completely determined by the number of larger magnitude. Thus, no arithmetic need be carried out in this case (see Exercises 24 and 25).

EXERCISE 22. Prove that in floating-point arithmetic $A + B = B + A$.

EXERCISE 22. In three-digit-mantissa arithmetic find

    (a) $.182 - 12.3$ (b) $185 - .01$

    (c) $185 + .01$ (d) $1.89 \times 10^1 - 18.9$

    (e) $-14.8 - 2.75$

EXERCISE 24. Suppose that $A = a_1 \times 10^{b_1}$, $B = a_2 \times 10^{b_2}$, and $b_1 - b_2 > t$. Prove that, if t-digit-mantissa arithmetic is used, then

$$A + B = A \qquad \text{and} \qquad A - B = A$$

Here A and B are assumed to be in standard form; so $a_1 \neq 0$, $a_2 \neq 0$.

EXERCISE 25. Suppose that zero is represented with a zero mantissa and the largest possible negative exponent. Prove that the results of Exercise 24 hold independent of what b_1 is if B is zero.

EXERCISE 26. Give an example using $t = 3$ floating-point arithmetic in which $A + (B + C) \neq (A + B) + C$. Here A, B, C are to be given in standard form.

EXERCISE 27. Determine whether or not it is true in general, for floating-point arithmetic, that $A(B + C) = AB + AC$. Here A, B, C are assumed to be in standard form.

Finally we come to floating-point division. Here the magnitude of the mantissa of the dividend is compared with the magnitude of the mantissa of the divisor. The dividend is adjusted, by shifting one place if necessary, so that it has a mantissa of smaller magnitude than that of the divisor. The quotient of the mantissas is determined to $t + 1$ digits, and the exponent of the divisor is subtracted from the exponent of the dividend. The result is then written in standard form. The following examples using $t = 3$ illustrate the basic ideas of division:

EXAMPLE 5-6.

    (a) $(.100 \times 10^1) / (.300 \times 10^1)$. Here the divisor is $.300 \times 10^1$, and the dividend is $.100 \times 10^1$. Because $.100 < .300$, no shifting is necessary.

$$
\begin{array}{r}
.3333 \\
\hline
.300\,\overline{)\,.100000} \\
900 \\
\hline
1000 \\
900 \\
\hline
100
\end{array}
$$

The result is $.3333 \times 10^0$. This is written in standard form to give $(1.000 \times 10^1) \div (.300 \times 10^1) = .333 \times 10^0$.

(b) $(.700 \times 10^1) \div (.300 \times 10^3)$. This is rewritten as $(.0700 \times 10^2) \div (.300 \times 10^3)$.

$$
\begin{array}{r}
.2333 \\
\hline
.300\overline{|.0700000} \\
600 \\
\hline
1000 \\
900 \\
\hline
1000
\end{array}
$$

Hence $(0.700 \times 10^1) \div (.300 \times 10^3) = .2333 \times 10^{-1}$
$$= .233 \times 10^{-1}$$

(c) $(.100 \times 10^2) \div (.600 \times 10^1)$

$$
\begin{array}{r}
.1666 \\
\hline
.600\overline{|.1000000}
\end{array}
$$

$(.100 \times 10^2) \div (.600 \times 10^1) = .1666 \times 10^1 = .167 \times 10^1$

EXERCISE 28. Using $t = 3$ floating-point arithmetic, evaluate 10. / 9., 3. / 9., 5. / 9..

EXERCISE 29. Using $t = 4$ floating-point arithmetic, find 9. / 5. and 5. / 9., and then find the product of the two quotients.

5-4   EXAMPLES FROM NEWTON-RAPHSON ITERATION

In the remainder of this chapter we shall illustrate certain peculiarities of floating-point arithmetic. Most of the examples used are quite simple in order to allow ease of hand computation.

EXAMPLE 5-7. Find $2^{1/2}$, using the Newton-Raphson procedure with $t = 2$ floating-point arithmetic.

Solution. We shall start with $p_0 = .20 \times 10^1$ and compute $p_1, p_2, \ldots$ by the algorithm $p_{i+1} = .5(p_i + 2/p_i)$.

$$p_1 = \left(.50 \times 10^0\right)\left(.20 \times 10^1 + \frac{.20 \times 10^1}{.20 \times 10^1}\right)$$

$$= \left(.50 \times 10^0\right)\left(.30 \times 10^1\right) = .15 \times 10^1$$

$$p_2 = \left(.50 \times 10^0\right)\left(.15 \times 10^1 + \frac{.20 \times 10^1}{.15 \times 10^1}\right)$$

$$= \left(.50 \times 10^0\right)\left(.28 \times 10^1\right) = .14 \times 10^1$$

It turns out that $p_3 = p_2$ so that all subsequent terms of the sequence are equal. The results of the computation are summarized in Table 5-1 (with $t = 2$ floating-point arithmetic used).

Tabel 5-1

n	p_n	$\left| p_n{}^2 - 2. \right|$
0	$.20 \times 10^1$	20×10^1
1	15×10^1	30×10^0
2	$.14 \times 10^1$	00×10^0
3	$.14 \times 10^1$	00×10^0

From Table 5-1 we have $2^{1/2} = .14 \times 10^1$. Moreover we see that in $t = 2$ floating-point arithmetic $|p_2{}^2 - 2.| = 0$. We know, however, that $.14 \times 10^1$ is only an approximation to $2^{1/2}$. The point to be made, then, is that it is incorrect to assume that an exact solution to an equation $f(x) = 0$ has been found merely because one has a p such that $f(p) = 0$ in floating-point arithmetic.

EXAMPLE 5-8. Find $2^{1/2}$ using $t = 3$ floating-point arithmetic.

Solution. Let $p_0 = .140 \times 10^1$. Then

$$p_1 = \left(.500 \times 10^0\right)\left(.140 \times 10^1 + \frac{.200 \times 10^1}{.140 \times 10^1}\right)$$

$$= \left(.500 \times 10^0\right)\left(.140 \times 10^1 + .143 \times 10^1\right)$$

$$= \left(.500 \times 10^0\right)\left(.283 \times 10^1\right) = .142 \times 10^1$$

$$p_2 = \left(.500 \times 10^0\right)\left(.142 \times 10^1 + \frac{.200 \times 10^1}{.142 \times 10^1}\right)$$

$$= \left(.500 \times 10^0\right)\left(.142 \times 10^1 + .141 \times 10^1\right)$$

$$= \left(.500 \times 10^0\right)\left(.283 \times 10^1\right) = .142 \times 10^1$$

Since $p_2 = p_1$, all subsequent terms have this value and the result of the computation is $2^{1/2} = .142 \times 10^1$. Notice that this result is more accurate than that achieved by using a two-digit mantissa in the calculation. However, in two-digit arithmetic

$$\left| \left(.14 \times 10^1\right)^2 - .20 \times 10^1 \right| = .00 \times 10^0$$

while in three-digit arithmatic

$$\left| \left(.142 \times 10^1\right)^2 - .200 \times 10^1 \right| = .200 \times 10^{-1}$$

Observe that $.141 \times 10^1$ is a still better three-digit answer and that

$$\left| \left(.141 \times 10^1\right)^2 - .200 \times 10^1 \right| = .100 \times 10^{-1}$$

We are prevented from getting this better result by the nature of the **Newton-Raphson** procedure, which converges from above in this case.

EXERCISE 30. Find $2^{1/2}$, using the Newton-Raphson procedure with $t = 2$ floating-point arithmetic, starting at

(a) $p_0 = 1.3$ (b) $p_0 = 1.6$ (c) $p_0 = 1.8$

EXERCISE 31. It seems likely in the above problem of finding $2^{1/2}$ that the result will be $.14 \times 10^1$ regardless of the starting value, provided only that $p_0 > 0$. Prove or disprove this assertion. (Assume that exponents stay within an allowable range.)

EXAMPLE 5-9. The way a problem is stated can have a significant effect upon the accuracy of the computed result. Suppose that we still wish to find $2^{1/2}$ but that we formulate the problem as that of finding $2^{1/2} - 1$. After this result is found, 1 is to be added on to give $2^{1/2}$. Using two-digit arithmetic, we find $2^{1/2} - 1$ as follows:

$$x = 2^{1/2} - 1$$

$$x + 1 = 2^{1/2}$$

$$(x+1)^2 = 2$$

$$x^2 + 2x - 1 = 0$$

Let $f(x) = x^2 + 2x - 1$. Find the positive solution to $f(x) = 0$, using Newton-Raphson iteration.

Solution. For simplicity, we start with the initial guess $x_0 = .40$. The general formula to be used is

$$x_{n+1} = x_n - \frac{x_n^2 + 2x_n - 1}{2x_n + 2}$$

Hence

$$x_1 = .40 \times 10^0 - \frac{(.40 \times 10^0)^2 + (.80 \times 10^0) - .10 \times 10^1}{.80 \times 10^0 + .20 \times 10^1}$$

$$= .40 \times 10^0 + \frac{.40 \times 10^{-1}}{.28 \times 10^{-1}}$$

$$= .41 \times 10^0$$

A similar calculation gives $x_2 = .41$, so that the solution is $x = .41$. Adding 1, we get $2^{1/2} = 1.41 = .141 \times 10^1$. This answer is better than the result of the previous three-digit calculation of $2^{1/2}$ and was found with only two-digit arithmetic used except in the last addition.

Basically, the error in the solution to a problem found by fixed-point iteration will depend upon the complexity of the function and the number of digits used in the calculation. The accuracy will usually increase with an increase in the number of digits carried in the calculation. If the iteration is continued until full machine accuracy seems to have been obtained, the error will be essentially that due to round-off in just one (the last) iteration. Thus, if a relatively small number of arithmetic operations are needed to perform one iteration, all but the last one or two digits of the result will usually be correct.

EXERCISE 32. In Example 5-9 show that $x_2 = .41 \times 10^0$.

EXERCISE 33. Using the Newton-Raphson iteration, find $2^{1/2}$ in four-digit floating point arithmetic. Use $x_0 = 1.4$.

EXERCISE 34. Find the value of $2^{1/2} - 1.4$ in two-digit arithmetic. Discuss the result. (Hint: Solve $x^2 + 2.8x - .04 = 0$. Use $x_0 = .01$.)

EXERCISE 35. Find the value of $2^{1/2} - 1$ in three-digit arithmetic. Discuss the result. Use $x_0 = .41$.

5-5   LINEAR INTERPOLATION EXAMPLE

EXAMPLE 5-10. Using $t = 3$ floating-point arithmetic, find an approximation to $f(2.05)$ by linear interpolation.

x	$f(x)$
2.00	7.39
2.10	8.17

Solution. Let $x_0 = .200 \times 10^1$. $x_1 = .210 \times 10^1$. We shall use the formula

$$P(x) = \frac{x - x_1}{x_0 - x_1} f(x_0) + \frac{x - x_0}{x_1 - x_0} f(x_1)$$

and shall perform the calculation in the order indicated by the formula.

$$P(.205 \times 10^1) = \frac{.205 \times 10^1 - .210 \times 10^1}{.200 \times 10^1 - .210 \times 10}(.739 \times 10^1)$$

$$+ \frac{.205 \times 10^1 - .200 \times 10^1}{.210 \times 10^1 - .200 \times 10}(.817 \times 10^1)$$

$$P(.205 \times 10^1) = (.500 \times 10^0)(.739 \times 10^1) + (.500 \times 10^0)(.817 \times 10^1)$$

$$= .370 \times 10^1 + .409 \times 10^1$$

$$= .779 \times 10^1$$

Observe that round-off error occurred in the calculation. The exact value of $P(2.05)$ is 7.78, whereas we obtained the value 7.79. The study of the effects due to round-off error is one of the more difficult problems of numerical analysis. One standard approach is to show that the numerical answer which results when round-off error has occurred is the exact answer to a problem similar to the original problem, but with slightly different parameters, coefficients, or data.

In our example suppose that $g(x)$ is given by

x	$g(x)$
2.00	7.40
2.10	8.18

Then, if $g(2.03)$ is approximated by linear interpolation, the exact answer is 7.79. That is, the exact solution to the problem of interpolating on $g(x)$ is the solution including round-off error in performing linear interpolation on $f(x)$.

The purpose of this approach is that it allows one to study the effect due to round-off error by studying inherent error. The values of $g(x)$ may be thought of as values of $f(x)$ plus inherent error. In general the effects of inherent error are easier to study than the effects of round-off error. A general study of this method is beyond the scope of this text, but Wilkinson shows that the study of round-off error in many numerical-analysis problems can be approached by studying the effects of inherent error. In particular, if the solution to a problem is sensitive to inherent error, then it will be sensitive to round-off error, and vice versa. The examples in the next section illustrate this point.

5-6   INVERSION OF AN ILL-CONDITIONED MATRIX

In this section we shall discuss an example of one of the more difficult aspects of numerical analysis. Many problems have the property that a very small change in the coefficients or the parameters of the problem produces a very large change in the solution or answer to the problem. Such a problem is called **ill-conditioned**.

At the end of Chap. 2 the reader was asked to find the inverse of the **Hilbert matrix**

$$H_n = \begin{bmatrix} 1 & \dfrac{1}{2} & \dfrac{1}{3} & \cdots & \dfrac{1}{n} \\[2mm] \dfrac{1}{2} & \dfrac{1}{3} & \dfrac{1}{4} & \cdots & \dfrac{1}{n+1} \\[2mm] \cdots & \cdots & \cdots & \cdots & \cdots \\[2mm] \dfrac{1}{n} & \dfrac{1}{n+1} & \dfrac{1}{n+2} & \cdots & \dfrac{1}{2n-1} \end{bmatrix}$$

for various n. The poor results which one obtains in performing the inversion in floating-point arithmetic are due to the fact that this is an ill-conditioned problem. Because of this it is frequently used as a test matrix in testing algorithms for the inversion of a matrix.

For ease of computation we shall restrict our attention to the case $n = 2$. Let

$$H = \begin{bmatrix} 1 & 1/2 \\ 1/2 & 1/3 \end{bmatrix}$$

To give an idea of what we mean by saying that the problem of finding H^{-1} is an ill-conditioned problem, let

$$H_\delta = \begin{bmatrix} 1 & 1/2 \\ 1/2 & 1/3+\delta \end{bmatrix}$$

Then, if $\delta \neq -\frac{1}{2}$

$$H_\delta^{-1} = \frac{1}{1+12\delta} \begin{bmatrix} 4+12\delta & -6 \\ -6 & 12 \end{bmatrix}$$

The effect of a small change in just one element of H can now be seen. This change affects all the elements of the inverse, producing much larger changes in these elements than the magnitude of δ itself. Note that when $\delta = 0$, we get

$$H^{-1} = \begin{bmatrix} 4 & -6 \\ -6 & 12 \end{bmatrix}$$

Now let us consider the problem of calculating H^{-1} by using $t = 2$ floating-point arithmetic. An immediate difficulty occurs in the computer representation of the problem. In base 10 arithmetic,

$$1/3 = .333...$$

(In base 2 arithmetic $1/3 = .01010101 \ldots .$) Thus we have an immediate round-off error. We can also think of this as an inherent error. In any case, with base 10 arithmetic used, the problem to be solved is that of finding H_1^{-1}, where

$$H_1 = \begin{bmatrix} .10\times10^1 & .50\times10^0 \\ .50\times10^0 & .33\times10^0 \end{bmatrix}$$

The matrix H_1 is itself still ill-conditioned with respect to the problem of finding its inverse. For example, if

$$H_{1\delta} = \begin{bmatrix} 1 & 1/2 \\ 1/2 & .33+\delta \end{bmatrix}$$

then the exact inverse is

$$H_\delta^{-1} = \frac{1}{.08+\delta} \begin{bmatrix} .33+\delta & -1/2 \\ -1/2 & 1 \end{bmatrix}$$

If a problem is ill-conditioned, then it will usually still be ill-conditioned after the first few steps of the solution process. If in performing these steps some round-off error occurs, then the resulting effects upon the final computed result may be large. We shall see this in the problem of finding H_1^{-1}.

Following the procedure of reducing H_1 to the identity matrix using the allowable transformations discussed in Chap. 2, we have

$$\begin{bmatrix} .10\times10^1 & .50\times10^0 & . & .10\times10^1 & 0 \\ .50\times10^0 & .33\times10^0 & . & 0 & 10\times10^1 \end{bmatrix}$$
$$\begin{bmatrix} .10\times10^1 & .50\times10^0 & . & .10\times10^1 & 0 \\ 0 & .80\times10^{-1} & . & -.50\times10^1 & .10\times10^1 \end{bmatrix}$$

$$\begin{bmatrix} .10\times10^1 & .50\times10^0 & . & .10\times10^1 & 0 \\ 0 & .10\times10^1 & . & -.63\times10^1 & .13\times10^2 \end{bmatrix}$$

$$\begin{bmatrix} .10\times10^1 & 0 & . & .42\times10^1 & -.65\times10^1 \\ 0 & .10\times10^1 & . & -.63\times10^1 & .13\times10^2 \end{bmatrix}$$

As can be seen by following through the calculation, some round-off error are made in the calculation. The result

$$\bar{H}_1^{-1} = \begin{bmatrix} 4.2 & -6.5 \\ -6.3 & 13 \end{bmatrix}$$

can be compared with the exact values of H^{-1} and H_1^{-1}, which are

$$H^{-1} = \begin{bmatrix} 4 & -6 \\ -6 & 12 \end{bmatrix} \qquad H_1^{-1} = \begin{bmatrix} 4.125 & -6.25 \\ -6.25 & 12.5 \end{bmatrix}$$

Thus, the approximate value H_1^{-1} is considerably different from the exact value of H_1^{-1}. This in turn differs considerably from H^{-1}.

We shall now discuss an important point. Suppose that, although the theoretical problem was to compute the inverse of H, the actual problem presented to the numerical analyst was to find H_1^{-1}. That is, the problem contains inherent error. Now the best that can be expected is that the numerical analyst will find the exact value of H_1^{-1}. Using two-digit arithmetic, he/she fails to do so. In three-digit arithmetic we have

$$\begin{bmatrix} .100\times10^1 & .500\times10^0 & \cdot & .100\times10^1 & 0 \\ .500\times10^0 & .330\times10^0 & \cdot & 0 & .100\times10^1 \end{bmatrix}$$

$$\begin{bmatrix} .100\times10^1 & .500\times10^0 & \cdot & .100\times10^1 & 0 \\ 0 & .800\times10^{-1} & \cdot & -.500\times10^0 & .100\times10^1 \end{bmatrix}$$

$$\begin{bmatrix} .100\times10^1 & .500\times10^0 & \cdot & .100\times10^1 & 0 \\ 0 & .100\times10^1 & \cdot & -.625\times10^1 & .125\times10^2 \end{bmatrix}$$

$$\begin{bmatrix} .100\times10^1 & 0 & \cdot & .413\times10^1 & -.625\times10^1 \\ 0 & .100\times10^1 & \cdot & -.625\times10^1 & .125\times10^2 \end{bmatrix}$$

Thus the result is

$$\begin{bmatrix} .413\times10^1 & -.625\times10^1 \\ -.625\times10^1 & .125\times10^2 \end{bmatrix}$$

It is easily seen that this is very close to the exact value of H_1^{-1}. Moreover, it is a better approximation to H^{-1} than was achieved by using only two digit arithmetic in the calculation. Thus, even though the data were correct to only two digits, a more accurate result was achieved by carrying three digits in the calculation. This is not an isolated result but, rather, is close to a general rule. By carrying more digits in the calculation, the **computational error** is reduced. In general, if a sufficient number of digits are carried in the calculation, the computational error can be made sufficiently smaller than the inherent error so that the computational error is negligible. This is often highly desirable.

EXERCISE 36. Find H_1^{-1}, using four-digit arithmetic.

EXERCISE 37. Solve the following system of equations, using $t = 2$ floating-point arithmetic, and compare with the exact answers:

$$x + y/2 = 2, \qquad x/2 + y/3 = 7/6$$

EXERCISE 38. Do Exercise 37, using the same procedure but $t = 3$ floating-point arithmetic.

5-7   EXAMPLES FROM TAYLOR'S SERIES

EXAMPLE 5-11. Find sin(.1), using Taylor's series and $t = 2$ floating-point arithmetic.

Solution.

$$\sin(x) = x - \frac{x^3}{3!} + \frac{x^5}{5!} - \frac{x^7}{7!} \cdots$$

Hence

$$\sin(.1) = .1 - \frac{.1^3}{6} + \frac{.1^5}{120} - \frac{.1^7}{5,040} \cdots$$

Now in floating-point arithmetic

$$\frac{.1^3}{6} = \frac{(.10 \times 10^0)(.10 \times 10^0)(.10 \times 10^0)}{.60 \times 10^1}$$

$$= \frac{.10 \times 10^{-2}}{.60 \times 10^1} = .17 \times 10^{-3}$$

The sum of the first two terms of the series is

$$.10 \times 10^0 - .17 \times 10^{-3} = .10 \times 10^0$$

Thus, the second term and all subsequent terms of the series do not affect the result.

EXAMPLE 5-12. It is desired to compute a table of sin(x) for $0 \le x \le .1$ in steps of .01, using $t = 4$ floating-point arithmetic. Discuss some of the computational aspects of this problem.

Solution. The expansion

$$\sin(x) = x - \frac{x^3}{3!} + \frac{x^5}{5!} - \frac{x^7}{7!} \cdots$$

will be used. For $0 \le x \le .1$ we know that sin(x) is approximately x. From this we can determine how many terms of the series to carry. We consider the ratio of successive terms of the series to the approximate answer x. Thus

$$\frac{x^3/6}{x} = \frac{x^2}{6}$$

$$\frac{x^5/120}{x} = \frac{x^4}{120}$$

In the range $0 \le x \le .1$ the maximum value of the first ratio is bounded by .0017 and the second by .00000084. It should be evident that when using $t = 4$ digits in the mantissas the term $x^5/5!$ and subsequent terms will not affect the result. Thus, the table should be computed using the approximation

$$\sin(x) \approx x - \frac{x^3}{6} \tag{5-2}$$

The analytic error (truncation error) does not exceed $x^5/5!$, which in this case is less than 0.84×10^{-7}.

Suppose now that $\sin(x)$ is computed by use of Eq. (5-2) as follows: First one computes x^3. This result is divided by 6 and then subtracted from x. For $x = .00, .01, .02, \ldots, 10$ the calculation of x^3 with $t = 4$ floating-point arithmetic involves no round-off error. Thus there is a chance for round-off error only in the division and the subtraction. The round-off error in computing $x^3/6$ does not exceed 1/2 in the fourth digit of the mantissa. Because $x^3/6$ is quite small relative to x in [0, .1], this round-off error is unlikely to affect the final outcome of the calculation. Thus one would expect that the table computed from Eq. (5-2) will have its errors bounded by 1/2 in the fourth digit of the mantissa.

EXERCISE 39. Discuss the problem of finding $\cos(.1)$ in $t = 2$ floating-point arithmetic, using the expansion

$$\cos(x) = 1 - \frac{x^2}{2!} + \frac{x^4}{4!} - \frac{x^6}{6!} \cdots$$

EXERCISE 40. Do Exercise 39 using $t = 3$ and $t = 4$.

5-8   EXAMPLES OF AN INFINITE SUM

The solution to a problem can often be written as a sum of the terms of a sequence. As the following examples illustrate, the order of summation can strongly affect the result.

EXAMPLE 5-13. Find the sums

$$1 + 1/2 + 1/3 + \cdots + 1/20$$
$$1/20 + 1/19 + 1/18 + \cdots + 1$$

using $t = 2$ floating-point arithmetic. Perform the operations working from left to right.

Solution.

$$
\begin{aligned}
1 &= .10\times10^1 & 1/11 &= .91\times10^{-1} \\
1/2 &= .50\times10^0 & 1/12 &= .83\times10^{-1} \\
1/3 &= .33\times10^0 & 1/13 &= .77\times10^{-1} \\
1/4 &= .25\times10^0 & 1/14 &= .71\times10^{-1} \\
1/5 &= .20\times10^0 & 1/15 &= .67\times10^{-1} \\
1/6 &= .17\times10^0 & 1/16 &= .63\times10^{-1} \\
1/7 &= .14\times10^0 & 1/17 &= .59\times10^{-1} \\
1/8 &= .13\times10^0 & 1/18 &= .56\times10^{-1} \\
1/9 &= .11\times10^0 & 1/19 &= .53\times10^{-1} \\
1/10 &= .10\times10^0 & 1/20 &= .50\times10^{-1}
\end{aligned}
$$

The partial sums for $1 + 1/2 + 1/3 + \cdots$ are

$$
\begin{aligned}
s_1 &= .10\times10^1 & s_{11} &= .30\times10^1 \\
s_2 &= .15\times10^1 & s_{12} &= .31\times10^1 \\
s_3 &= .18\times10^1 & s_{13} &= .32\times10^1 \\
s_4 &= .21\times10^1 & s_{14} &= .33\times10^1 \\
s_5 &= .23\times10^1 & s_{15} &= .34\times10^1 \\
s_6 &= .25\times10^1 & s_{16} &= .35\times10^1 \\
s_7 &= .26\times10^1 & s_{17} &= .36\times10^1 \\
s_8 &= .27\times10^1 & s_{18} &= .37\times10^1 \\
s_9 &= .28\times10^1 & s_{19} &= .38\times10^1 \\
s_{10} &= .29\times10^1 & s_{20} &= .39\times10^1
\end{aligned}
$$

Observe that, if we add $1/21$ to the sum, the result remains unchanged. From this it follows that the infinite sum $1 + 1/2 + 1/3 \ldots$ is $.39 \times 10^1$ in $t = 2$ floating-point arithmetic. This is a rather small result considering that the sequence of exact partial sums diverges to $+ \infty$.

Adding in the order $1/20 + 1/10 + \cdots + 1$, the partial sums are

$$
\begin{aligned}
s_1 &= .50\times10^{-1} & s_{11} &= .77\times10^0 \\
s_2 &= .10\times10^0 & s_{12} &= .88\times10^0 \\
s_3 &= .16\times10^0 & s_{13} &= .10\times10^1 \\
s_4 &= .22\times10^0 & s_{14} &= .11\times10^1
\end{aligned}
$$

$$s_5 = .28 \times 10^0 \quad s_{15} = .13 \times 10^1$$
$$s_6 = .35 \times 10^0 \quad s_{16} = .15 \times 10^1$$
$$s_7 = .42 \times 10^0 \quad s_{17} = .18 \times 10^1$$
$$s_8 = .50 \times 10^0 \quad s_{18} = .21 \times 10^1$$
$$s_9 = .58 \times 10^0 \quad s_{19} = .26 \times 10^1$$
$$s_{10} = .67 \times 10^0 \quad s_{20} = .36 \times 10^1$$

The foregoing example is of interest for several reasons. Notice that the sum depends upon the order in which the numbers are added. Notice also that if the infinite sum $1 + 1/2 + 1/3 \ldots$ is to be formed in this order only a finite number (20) terms enter into the final result. Thus, it would be quite silly to instruct the computer to find

$$\sum_{i=1}^{5,000} \frac{1}{i}$$

using two-digit-mantissa floating-point arithmetic because most of the computational time would be wasted and the result would be hopelessly incorrect.

EXERCISE 41. Find the sum $1/25 + 1/24 + \cdots + 1$, using $t = 2$ floating-point arithmetic.

EXERCISE 42. The series $1 + 1/2 + 1/3 + \cdots$ is to be summed by using $t = 3$ floating-point arithmetic. Determine the number of terms which need to be added before the partial sums no longer increase.

EXERCISE 43. The exact value of $\pi/4$ is given by

$$\frac{\pi}{4} = 1 - \frac{1}{3} + \frac{1}{5} - \frac{1}{7} + \frac{1}{9} - \frac{1}{11} \cdots$$

Determine the number of terms which need to be added before the partial sums no longer change, using $t = 2$ floating-point arithmetic.

EXERCISE 44. Repeat Exercise 43, this time using the expression

$$\frac{\pi}{4} = \left(1 - \frac{1}{3}\right) + \left(\frac{1}{5} - \frac{1}{7}\right) + \left(\frac{1}{9} - \frac{1}{11}\right) + \cdots$$

Here $1 - 1/3$ is the first term of the series, $1/5 - 1/7$ is the second term, etc.

EXERCISE 45. Find an approximation to $\pi/4$ by summing the series of Exercise 44, using $t = 2$ floating-point arithmetic.

Numerical Differentiation and Integration

6-1 INTRODUCTION

This chapter is devoted to the study of the problems of **numerical differentiation** and **numerical integration**. Much of our development of these two topics is based upon the Lagrange interpolation formula. However, in the later sections of this chapter we show how to derive certain formulas by a process known as **Richardson's extrapolation**. This leads to some very useful algorithms for numerical differentiation and integration.

6-2 NUMERICAL DIFFERENTIATION

In Chap. 4 we showed that if $f \in C^{n+1}[a, b]$, $x_0, x_1, \ldots, x_n$, are distinct points in $[a,b]$ and $x \in [a,b]$, then

$$f(x) = \sum_{i=0}^{n} L_i(x) f(x_i) + R(x) \tag{6-1}$$

where

$$R(x) = \frac{\prod_{j=0}^{n}(x - x_j)}{(n+1)!} f^{(n+1)}(\xi(x))$$

and min $(x_0, x_1, \ldots, x_n, x) < \xi(x) < $ max $(x_0, x_1, \ldots, x_n, x)$. We shall often use the notation f_i for $f(x_i)$. Differentiation of both sides of Eq. (6-1) with respect to x gives

$$f'(x) = \sum_{i=0}^{n} L_i'(x) f_i + R'(x) \tag{6-2}$$

The term $R'(x)$ in Eq. (6-2) can be written as the sum of two expressions.

$$R'(x) = \left(\frac{d}{dx} \prod_{j=0}^{n} (x - x_j) \right) \frac{f^{(n+1)}(\xi(x))}{(n+1)!} + \prod_{j=0}^{n} (x - x_j) \frac{d}{dx} \left(\frac{f^{(n+1)}(\xi(x))}{(n+1)!} \right)$$

Because of the unknown function $\xi(x)$ the second expression is not easily simplified. However, we are most often interested in the value of $f'(x)$ at one of the **interpolation** points x_k. Then it is sufficient to know only that

$$\frac{d}{dx} f^{(n+1)}(\xi(x)) \Big|_{x=x_k}$$

is bounded to conclude

$$\prod_{j=0}^{n} (x - x_j) \frac{d}{dx} \frac{f^{(n+1)}(\xi(x))}{(n+1)!} \Big|_{x=x_k} = 0$$

With this assumption, the remainder takes the simpler from

$$R'(x_k) = \left(\frac{d}{dx} \prod_{j=0}^{n} (x - x_j) \right) \frac{f^{(n+1)}(\xi(x))}{(n+1)!} \Big|_{x=x_k} \tag{6-3}$$

Let us consider the expression

$$\frac{d}{dx} \prod_{j=0}^{n} (x - x_j) \Big|_{x=x_k}$$

for a moment. For simplicity, consider the case $n = 2$.

$$\frac{d}{dx} (x - x_0)(x - x_1)(x - x_2) = (x - x_1)(x - x_2)$$

$$+ (x - x_0)(x - x_2) + (x - x_0)(x - x_1)$$

Evaluating this at x_k gives us

$$(x_k - x_1)(x_k - x_2) + (x_k - x_0)(x_k - x_2) + (x_k - x_0)(x_k - x_1)$$

But k is 0, 1, or 2. Thus two terms in the above expression will be zero for a given k. The result is summarized by

$$\frac{d}{dx} \prod_{j=0}^{2} (x - x_j) \Big|^{x=x_k} = \prod_{\substack{j=0 \\ j \neq k}}^{2} (x_k - x_j)$$

This argument can be generalized to give

$$\frac{d}{dx} \prod_{j=0}^{n} (x - x_j) \Big|^{x=x_k} = \prod_{\substack{j=0 \\ j \neq k}}^{n} (x_k - x_j) \tag{6-4}$$

The proof is left as Exercise 5.

The assumption that $(d/dx) f^{(n+1)}(\xi(x))$ be bounded can be eliminated, but the argument is beyond the scope of this text (see Hildebrand). Thus, we have the following theorem.

THEOREM 6-2-1. If $f \in C^{n+1}[a, b]$ and $x_0, x_1, \cdots, x_n$ are distinct points of $[a, b]$, then

$$f'(x_k) = \sum_{\substack{j=0}}^{n} L'_j(x_k) f(x_j) + \prod_{\substack{j=0 \\ j \neq k}}^{n} (x_k - x_j) \frac{f^{(n+1)}(\xi_k)}{(n+1)!}$$

for $k = 0, 1, \ldots, n$, where

$$\min(x_0, x_1, \ldots, x_n) < \xi_k < \max(x_0, x_1, \ldots, x_n)$$

EXAMPLE 6-1. Let $x_i = x_0 + ih$ for $i = 0, 1, 2$. Here x_0 and h are specified real numbers. Let $f \in C^3[x_0, x_2]$. Determine differentiation formulas for $f'(x_0)$, $f'(x_1)$, and $f'(x_2)$.

Solution. By Theorem 6-2-1 we have

$$f'(x_k) = L'_0(x_k) f_0 + L'_1(x_k) f_1 + L'_2(x_k) f_2 + R'(x_k)$$

for $k = 0, 1, 2$. The Lagrange coefficient polynomials on the three points are

$$L_0(x) = \frac{(x-x_1)(x-x_2)}{(x_0-x_1)(x_0-x_2)} = \frac{[(x-x_0)-h][(x-x_0)-2h]}{2h^2}$$

$$L_1(x) = \frac{(x-x_0)(x-x_2)}{(x_1-x_0)(x_1-x_2)} = \frac{[(x-x_0)][(x-x_0)-2h]}{-h^2}$$

$$L_2(x) = \frac{(x-x_0)(x-x_1)}{(x_2-x_0)(x_2-x_1)} = \frac{[(x-x_0)][(x-x_0)-h]}{2h^2}$$

Thus we have

$$L'_0(x) = \frac{1}{2h^2}\{[(x-x_0)-2h] + [(x-x_0)-h]\}$$

$$L'_1(x) = -\frac{1}{h^2}\{[(x-x_0)-2h] + (x-x_0)\}$$

$$L'_2(x) = \frac{1}{2h^2}\{[(x-x_0)-h] + (x-x_0)\}$$

This gives, for $k = 0$,

$$f'(x_0) = -\frac{3}{2h} f_0 + \frac{2}{h} f_1 - \frac{1}{2h} f_2 + R'(x_0) \tag{6-5}$$

where

$$R'(x_0) = \frac{(x_0-x_1)(x_0-x_2)}{3!} f^{(3)}(\xi_0) = \frac{2h^2}{3!} f^{(3)}(\xi_0) \tag{6-6}$$

A better form for Eq. (6-5) is

$$f'(x_0) = \frac{1}{2h}(-3f_0 + 4f_1 - f_2) + \frac{h^2}{3} f^{(3)}(\xi_0) \tag{6-7}$$

In a similar manner one obtains the formulas for $f'(x_1)$ and $f'(x_2)$.

$$f'(x_1) = \frac{1}{2h}(-f_0 + f_2) - \frac{h^2}{6}f^{(3)}(\xi_1) \tag{6-8}$$

$$f'(x_2) = \frac{1}{2h}(f_0 - 4f_1 + 3f_2) + \frac{h^2}{3}f^{(3)}(\xi_2) \tag{6-9}$$

Usually in numerical differentiation the function $f(x)$ will be tabulated or available at a number of equally spaced points. Then, for a given point any one of the formulas (6-7) to (6-9) could be used. The most commonly used of the three formulas is

$$f'(x_1) = \frac{1}{2h}(-f_0 + f_2) - \frac{h^2}{6}f^{(3)}(\xi_1)$$

This formula requires only two evaluations of $f(x)$. In addition, if $f^{(3)}(x)$ is nearly constant in the interval $[x_0, x_2]$, then the truncation error using this formula is about half that for the other formulas.

EXAMPLE 6-2. Find an approximate value for $f'(1.1)$ if the following table gives $f(x)$ exactly at the specified points.

TABLE 6-1

x	$f(x)$
1.0	−1.000
1.1	−.869
1.2	−.672
1.3	−.403

Solution. Using Eq. (6-8) with $x_0 = 1.0$, $x_1 = 1.1$, $x_2 = 1.2$, we get

$$f'(1.1) \approx \frac{1}{.2}(+1.000 - .672) = 1.64$$

Using Eq. (6-8) with $x_0 = 1.1$, $x_1 = 1.2$, $x_2 = 1.3$, we get

$$f'(1.1) \approx \frac{1}{.2}(2.607 - 2.688 + .403) = 1.61$$

Notice that we have no way of knowing for sure which of the two approximations to $f'(1.1)$ is more accurate. This particular example was constructed by using $f(x) = x^3 - 2x$. Thus, $f'(x) = 3x^2 - 2$ and $f'(1.1) = 1.63$. Notice that in this example, the magnitude of the error from using Eq. (6-7), is exactly twice that from using Eq. (6-8). This is to be expected, because $f''(x)$ is a constant.

EXERCISE 1. Table 6-2 gives values of $\ln(x)$ for various x. Find approximations to the **derivative** of $\ln(x)$ at $x = 2.4$, using each of the three formulas (6-7) to (6-9); obtain bounds on the truncation error, and compare your answers with the exact solution.

TABLE 6-2

x	$\ln x$
2.0	.69315
2.2	.78846
2.4	.87547
2.6	.95551
2.8	1.02962

EXERCISE 2. Using Table 6-2, approximate the derivative of $\ln(x)$ at $x = 2.0$, using $h = .2$ and $h = .4$. Compare the truncation-error bounds and the actual errors.

EXERCISE 3. Derive formula (6-8).

EXERCISE 4. Derive formula (6-9).

EXERCISE 5. Prove that for $n \geq 1$, $0 \leq k \leq n$,

$$\frac{d}{dx}\prod_{j=0}^{n}(x-x_j)\Bigg|_{x=x_k} = \prod_{\substack{j=0 \\ j\neq k}}^{n}(x_k - x_j)$$

EXERCISE 6. Derive a formula with remainder for approximating $f'(x_1)$ in terms of $f(x_0), f(x_1), f(x_2)$ where $x_1 = x_0 + h$, $x_2 = x_0 + 4h$

EXERCISE 7. Apply the formula derived in Exercise 6 to the problem of approximating the **derivative** of $\ln(x)$ at 2.2. Use $h = .2$ and Table 6-2. Find a bound on the truncation error, and find the actual error.

EXERCISE 8. Suppose that $P(x)$ is the Lagrange interpolating polynomial of degree not exceeding 1 which interpolates $f(x)$ at x_0 and $x_2 = x_0 + 2h$. By differentiation obtain a formula for approximating $f'(x_1)$, where $x_1 = x_0 + h$. Compare this formula and its remainder with Eq. (6-8).

Example (6-2) was somewhat atypical. Usually the values of $f(x)$ are given by a table which is not exact owing to experimental error and/or round-off errors. That is, the data in numerical differentiation will usually contain inherent error. Let us analyze Eq. (6-8) in such a case. Suppose that

$$f(x_0) = y_0 + \varepsilon_0$$
$$f(x_2) = y_2 + \varepsilon_2$$

where y_0, y_2 are the tabulated values and ε_0, ε_2 represent errors in these values. The approximate value of $f'(x_1)$ is given by

$$f'(x_1) = \frac{1}{2h}(-y_0 + y_2) \tag{6-10}$$

while the exact value is

$$f'(x_1) = \frac{1}{2h}(-y_0 - \varepsilon_0 + y_2 + \varepsilon_2) - \frac{h^2}{6} f^{(3)}(\xi_1) \qquad (6\text{-}11)$$

Thus,

$$|f'(x_1) - f'(x_1)| = \left| \frac{1}{2h}(-\varepsilon_0 + \varepsilon_2) - \frac{h^2}{6} f^{(3)}(\xi_1) \right| \qquad (6\text{-}12)$$

for some $\xi_1 \in (x_0, x_2)$. The effect of the inherent error is $(-\varepsilon_0 + \varepsilon_2)/2h$. As usual, for convenience we use the terminology inherent error for the effect of the inherent error.

In a typical problem some bound on $|\varepsilon_0|$ and $|\varepsilon_2|$ is available. Suppose that ε is such a bound. Then the inherent error is bounded by

$$\left| \frac{1}{2h}(-\varepsilon_0 + \varepsilon_2) \right| \le \frac{|\varepsilon_0| + |\varepsilon_2|}{2h} \le \frac{\varepsilon}{h} \qquad (6\text{-}13)$$

Notice that the bound on the **inherent error** increases as h decreases. Thus, numerical differentiation leads to a very difficult situation. To ensure an accurate result, one must take h small enough so that the magnitude of the truncation error

$$\frac{h^2}{6}|f^{(3)}(\xi)|$$

will be small. But if h is small relative to the errors inherent in the data, then the inherent error in the calculated result may be quite large.

EXAMPLE 6-3. Table 6-3 gives the values of $f(x) = e^x$ to four decimal places for certain x's in $[1.0, 1.2]$. Using Eq. (6-8), approximate $(d/dx)e^x$ at $x = 1.1$, with $h = 0.1, 0.05$, and 0.01. Analyze the error in using this approximation for each h.

TABLE 6-3

x	$f(x)$
1.00	2.7183
1.05	2.8577
1.09	2.9743
1.10	3.0042
1.11	3.0344
1.15	3.1582
1.20	3.3201

Solution. Since Table 6-3 gives the values of e^x to four decimal places, it contains rounding errors. A reasonable assumption is that these errors are bounded by $\varepsilon = .00005$. The approximations to $f'(1.1)$ are given in Table 6-4.

TABLE 6-4

h	Approximation to $f'(1.1)$	Bound on inherent error	Bound on truncation error
.10	3.009	.0005	$.00167f''(\xi_1)$
.05	3.005	.0010	$.00043f''(\xi_2)$
.01	3.005	.0050	$.00002f''(\xi_3)$

From the last two entries in the table one would probably conclude that 3.005 was a reasonable approximation to $f'(x)$. The exact **derivative** of $f(x) = e^x$ is $f'(x) = e^x$. Thus, from the table of e^x, the correct value of $f'(1.1)$ is 3.0042 to four decimal places. The actual error is considerably smaller than the error bound.

EXAMPLE 6-4. Using a table of $f(x) = e^x$ correct to four decimal places, find the approximate value of $f'(1.15)$ for various h and discuss the results.

Solution. The exact value of $f'(1.15)$ to four decimal places is 3.1582. For $h = .01, .02, \ldots, 08$ and by using Eq. (6-8) the approximations of Table 6-5 were found. Observe that in this table the total error decreases in magnitude as h decreases to .03. Then the error increases as h decreases.

TABLE 6-5

h	Approximation to $f'(1.15)$	Error
.08	3.1613	−.0031
.07	3.1607	−.0025
.06	3.1600	−.0018
.05	3.1590	−.0008
.04	3.1588	−.0006
.03	3.1583	−.0001
.02	3.1575	.0008
.01	3.1550	.0032

Numerical differentiation is particularly useful when the function $f(x)$ involved can be evaluated quite accurately. In such a case the truncation error can be made small without introducing a large inherent error. For an application, we shall reconsider the **Newton-Raphson** procedure for solving a nonlinear system of equations given by Algorithm 3-2. Consider the system

$$f_1(x_1, x_2, \ldots, x_n) = 0$$
$$f_2(x_1, x_2, \ldots, x_n) = 0$$
$$\cdots\cdots\cdots\cdots\cdots\cdots\cdots$$
$$f_n(x_1, x_2, \ldots, x_n) = 0$$

Each iteration of the procedure requires the calculation of the Jacobian matrix

$$\frac{\partial f_i}{\partial x_j, n \times n}$$

at the point most recently computed. In Chap. 3 we found it necessary to write n^2 functions to compute the partial derivatives. While the case 2×2 was not over involved, the value n^2 increases rapidly for larger n.

Partial derivatives can be approximated by numerical differentiation. We shall illustrate this by using the case $n = 3$. For the function $f(x_1, x_2, x_3)$

$$\frac{\partial f}{\partial x_1} = \frac{1}{2h}\left[-f\left(x_1-h\right),x_2,x_3\right)+f\left(x_1+h, x_2,x_3\right)\right]-\frac{h^2}{6}\frac{\partial^3 f}{\partial x_1^3}\left(\xi_1,x_2,x_3\right) \quad (6\text{-}14)$$

where $x_1 - h < \xi_1 < x_1 + h$
Similarly

$$\frac{\partial f}{\partial x_2} = \frac{1}{2h}\left[-f\left(x_1, x_2-h, x_3\right)+f(x_1, x_2,+h, x_3)\right]-\frac{h^2}{6}\frac{\partial^3 f}{\partial x_2^3}(x_1, \xi_2, x_3)$$

where $x_2 - h < \xi_2 < x_2 + h$

$$\frac{\partial f}{\partial x_3} = \frac{1}{2h}\left[-f\left(x_1, x_2, x_3-h\right)+f(x_1, x_2, x_3+h)\right]-\frac{h^2}{6}\frac{\partial^3 f}{\partial x_3^3}(x_1, x_2, \xi_3)$$

where $x_3 - h < \xi_3 < x_3 + h$
The key to using numerical differentiation in the Newton-Raphson iteration is the writing of the functions $f_1(X), f_2(X), \ldots, f_n(X)$ in an indexed form.

To illustrate, suppose that the system to be solved is

$$x_1^2 + x_1 x_2 + x_2^2 + x_3 - 4 = 0$$
$$2x_1 + x_1 x_2 + x_3^2 - 4 \quad = 0$$
$$x_1^2 + x_2^2 + x_3^2 - 3 \quad = 0$$

The functions are evaluated by the following function subprogram.

```
float fun(int i, float x[ ] )
{
  if( i == 1)
    {
    f = x[1]*x[1] + x[1]*x[2]  + x[2] * x[2]  + x[3]  - 4 ;
    return(f);
    }
  if( i == 2)
    {
    f =  2.*x[1] + x[1]*x[2] + x[3]*x[3] - 4 ;
    return(f);
    }
  if( i == 3)
    {
```

```
    f = x[1]*x[1] + x[2]*x[2]  + x[3] * x[3]  - 3 ;
    return(f);
    }
}
```

Now the calculation of the partial derivatives

$$A = \frac{\partial f_i}{\partial x_j}, \; n \times n$$

can be accomplished with a computer program using a number of loops.

Listed below are some of the standard formulas for numerical differentiation.

Three-point formulas for $f'(x)$:

$$f_0' = \frac{1}{2h}(-3f_0 + 4f_1 - f_2) + \frac{h^2}{3} f^{(3)}(\xi_0)$$

$$f_1' = \frac{1}{2h}(-f_0 + f_2) - \frac{h^2}{6} f^{(3)}(\xi_1)$$

$$f_2' = \frac{1}{2h}(f_0 - 4f_1 + 3f_2) + \frac{h^2}{3} f^{(3)}(\xi_2)$$

$$(6\text{-}15)$$

Four-point formula for $f'(x)$:

$$f_0' = \frac{1}{6h}(-11f_0 + 18f_1 - 9f_2 + 2f_3) - \frac{h^3}{4} f^{(4)}(\xi_0)$$

$$f_1' = \frac{1}{6h}(-2f_0 - 3f_1 + 6f_2 - f_3) + \frac{h^3}{12} f^{(4)}(\xi_1)$$

$$f_2' = \frac{1}{6h}(f_0 - 6f_1 + 3f_2 + 2f_3) - \frac{h^3}{12} f^{(4)}(\xi_2)$$

$$(6\text{-}16)$$

$$f_3' = \frac{1}{6h}(-2f_0 + 9f_1 - 18f_2 + 11f_3) + \frac{h^3}{4} f^{(4)}(\xi_3)$$

Five-point formulas for $f'(x)$:

$$f_0' = \frac{1}{12h}(-25f_0 + 48f_1 - 36f_2 + 16f_3 - 3f_4) + \frac{h^4}{5} f^{(5)}(\xi_0)$$

$$f_1' = \frac{1}{12h}(-3f_0 - 10f_1 + 18f_2 - 6f_3 + f_4) - \frac{h^4}{20} f^{(5)}(\xi_1)$$

$$f_2' = \frac{1}{12h}(f_0 - 8f_1 + 8f_3 - f_4) + \frac{h^4}{30} f^{(5)}(\xi_2)$$

$$(6\text{-}17)$$

$$f_3' = \frac{1}{12h}(-f_0 + 6f_1 - 18f_2 + 10f_3 + 3f_4) + \frac{h^4}{20} f^{(5)}(\xi_3)$$

$$f_4' = \frac{1}{12h}(3f_0 - 16f_1 + 36f_2 - 48f_3 + 25f_4) + \frac{h^4}{5} f^{(5)}(\xi_4)$$

EXERCISE 9. Using the Lagrange interpolation formula for $n = 4$ with equally spaced points $x_i = x_0 + ih$, $i = 0, 1, 2, 3, 4$, derive

$$f'(x_2) = \frac{1}{12h}[f(x_0) - 8f(x_1) + 8f(x_3) - f(x_4)] + \frac{h^4}{30} f^{(5)}(\xi)$$

where $\xi \in [x_0, x_4]$.

EXERCISE 10. Table 6-6 gives arctan x for $.475 \leq x \leq .525$ in intervals of .005.

TABLE 6-6

x	arctan x
.475	.4434483
.480	.4475200
.485	.4515758
.490	.4556157
.495	.4596396
.500	.4636476
.505	.4676396
.510	.4716156
.515	.4755755
.520	.4795193
.525	.4834470

(a) For $f(x) =$ arctan x approximate $f'(x)$ at $x = 0.5$, using Eq. (6-8) and $h = .005, .01, .02, .025$.
(b) Carry out an error analysis for each h in part (a).
(c) Compare the error bounds found in (b) with the actual errors.

EXERCISE 11. Obtain an expression for the inherent error in numerical differentiation when using Eq. (6-7) and also when using Eq. (6-9).

EXERCISE 12. Using a five-place table of $f(x) = \sin(x)$, calculate $f'(.3)$ with $h = -1$ in Eq. (6-8) and in the third formula of Eqs. (6-17). Compare the results with the exact value $\cos(.3)$.

EXERCISE 13. Obtain an expression for the inherent error in calculating $f(x_2)$ from the third formula in Eqs. (6-17). Discuss the advantages and disadvantages of using this higher-order formula over formula (6-8).

EXERCISE 14. Suppose that the nth **derivative** of $f(x)$ is known to satisfy $|f^{(n)}(x)| < 10^n$. Further suppose that tabulated values of $f(x)$ are available for any desired **step size** but that they contain inherent errors which are bounded by .02. For formulas (6-15) to (6-17) determine the formula and step size h which makes the error bound for the total error in numerical differentiation of $f(x)$ as small as possible.

6-3 NUMERICAL INTEGRATION

To obtain numerical integration formulas, we again begin with the Lagrange interpolation formula. Suppose that $f \in C^{n+1}[c, d]$ and $x_0, x_1, \ldots, x_n$ are distinct points of $[c, d]$. Then

$$f(x) = \sum_{i=0}^{n} L_i(x) f(x_i) + R(x) \qquad (6\text{-}18)$$

where

$$R(x) = \frac{\prod_{j=0}^{n}(x - x_j)}{(n-1)!} f^{(n+1)}(\xi(x))$$

Now suppose that $c \le a \le b \le d$. Integrated both sides of Eq. (6-18) from a to b gives

$$\int_a^b f(x)\,dx = \int_a^b \sum_{i=0}^{n} L_i(x) f(x_i)\,dx + \int_a^b R(x)\,dx \qquad (6\text{-}19)$$

This simplifies to

$$\int_a^b f(x)\,dx = \sum_{i=0}^{n} f(x_i) \int_a^b L_i(x)\,dx + \int_a^b R(x)\,dx$$

which we rewrite as

$$\int_a^b f(x)\,dx = \sum_{i=0}^{n} C_i f(x_i) + \int_a^b R(x)\,dx \qquad (6\text{-}20)$$

where

$$C_i = \int_a^b L_i(x)\,dx$$

The actual **integration** formula, frequently called a **quadrature formula**, is

$$\int_a^b f(x)\,dx \approx \sum_{i=0}^{n} C_i f(x_i) \qquad (6\text{-}21)$$

The remainder, or truncation error, is

$$\int_a^b R(x)\,dx = \int_a^b \prod_{j=0}^{n}(x - x_j) \frac{f^{(n+1)}(\xi(x))}{(n+1)!}\,dx$$

The truncation error in **numerical integration** depends upon the function being integrated. To indicate this dependence, we introduce the notation $E(f)$ to denote the error in integrating a particular $f(x)$. Thus from Eq. (6-20)

$$E(f) = \int_a^b f(x)\,dx - \sum_{i=0}^{n} C_i f(x_i) \qquad (6\text{-}22)$$

A detailed study of error terms will be made in Sec. 6-5. However, we can observe an important fact here. There are certain functions $f(x)$ for which the truncation error in Eq. (6-20) will be zero.

THEOREM 6-3-1. Let $x_0, x_1, \ldots, x_n$ be arbitrary distinct points, and let $[a, b]$ be an arbitrary interval. If $P(x)$ is a polynomial of degree not exceeding n, then

$$\int_a^b P(x) \, dx = \sum_{i=0}^n C_i P(x_i)$$

where

$$C_i = \int_a^b L_i(x) \, dx$$

Proof. A polynominal $P(x)$ has $n + 1$ continuous derivatives on any interval. Since the degree of $P(x)$ dose not exceed n, we have for all x

$$\frac{d^{n+1}}{dx^{n+1}} P(x) = 0$$

Since the remainder in Eq. (6-20) is

$$\int_a^b \prod_{j=0}^n (x - x_j) \frac{P^{(n+1)}(\xi(x))}{(n+1)!} \, dx$$

it follows that the truncation error is zero.

DEFINITION 6-1. If the truncation error in Eq. (6-20) is zero for a particular funtion $f(x)$, the integration formula is said to be *exact* for that $f(x)$.

EXAMPLE 6-5. Derive a formula for

$$\int_{x_0}^{x_1} f(x) \, dx$$

it terms of $f(x_0), f(x_1)$.

Solution. The appropriate Lagrange interpolation formula is

$$f(x) = \frac{x - x_1}{x_0 - x_1} f(x_0) + \frac{x - x_0}{x_1 - x_0} f(x_1) + \frac{(x - x_0)(x - x_1)}{2} f''(\xi(x))$$

Using Eq. (6-20) with $a = x_0$, $b = x_1$ gives

$$C_0 = \int_{x_0}^{x_1} \left(\frac{x - x_1}{x_0 - x_1} \right) dx = \frac{x_1 - x_0}{2}$$

$$C_1 = \int_{x_0}^{x_1} \left(\frac{x - x_0}{x_1 - x_0} \right) dx = \frac{x_1 - x_0}{2}$$

Hence

$$\int_{x_0}^{x_1} f(x) \, dx = \frac{x_1 - x_0}{2} [f(x_0) + f(x_1)]$$

$$+ \int_{x_0}^{x_1} (x - x_0)(x - x_1) \frac{f''(\xi(x))}{2} \, dx \qquad (6\text{-}23)$$

The formula

$$\int_{x_0}^{x_1} f(x)\, dx \approx \frac{x_1 - x_0}{2}[f(x_0) + f(x_1)] \tag{6-24}$$

is called the *trapezoidal rule*. It is exact for polynomials of degree not exceeding 1.

EXERCISE 15 Find the approximate value of

$$\int_0^1 x^2\, dx$$

by use of the trapezoidal rule. Find the exact error, and show that it is the same as is predicted by the remainder in Eq. (6-23).

6-4   FORMULAS FOR EQUALLY SPACED POINTS

In the remainder of this chapter we shall be concerned with **numerical-integration** formulas based upon equally spaced points. That is, the points can be represented in the form

$$x_1 = x_0 + ih \qquad i = 0, \ldots, n \tag{6-25}$$

In deriving such formulas, we shall find it computationally convenient to consider first the special case $x_0 = 0$, $h = 1$. The extension of the resulting formulas to the case of arbitrary x_0 and $h > 0$ is not difficult and is given later in this section.

If $x_0 = 0$, $x_1 = 1, \ldots, x_n = n$, then the Lagrange interpolation coefficients $L_i(x)$ have the form

$$
\begin{aligned}
L_i(x) &= \prod_{\substack{j=0 \\ j \neq i}}^{n} \frac{x - j}{i - j} \\
&= \frac{x(x-1)\, \cdots\, (x-i+1)(x-i-1)\, \cdots\, (x-n)}{i(i-1)\, \cdots\, (2)(1)(-1)(-2)\, \cdots\, (i-n)}
\end{aligned}
\tag{6-26}
$$

or equivalently

$$L_i(x) = \frac{(-1)^{n-i}}{i!(n-i)!} \prod_{\substack{j=0 \\ j \neq i}}^{n} (x - j) \tag{6-27}$$

The coefficients in Eq. (6-20) are then given by

$$C_i = \frac{(-1)^{n-i}}{i!(n-i)!} \int_a^b \prod_{\substack{j=0 \\ j \neq i}}^{n} (x - j)\, dx \tag{6-28}$$

for particular n, a, b.

EXAMPLE 6-6. Find the formula corresponding to the case $n = 2$, $a = 0$, $b = 2$.

Solution. We are seeking a formula of the form

$$\int_0^2 f(x)\, dx = C_0 f(0) + C_1 f(1) + C_2 f(2) + E(f)$$

From Eq. (6-28) we have

$$C_0 = \frac{1}{2}\int_0^2 (x-1)(x-2)\,dx = \frac{1}{3}$$

$$C_1 = -\int_0^2 x(x-2)\,dx = \frac{4}{3}$$

$$C_2 = \frac{1}{2}\int_0^2 x(x-1)\,dx = \frac{1}{3}$$

The resulting formula is

$$\int_0^2 f(x)\,dx = \frac{1}{3}[f(0)+4f(1)+f(2)]+E(f) \qquad (6\text{-}29)$$

It is known as *Simpson's rule.*

EXERCISE 16. Derive a formula for approximating $\int_0^1 f(x)\,dx$ in terms of $f(0), f(1)$.

EXERCISE 17. Derive a formula for approximating $\int_0^3 f(x)\,dx$ in terms of $f(0), f(1), f(2), f(3)$.

EXERCISE 18. Derive a formula for approximating $\int_2^3 f(x)\,dx$ in terms of $f(0), f(1), f(2)$.

EXERCISE 19. Using $n = 3$, derive a formula for approximating $\int_2^3 f(x)\,dx$ in terms of $f(0), f(1), f(2), f(3)$.

EXERCISE 20. Compute C_0, C_1, and C_2 so that

$$\int_0^2 f(x)\,dx \approx C_0 f(0)+C_1 f(1)+C_2 f(2)$$

is exact for $f(x) = 1$, x, and x^2. (*Hint*: Set up a system of three linear equations for the three unknowns C_0, C_1, and C_2.)

We shall now show how to find a quadrature formula for an arbitrary x_0 and $h > 0$, given a quadrature formula for $x_0 = 0$, $h = 1$. To illustrate the technique, suppose that we have the formula

$$\int_0^2 f(x)\,dx \approx \frac{1}{3}[f(0)+4f(1)+f(2)] \qquad (6\text{-}30)$$

derived in Eq. (6-29). The problem is to find a formula for

$$\int_{x_0}^{x_2} g(x)\,dx \qquad (6\text{-}31)$$

in terms of $g(x_0)$, $g(x_1)$, $g(x_2)$, where

$$x_i = x_0 + ih \qquad i = 0, 1, 2$$

Let $x = x_0 + sh$. Then $dx = h\,ds$, and Eq. (6-31) becomes

$$h\int_0^2 g(x_0 + sh)\,ds$$

Now, applying Eq. (6-30), we get

$$\int_{x_1}^{x_2} g(x)\, dx = h\int_0^2 g(x_0 + sh)\, ds$$

$$\approx \frac{h}{3}\left[g(x_0 + 0h) + 4g(x_0 + 1h) + g(x_0 + h)\right]$$

$$\int_{x_0}^{x_2} g(x)\, dx \approx \frac{h}{3}\left[g(x_0) + 4g(x_1) + g(x_2)\right] \tag{6-32}$$

This is the desired formula.

The procedure of employing a change of variable $x = x_0 + sh$ will work in the general case. It is also useful in the calculation of error terms, as we shall see later.

EXERCISE 21. Derive a formula for approximating $\int_{x_0}^{x_3} f(x)\, dx$ in terms of $f(x_0), f(x_1), f(x_2), f(x_3)$, where $x_i = x_0 + ih$, $i = 0, \ldots, 3$.

EXERCISE 22. Derive a formula for approximating $\int_{x_2}^{x_3} f(x)\, dx$ in terms of $f(x_0), f(x_1), f(x_2)$, where $x_i = x_0 + ih$, $i = 0, 1, 2, 3$.

EXERCISE 23. Derive a formula for approximating $\int_{x_2}^{x_3} f(x)\, dx$ in terms of $f(x_0), f(x_1), f(x_2), f(x_3)$, where $x_i = x_0 + ih$, $i = 0, \ldots, 3$.

EXERCISE 24. Formulate a general procedure for the immediate conversion of a formula

$$\int_a^b f(x)\, dx \approx C_0 f(0) + C_1 f(1) + \cdots + C_n f(n)$$

into a formula for

$$\int_{x_a}^{x_b} f(x)\, dx$$

in terms of $f(x_0)$, $f(x_1)$, $\ldots$, $f(x_n)$, where $x_i = x_0 + ih$, $i = 0, \ldots, n$, $x_a = x_0 + ah$, $x_b = x_0 + bh$.

EXERCISE 25. Using the table of e^x below (Table 6-7), find the approximate value of $\int_{2.10}^{2.20} e^x\, dx$:

(a) By means of the trapezoidal rule, with $x_0 = 2.10$, $x_1 = 2.20$.
(b) By means Simpson's rule, with $x_0 = 2.10$, $x_1 = 2.15$, $x_2 = 2.20$.
(c) Compare the above results with the exact answer.

TABLE 6-7

x	e^x
2.10	8.1662
2.15	8.5849
2.20	9.0250

6-5   REMAINDER TERMS

In this section we shall derive some remainder formulas for numerical integration. We assume that $x_0, x_1, \ldots, x_n$ are distinct points. Let I denote the smallest interval containing $x_0, \ldots, x_n$ and a, b. Then, if $f \in C^{n+1}(I)$,

$$\int_a^b f(x)dx = \sum_{i=0}^n C_i f(x_i) + \int_a^b \prod_{j=0}^n (x - x_j) \frac{f^{(n+1)}(\xi(x))}{(n+1)!} dx \qquad (6\text{-}33)$$

Here $\xi(x) \in I$ and the C_i are integrals of the appropriate Lagrange coefficients. The remainder above is not in an easily usable form because of the unknown function $\xi(x)$ in the integrand. Thus, we wish to find simpler expressions for

$$E(f) = \int_a^b \prod_{j=0}^n (x - x_j) \frac{f^{(n+1)}(\xi(x))}{(n+1)!} dx \qquad (6\text{-}34)$$

For certain choices of $[a, b]$ the problem is not too difficult. The most general problem, however, is still the subject of current research. To illustrate some of the simpler cases, we need the following **mean-value theorem** of integral calculus.

THEOREM 6-5-1. If $f(x) \in C[a, b]$ and $g(x)$ is integrable and does not change sign on $[a, b]$, then there exists a point $t \in [a, b]$ such that

$$\int_a^b f(x)g(x)dx = f(t)\int_a^b g(x)\, dx$$

Proof. The proof of this theorem is based upon the intermediate-value theorem. Let

$$m = \min_{x \in [a,b]} f(x) \qquad M = \max_{x \in [a,b]} f(x)$$

For simplicity we consider the case $g(x) \geq 0$ on $[a,b]$. Then

$$m\int_a^b g(x)\, dx \leq \int_a^b f(x)g(x)\, dx \leq M\int_a^b g(x)\, dx$$

The quantity $\int_a^b g(x)\, dx$ is just a constant. By the intermediate-value theorem, the function

$$f(x)\int_a^b g(x)\, dx$$

takes on all values between

$$m\int_a^b g(x)\, dx \equiv l \qquad \text{and} \qquad M\int_a^b g(x)\, dx \equiv u$$

as x varies over $[a, b]$. In particular, there exists a $t \in [a,b]$ such that

$$f(t) = \int_a^b g(x)\, dx = \int_a^b f(x)g(x)\, dx$$

because the integral on the right is a number in $[l,u]$. Since a similar argument can be given in case $g(x) \leq 0$ on $[a,b]$, the proof is complete.

Now suppose that the interval of integration $[a, b]$ in Eq. (6-20) is one in which the product

$$(x - x_0)(x - x_1) \cdots (x - x_n)$$

does not change sign. If $x_0 < x_1 < x_2 < \cdots < x_n$, then all such intervals are of the form $[a, b]$, where one of the following holds:

$$a < b \leq x_0 \tag{6-35a}$$

$$x_i \leq a < b \leq x_{i+1} \qquad i = 0, \ldots, n-1 \tag{6-35b}$$

$$x_n \leq a < b \tag{6-35c}$$

In these cases, if $f^{(n+1)}(\xi(x))$ is continuous on $[a, b]$, then there exists a point $t \in [a, b]$ such that

$$\int_a^b \prod_{j=0}^n (x - x_j) \frac{f^{(n+1)}(\xi(x))}{(n+1)!} \, dx = \frac{f^{(n+1)}(\xi(t))}{(n+1)!} \int_a^b \prod_{j=0}^n (x - x_j) \, dx \tag{6-36}$$

Since $\xi(x) \in I$, this can be simplified to

$$E(f) = \frac{f^{(n+1)}(\mu)}{(n+1)!} \int_a^b \prod_{j=0}^n (x - x_j) \, dx \tag{6-37}$$

for some $\mu \in I$. Finally, the integrand on the right is a polynomial; so for a given $x_0, \ldots, x_n, a, b$ it can be evaluated.

The hypothesis $f^{(n+1)}(\xi(x)) \in C[a, b]$ is actually redundant. It can be shown that it can be omitted. A proof of the following theorem is given in Henrici (1964, p. 188).

THEOREM 6-5-2. Suppose that $f \in C^{n+1}(I)$ and that $(x - x_0)(x - x_1)$ $\cdots (x - x_n)$ does not change sign on $[a, b]$. Then there exists a $\mu \in I$ such that

$$\int_a^b f(x) \, dx = \sum_{i=0}^n C_i f_i + \frac{f^{(n+1)}(\mu)}{(n+1)!} \int_a^b \prod_{j=0}^n (x - x_j) \, dx \tag{6-38}$$

where

$$C_i = \int_a^b \prod_{\substack{j=0 \\ j \neq i}}^n \left(\frac{x - x_j}{x_i - x_j} \right) dx$$

EXAMPLE 6-7. Find a simple expression for the remainder in the trapezoidal rule.

Solution. The trapezoidal rule is

$$\int_{x_0}^{x_1} f(x) \, dx = \frac{x_1 - x_0}{2} [f(x_0) + f(x_1)] + \int_{x_0}^{x_1} \frac{(x - x_0)(x - x_1) f''(\xi(x))}{2} \, dx$$

The product $(x - x_0)(x - x_1)$ does not change sign on $[x_0, x_1]$. Thus, by Theorem 6-5-2

$$\int_{x_0}^{x_1} (x - x_0)(x - x_1) \frac{f''(\xi(x))}{2} \, dx = \frac{f''(\mu)}{2} \int_{x_0}^{x_1} (x - x_0)(x - x_1) \, dx$$

for some μ satisfying $x_0 \le \mu \le x_1$. The integral on the right is more easily evaluated after the change of variable $x = x_0 + sh$, where $h = x_1 - x_0$. The result is

$$E(f) = \frac{h^3 f''(\mu)}{2} \int_0^1 s(s-1)\, ds = \frac{-h^3 f''(\mu)}{12}$$

Thus, the **trapezoidal rule** with remainder is

$$\int_{x_0}^{x_1} f(x)\, dx = \frac{h}{2}(f_0 + f_1) - \frac{h^3}{12} f''(\mu) \qquad x_0 \le \mu \le x_1 \qquad (6\text{-}39)$$

EXAMPLE 6-8. Find a bound on the error in evaluating

$$\int_{.1}^{.2} e^x dx$$

by use of the trapezoidal rule.

Solution. Let $f(x) = e^x$; then $f''(x) = e^x$. Using Eq. (6-39), with $x_0 = .1$, $x_2 = .2$, $h = .1$, we get

$$\int_{.1}^{.2} e^x dx = \frac{.1}{2}(e^{.1} + e^{.2}) - \frac{.1^3}{12} e^{\xi} \qquad .1 \le \xi \le .2$$

The error is bounded by

$$\max_{.1 \le \xi \le .2} \left| \frac{.1^3}{12} e^{\xi} \right| = \frac{.001}{12} e^{.2} \approx .00010$$

EXERCISE 26. Derive a formula, with remainder simplified as much as possible, for

$$\int_{x_1}^{x_2} f(x)\, dx$$

in terms of $f(x_0), f(x_1)$. Here $x_1 = x_0 + ih$ ($i = 0, 1, 2$).

The difficulty with Theorem 6-5-2 is that it does not apply in many simple cases, such as Simpson's rule. In **Simpson's rule** the points are $x_0, x_1 = x_0 + h$, $x_2 = x_0 + 2h$, with $a = x_0$, $b = x_2$. The function $(x - x_0)(x - x_1)(x - x_2)$ does change sign on $[a, b]$. The problem in this more general case has been studied for many years. It was established by Steffensen that a remainder formula similar in form to Eq. (6-38) exists for a class of quadrature formulas where the hypotheses of Theorem 6-5-2 are not satisfied. Subsequent research has extended that class. Daniell proposed that for a broad class of quadrature formulas it is reasonable to seek a remainder formula of the form

$$K f^{(m)}(\mu) \qquad (6\text{-}40)$$

where K and m are constants depending upon the integration formula and M is a point in the interval I containing $x_0, \ldots, x_n, a, b$. In the trapezoidal rule, for example,

$$K = \frac{-h^3}{12} \qquad m = 2$$

It should be emphasized that not all quadrature formulas have a remainder of this form. However, except for **Romberg integration** (Sec. 6-8), all the formulas considered in this text do have such a remainder. We shall illustrate how to find the appropriate K and m for the case of equally spaced points. The algebra is much simplified by again taking the points as $x_0 = 0, x_1 = 1, \ldots, x_n = n$. The extension to the general case will be given later.

For a given n we are trying to find K and m independent of f such that

$$\int_a^b f(x)\, dx = \sum_{i=0}^n C_i f(i) + K f^{(m)}(\xi)$$

where

$$C_i = \int_a^b L_i(x)\, dx \qquad \xi \in I$$

Here I is the smallest interval containing a, b, 0, n.

The trick is to apply the **quadrature formula** to various powers of x. Thus, if $f(x) = x^m$ for $m \le n$, we know that the error must be zero, because the quadrature formula is exact for these functions. Hence $m > n$. Trying $m = n + 1$, then $n + 2$, etc., one finds the smallest value of m for which $E(x^{m-1}) = 0$ and $E(x^m) \ne 0$. Then, because $(d^m/dx^m)x^m = m!$, it follows that

$$\int_a^b x^m\, dx = \sum_{i=0}^n C_i f(i) + K m! \tag{6-41}$$

This is solved for K.

$$K = \frac{1}{m!}\left[\int_a^b x^m\, dx - \sum_{i=0}^n C_i f(i) \right] = \frac{E(x^m)}{m!} \tag{6-42}$$

Thus we obtain K and m so that $E(f) = K f^{(m)}(\xi)$.

Example 6.9. Find a remainder formula for Simpson's rule (6-29),

$$\int_0^2 f(x)\, dx = \frac{1}{3}[f(0) + 4f(1) + f(2)] + E(f)$$

Solution. We know that $E(1) = E(x) = E(x^2) = 0$. Setting $f(x) = x^3$, we calculate

$$\int_0^2 x^3\, dx = 4$$
$$1/3[0^3 + (4)(1^3) + 2^3] = 4$$

Hence $E(x^3) = 0$.

Next setting $f(x) = x^4$, we calculate

$$\int_0^2 x^4\, dx = \frac{32}{5}$$
$$1/3[0^4 + (4)(1^4) + 2^4] = 20/3$$

This gives $E(x^4) = -4/15$, which means that $m = 4$ and

$$K = \frac{(-4/15)}{4!} = -\frac{1}{90}$$

Hence the error term of **Simpson's rule** is

$$E(f) = -1/90 \; f^{(4)}(\xi)$$

where $0 \leq \xi \leq 2$.

EXAMPLE 6-10. Find a remainder formula for the general Simpson's rule

$$\int_{x_0}^{x_2} f(x) \; dx = \frac{h}{3}[f(x_0) + 4f(x_1) + f(x_2)]$$

Solution. Let $x = x_0 + sh$. Upon using the notation $F(s) = f(x_0 + sh)$, the above formula becomes

$$\int_{x_0}^{x_2} f(x)dx = h\int_0^2 F(s)ds \approx \frac{h}{3}[F(0) + 4F(1) + F(2)]$$

From Example 6-9 (after multiplication by h)

$$h\int_0^2 F(s)ds = \frac{h}{3}[F(0) + 4F(1) + F(2)] - \frac{hF^{(4)}(\xi)}{90} \tag{6-43}$$

Now

$$F(s) = f(x_0 + sh)$$
$$F'(s) = hf'(x_0 + sh)$$
$$\cdots\cdots\cdots\cdots$$
$$F^{(4)}(s) = h^4 f^{(4)}(x_0 + sh)$$

Making the appropriate substitutions in Eq. (6-43), we get

$$\int_{x_0}^{x_2} f(x) \; dx = \frac{h}{3}[f(x_0) + 4f(x_1) + f(x_2)] - \frac{h^5}{90} f^{(4)}(\mu) \tag{6-44}$$

$$x_0 \leq \mu \leq x_2$$

Below are listed a few additional formulas with remainders. We shall use some of these formulas in our study of the numerical solution of ordinary **differential equations**. All these formulas can be derived by means of the techniques developed so far in this chapter.

The following are known as **closed-type formulas** because the interval of integration is the same as the interval determined by the interpolation points:

$$\int_{x_0}^{x_1} f(x) \; dx = \frac{h}{2}(f_0 + f_1) - \frac{h^3}{12} f''(\xi) \tag{6-45}$$

$$\int_{x_0}^{x_2} f(x) \; dx = \frac{h}{3}(f_0 + 4f_1 + f_2) - \frac{h^3}{90} f^{(4)}(\xi) \tag{6-46}$$

$$\int_{x_0}^{x_3} f(x) \; dx = \frac{3h}{8}(f_0 + 3f_1 + 3f_2 + f_3) - \frac{3h^5}{80} f^{(4)}(\xi) \tag{6-47}$$

$$\int_{x_0}^{x_4} f(x) \; dx = \frac{4h}{90}(7f_0 + 32f_1 + 12f_2 + 32f_3 + 7f_4) - \frac{8h^7}{945} f^{(6)}(\xi) \tag{6-48}$$

$$\int_{x_0}^{x_5} f(x)\,dx = \frac{6h}{840}(41f_0 + 216f_1 + 27f_2 + 272f_3 + 27f_4 + 216f_5$$

$$+ 41f_6) - \frac{9h^9}{1,400}f^{(8)}(\xi) \qquad (6\text{-}49)$$

The following are known as **open-type formulas** because the interpolation points are interior to the interval of integration:

$$\int_{x_{-1}}^{x_1} f(x)\,dx = 2hf_0 + \frac{h^3}{3}f''(\xi) \qquad (6\text{-}50)$$

$$\int_{x_{-1}}^{x_2} f(x)\,dx = \frac{3h}{2}(f_0 + f_1) + \frac{3h^3}{4}f''(\xi) \qquad (6\text{-}51)$$

$$\int_{x_{-1}}^{x_3} f(x)\,dx = \frac{4h}{3}(2f_0 - f_1 + 2f_2) + \frac{28h^5}{90}f^{(4)}(\xi) \qquad (6\text{-}52)$$

$$\int_{x_{-1}}^{x_6} f(x)\,dx = \frac{6h}{20}(11f_0 - 14f_1 + 26f_2 - 14f_3 + 11f_4) + \frac{41h^7}{140}f^{(6)}(\xi) \qquad (6\text{-}53)$$

The following formulas are useful as **predictors** in the **predictor-corrector methods** of solving ordinary differential equations considered in Chap. 8:

$$\int_{x_0}^{x_1} f(x)\,dx = hf_0 + \frac{h^2}{2}f'(\xi) \qquad (6\text{-}54)$$

$$\int_{x_1}^{x_2} f(x)\,dx = \frac{h}{2}(-f_0 + 3f_1) + \frac{5h^3}{12}f''(\xi) \qquad (6\text{-}55)$$

$$\int_{x_2}^{x_3} f(x)\,dx = \frac{h}{24}(f_0 - 5f_1 + 19f_2 + 9f_3) - \frac{19h^5}{720}f^{(4)}(\xi) \qquad (6\text{-}56)$$

$$\int_{x_3}^{x_4} f(x)\,dx = \frac{h}{24}(-9f_0 - 37f_1 + 59f_2 + 55f_3) + \frac{251h^5}{720}f^{(4)}(\xi) \qquad (6\text{-}57)$$

EXERCISE 27. Find a remainder for the formula

$$\int_0^3 f(x)\,dx \approx \frac{3}{8}[f(0) + 3f(1) + 3f(2) + f(3)]$$

This is sometimes called *Simpson's three-eighths rule.*

EXERCISE 28. Using the results of Exercise 27, find a remainder for the formula.

$$\int_{x_0}^{x_3} f(x)\,dx \approx \frac{3h}{8}[f(x_0) + 3f(x_1) + 3f(x_2) + f(x_3)]$$

for equally spaced points $x_0, \ldots, x_3$.

EXERCISE 29. Derive the following formula with remainder:

$$\int_{x_{-1}}^{x_2} f(x)\,dx \approx \frac{3h}{2}(f_0 + f_1) + \frac{3h^3}{4}f''(\xi)$$

[*Hint:* Use an interpolating polynomial of degree 1 to $f(x)$ for $x = x_0$, $x = x_1$.]

EXERCISE 30. Find a formula with remainder for $\int_{x_2}^{x_3} f(x)\,dx$ in terms of $f(x_0)$, $f(x_1), f(x_2)$, where $x_i = x_0 + ih$ $(i = 0, 1, 2, 3)$.

EXERCISE 31. Approximate $\int_0^1 dx/(1+x)$ by use of Simpson's rule. Find a bound on the error.

EXERCISE 32. Using Simpson's three-eighths rule (see Exercise 28) with remainder, find an approximation to the integral of Exercise 31, and find a bound on the error.

6-6   INTEGRATION OVER LARGE INTERVALS

Let us consider the problem of approximating $\int_0^\pi \sin x\,dx$, using Simpson's rule with $h = \pi/2$.

$$\int_0^\pi \sin x\,dx = \frac{\pi}{6}\left(\sin 0 + 4\sin\frac{\pi}{2} + \sin \pi\right) - \left(\frac{\pi}{2}\right)^5 \frac{\sin \xi}{90} \qquad (6\text{-}58)$$

The approximate value of the integral is $2\pi/3$, and the absolute value of the remainder is

$$\left|\int_0^\pi \sin x\,dx = 2\pi/3\right| = \frac{\pi^5}{2{,}880}|\sin \xi| \le \frac{\pi^5}{2{,}880} \approx 0.1 \qquad (6\text{-}59)$$

The actual error is $2 - 2\pi/3 \approx -.094$. In normal situations this error is much larger than can be tolerated.

Using the fact that

$$\int_0^\pi \sin x\,dx = \int_0^{\pi/2} \sin x\,dx + \int_{\pi/2}^\pi \sin x\,dx \qquad (6\text{-}60)$$

we can approximate $\int_0^{\pi/2}\sin x\,dx$ and $\int_{\pi/2}^\pi \sin x\,dx$ separately, using Simpson's rule, and we can add the two results to give an approximation to $\int_0^x \sin x\,dx$. This will decrease the error, as will now be shown.

$$\int_0^{\pi/2}\sin x\,dx = \frac{\pi}{12}\left(\sin 0 + 4\sin\frac{\pi}{4} + \sin\frac{\pi}{2}\right) - \frac{\pi^5}{(4^5)(90)}\sin \xi_1$$

The approximate value of the integral is $[(2)(2^{1/2}) + 1]\,(\pi/12)$ and

$$\left|\int_0^{\pi/2}\sin x\,dx - [(2)(2^{1/2}) + 1]\frac{\pi}{12}\right| = \frac{\pi^5}{(4^5)(90)}|\sin \xi_1| \le .003$$

Similarly

$$\int_{\pi/2}^\pi \sin x\,dx = \frac{\pi}{12}\left(\sin\frac{\pi}{2} + 4\sin\frac{3\pi}{4} + \sin \pi\right) - \frac{\pi^5}{(4^5)(90)}\sin \xi_2$$

The approximate value of the integral is $[(2)(2^{1/2}) + 1](\pi/12)$, and

$$\left|\int_{\pi/2}^\pi \sin x\,dx - [(2)(2^{1/2}) + 1]\frac{\pi}{12}\right| \le .003$$

By adding the above results, we get $\int_0^\pi \sin x \, dx \approx [(2)(2^{1/2}) + 1](\pi/6)$ and $\left| \int_0^\pi \sin x \, dx - [(2)(2)^{1/2} + 1](\pi/6) \right| \leq .006$. The magnitude of the actual error is approximately .033.

Going now to the general case, suppose that we have an interval $[a, b]$ and that we divide it into $2m$ subintervals by the points

$$a = x_0 < x_1 < x_2 < \cdots < x_{2m} = b$$

where $x_i = x_0 + ih$ and $h = (b - a)/2m$. Then we may write

$$\int_a^b f(x) \, dx = \sum_{j=1}^m \int_{x_{2j-1}}^{x_{2j}} f(x) \, dx \tag{6-61}$$

Assuming that $f^{(4)}(x)$ is continuous on $[x_0, x_{2m}]$, we apply Simpson's rule with remainder to each integral.

$$\int_{x_{2j-3}}^{x_{2j}} f(x) \, dx = \frac{h}{3}(f_{2j-2} + 4f_{2j-1} + f_{2j}) - \frac{h^5}{90} f^{(4)}(\xi_j) \tag{6-62}$$

where $f_i = f(x_i)$ and $x_{2j-2} \leq \xi_j \leq x_{2j}$. Using Eqs. (6-62) and (6-61), we get

$$\int_a^b f(x) \, dx = \frac{h}{3} \sum_{j=1}^m (f_{2j-2} + 4f_{2j-1} + f_{2j}) - \frac{h^5}{90} \sum_{j=1}^m f^{(4)}(\xi_j) \tag{6-63}$$

The first sum in Eq. (6-63) represents the approximation to the integral; the second sum is the remainder.

The remainder may be simplified by use of the intermediate-value theorem. Notice that

$$\sum_{j=1}^m f^{(4)}(\xi_j) \leq m[\max_{1 \leq j \leq m} f^{(4)}(\xi_j)]$$

and

$$\sum_{j=1}^m f^{(4)}(\xi_j) \geq m[\min_{1 \leq j \leq m} f^{(4)}(\xi_j)]$$

Thus, we get

$$\min_{1 \leq j \leq m} f^{(4)}(\xi_j) \leq \frac{1}{m} \sum_{j=1}^m f^{(4)}(\xi_j) \leq \max_{1 \leq j \leq m} f^{(4)}(\xi_j) \tag{6-64}$$

Finally, because $f^{(4)}(x)$ is continuous on $[a, b]$, it takes on all values between its minimum and maximum. Thus, there exists a point $\mu \in [a, b]$ such that

$$f^{(4)}(\mu) = \frac{1}{m} \sum_{j=1}^m f^{(4)}(\xi_j) \tag{6-65}$$

Using Eq. (6-65) in (6-63) gives the following theorem.

THEOREM 6-6-1. Simpson's Rule for $2m$ Intervals. Let $m > 0$ be an integer, and let $h = (b - a)/2m$. Let $x_i = a + ih$ for $i = 0, 1, \ldots, 2m$. Then, if $f \in C^4 [a, b]$, there exists a $\mu \in [a, b]$ such that

$$\int_a^b f(x)dx = \frac{h}{3}\left[f(a) + f(b) + 2\sum_{i=1}^{m-1} f(x_{2i}) \right.$$

$$\left. + 4\sum_{i=1}^{m} f(x_{2i-1}) \right] - \frac{mh^5}{90} f^{(4)}(\mu) \tag{6-66}$$

Often the remainder is written in the alternative form

$$-\frac{mh^5}{90} f^{(4)}(\mu) = -\frac{(b-a)h^4}{180} f^{(4)}(\mu) \tag{6-67}$$

EXAMPLE 6-11. Determine m, and hance h, so that the absolute value of the error in using Simpson's rule for $2m$ intervals to approximate $\int_0^2 dx/(x+1)$ is less than 5×10^{-5}.

Solution. By Eq. (6-66), we know that

$$\left| \int_a^b f(x)\, dx - \frac{h}{3}\left(f_0 + f_{2m} + 2\sum_{j=1}^{m-1} f_{2j} + 4\sum_{j=1}^{m} f_{2j-1} \right) \right| \le \frac{mh^5}{90} \max_{x\in[a,b]} \left| f^{(4)}(x) \right|$$

Thus we seek an m such that

$$\frac{mh^5}{90} \max_{x\in[a,b]} \left| f^{(4)}(x) \right| \le 5 \times 10^{-5}$$

Because $f(x) = 1/(x + 1)$, we have

$$f^{(4)}(x) = \frac{4!}{(x+1)^5}$$

This means that $\max_{x\in[0,2]} \left| f^{(4)}(x) \right| = 4!$. Thus

$$\left| \int_0^2 \frac{dx}{x+1} - (\text{Simpon's rule approximation}) \right| \le \frac{mh^5}{90} 4!$$

Because $h = (2 - 0)/2m = 1/m$, m must satisfy

$$\frac{4!}{m^4(90)} \le 5 \times 10^{-5} \quad \text{or} \quad m^4 \ge \frac{4}{75} \times 10^5$$

Because m is to be an integer, we conclude any integer $m > 9$ provides the desired accuracy.

The derivation of a **trapezoidal rule** for m intervals follows the same ideas as the derivation of **Simpson's rule** for $2m$ intervals. The details are left as Exercise 35.

THEOREM 6-6-2. Trapezoidal Rule for m Intervals. Let $m > 0$ be an integer, and let $h = (b - a)/m$. Let $x_i = a + ih$ for $i = 0, 1, \ldots, m$. Then, if $f \in C^2[a, b]$, there exists a $\mu \in [a, b]$ such that

$$\int_a^b f(x)\, dx = \frac{h}{2}\left[f(a) + f(b) + 2\sum_{i=1}^{m-1} f(x_i) \right] - \frac{mh^3}{12} f''(\mu) \tag{6-68}$$

The remainder may also be written as

$$-\frac{mh^3}{12} f''(\mu) = -\frac{(b-a)h^2}{12} f''(\mu) \tag{6-69}$$

We have not yet discussed the problem of inherent error in numerical integration. We shall illustrate some of the basic ideas by considering the trapezoidal rule for m intervals. Using the notation of Theorem 6-6-2 suppose that

$$f(x_i) = y_i + \varepsilon_i$$

and that the integration is to be performed by means of the values y_i as approximations to $f(x_i)$. A bound on the effect due to the inherent error is given by

$$\left| \frac{h}{2} \left\{ [f(x_0) - y_0] + [f(x_m) - y_m] + 2 \sum_{i=1}^{m-1} [f(x_i) - y_i] \right\} \right|$$

This simplifies to

$$\left| \frac{h}{2} \left(\varepsilon_0 + \varepsilon_m + 2 \sum_{i=1}^{m-1} \varepsilon_i \right) \right|$$

In the worst possible CAse all the errors ε_i will be of the same sign. Let $\varepsilon = \max_{1 \le i \le m} |\varepsilon_i|$. Then the effect due to inherent error is bounded by

$$\left| \frac{h}{2} \left(\varepsilon + \varepsilon + 2 \sum_{i=1}^{m-1} \varepsilon \right) \right| = (b - a)\varepsilon$$

This bound does not depend upon the number of subintervals into which $[a, b]$ is divided. In actual practice one would expect some of the ε_i to be positive and some to be negative. In such a case cancellation will occur, and the actual error due to inherent error will be considerably smaller than the error bound.

The problem of inherent error for **Simpson's rule** is left as Exercise 37. The results are similar to the trapezoidal rule result derived above.

Exercise 33.
   (a) Write a COmputer program for Simpson's rule for $2m$ intervals.
   (b) Run your program to approximate $\int_0^2 dx/(x+1)$. Take $m = 9$ and $m = 18$.

Exercise 34.
   (a) Determine the m, and hence also the h, needed to approximate, with an error less than 10^{-6}, the integral $\int_0^3 e^x \cos(x)\, dx$, using **Simpson's rule** for $2m$ intervals.
   (b) Carry out this integration using the program of Exercise 33.

Exercise 35.
   (a) Show that the trapezoidal rule for m intervals is

$$\int_a^b f(x)\, dx = \frac{h}{2} \left[f(x_0) + f(x_m) + 2 \sum_{i=1}^{m-1} f(x_i) \right] - \frac{mh^3}{12} f''(\mu)$$

where $x_i = a + ih$, $h = (b - a)/m$, and $\mu \in [a, b]$.

(b) Determine the m, and hence also the h, needed to approximate, with an error less than 5×10^{-5}, the integral $\int_0^2 dx/(x+1)$, using the trapezoidal rule for m intervals.

EXERCISE 36.

(a) Show that three-eighths rule for $3m$ intervals is

$$\int_a^b f(x)dx = \frac{3h}{8}\left\{ f(x_0) + f(x_{3m}) + 2\sum_{i=1}^{m-1} f(x_{3i}) \right.$$

$$\left. + 3\sum_{i=1}^m [f(x_{3i-2}) + f(x_{3i-1})] \right\} - \frac{3mh^5}{80} f^{(4)}(\mu)$$

where $x_i = a + ih$, $h = (b - a)/3m$, and $\mu \in [a, b]$.

(b) Draw a flow chart for the method in (a).

EXERCISE 37. Derive an expression for the effect of the inherent error in Simpson's rule for $2m$ intervals. Find a bound on this error which is independent of m.

6-7   RICHARDSON'S EXTRAPOLATION

In this section, we shall illustrate a technique which has wide application in numerical differentiation, integration, and solution of **differential equations**. To begin with, we shall re-derive the differentiation formula

$$f'(x_0) = \frac{f(x_0 + h) - f(x_0 - h)}{2h} + \text{remainder} \tag{6-70}$$

The technique used will yield a different form of the remainder from that given previously.

We shall assume that $f(x)$ can be expanded in a Taylor's series about x_0. We terminate the series after the fourth-degree term.

$$f(x) = f(x_0) + (x - x_0)f'(x_0) + \cdots + \frac{(x - x_0)^4}{4!} f^{(4)}(x_0)$$

$$+ \frac{(x - x_0)^5}{5!} f^{(5)}(\xi) \tag{6-71}$$

Now set $x = x_0 - h$ and then $x = x_0 + h$ in Eq. (6-71). Using the notation

$$x_i = x_0 + ih \qquad i = 0, \pm 1, \pm 2, \ldots$$

$$f_i = f(x_i)$$

$$f_0^{(n)} = f^{(n)}(x_0) \qquad n = 1, 2, \ldots$$

we get

$$f_{-1} = f_0 - hf_0' + \frac{h^2}{2!} f_0'' - \frac{h^3}{3!} f_0^{(3)} + \frac{h^4}{4!} f_0^{(4)} - \frac{h^5}{5!} f^{(5)}(\xi_{-1})$$

$$f_1 = f_0 + hf_0' + \frac{h^2}{2!} f_0'' + \frac{h^3}{3!} f_0^{(3)} + \frac{h^4}{4!} f_0^{(4)} + \frac{h^5}{5!} f^{(5)}(\xi_1) \tag{6-72}$$

Finally, we subtract the first of these two expressions from the second and divide the result by $2h$. The result is a formula similar to Eq. (6-70).

$$\frac{f_1 - f_{-1}}{2h} = f_0' + \frac{h^2}{6} f_0^{(3)} + \frac{h^4}{240}\left[f^{(5)}(\xi_{-1}) + f^{(5)}(\xi_1)\right] \qquad (6\text{-}73)$$

If $f^{(5)}(x)$ is continuous on $[x_{-1}, x_1]$, then the remainder can be simplified by using the intermediate-value theorem. There exists an $\xi \in (x_{-1}, x_1)$ such that

$$\frac{f_1 - f_{-1}}{2h} = f_0' + \frac{h^2}{6} f_0^{(3)} + \frac{h^4}{120} f^{(5)}(\xi) \qquad (6\text{-}74)$$

Now consider repeating the whole procedure, this time taking $x = x_0 - 2h$ and $x = x_0 + 2h$. Alternatively we note that the result will be similar to Eq. (6-74) with h replaced by $2h$. If $f^{(5)}(x)$ is continuous on $[x_{-2}, x_2]$, the result is

$$\frac{f_2 - f_{-2}}{4h} = f_0' + \frac{4h^2}{6} f_0^{(3)} + \frac{16h^4}{120} f^{(5)}(\mu) \quad x_{-2} < \mu < x_2 \qquad (6\text{-}75)$$

This gives us another formula for $f'(x_0)$.

Richardson's extrapolation procedure consists in combining the formulas (6-74) and (6-75) in a way which eliminates the terms involving h^2. This can be done by multiplying formula (6-74) by 4 and subtracting formula (6-75). This gives

$$\frac{f_{-2} - 8f_{-1} + 8f_1 - f_2}{4h} = 3f_0' - \frac{h^4}{30}\left[4f^{(5)}(\mu) - f^{(5)}(\xi)\right] \qquad (6\text{-}76)$$

This can be solved for $f'(x_0)$, giving

$$f_0' = \frac{f_{-2} - 8f_{-1} + 8f_1 - f_2}{12h} + \frac{h^4}{90}\left[4f^{(5)}(\mu) - f^{(5)}(\xi)\right] \qquad (6\text{-}77)$$

The result is a new formula for numerical differentiation. In this formula the truncation error involves a factor h^4. [Compare this with the third formula of Eqs. (6-17).]

The notation $O(h^n)$ is frequently used in talking about error terms. This is read "Oh of h^n." For example, we write

$$f_0' = \frac{f_1 - f_{-1}}{2h} + 0(h^2) \qquad (6\text{-}78)$$

The notation $O(h^2)$ in this case can be thought of as a shorthand for the remainder in Eq. (6-8).

In general, an error expression $E(h)$ is said to be $O(h^n)$ if there exists a constant K which does not depend upon h such that for small values of h

$$|E(h)| \le Kh^n$$

A formula for numerical differentiation, integration, approximation of functions, etc., is said to be of order n if the truncation error (or remainder) is $O(h^n)$.

EXAMPLE 6-12. Suppose that $f^{(5)}(x)$ is continuous on an interval $[a, b]$ containing $[x_{-2}, x_2]$. Show that

$$f_0' = \frac{f_{-2} - 8f_{-1} + 8f_1 - f_{+2}}{12h} + O(h^4)$$

That is, show that formula (6-77) is a fourth-order method for numerical differentiation.

Solution. From Eq. (6-77) the remainder is given by

$$\frac{h^4}{90}\left[4f^{(5)}(\mu) - f^{(5)}(\xi)\right] \quad x_{-1} < \xi < x_1, \; x_{-2} < \mu < x_2$$

Let

$$M = \max_{x \in [a,b]}\left|f^{(5)}(x)\right|$$

Then

$$\left|\frac{4f^{(5)}(\mu) - f^{(5)}(\xi)}{90}\right| \leq \frac{\left|4f^5(\mu)\right| + \left|f^{(5)}(\xi)\right|}{90} \leq \frac{M}{18}$$

Hence

$$\left|\frac{h^4}{90}\left[4f^{(5)}(\mu) - f^{(6)}(\xi)\right]\right| \leq \frac{M}{18}h^4$$

EXAMPLE 6-13. Determine the order of Simpson's rule for $2m$ intervals

Solution. If $f^{(4)}(x)$ is continuos on $[a, b]$ then from the equation in Example 6-8 and Eq. (6-40)

$$\int_a^b f(x)\,dx = \frac{h}{3}\left(f_0 + f_{2m} + 2\sum_{j=1}^{m-1} f_{2j} + 4\sum_{j=1}^{m} f_{2j-1}\right) - \frac{(b-a)h^4 f^{(4)}(\mu)}{180}$$

Let

$$K = \frac{b-a}{180}\max_{a \leq x \leq b}\left|f^{(x)}(x)\right|$$

Then the remainder is bounded by Kh^4, so the remainder is $O(h^4)$.

We can express Richardson's extrapolation in general terms. Suppose that MP denotes a mathematical procedure, while $NP(h)$ denotes a numerical procedure which approximates MP and depends upon a **step size** h. Finally, suppose that there exist constants c, n, m such that

$$MP = NP(h) + ch^n + O(h^m) \tag{6-79}$$

where $m > n$. If the same numerical procedure is applied with a different step size αh, where $\alpha > 0$ and $\alpha \neq 1$, then

$$MP = NP(\alpha h) + c(\alpha h)^n + O(\alpha^m h^m) \tag{6-80}$$

Now multiply Eq. (6-79) by α^n, subtract (6-80), and divide the result by $\alpha^n - 1$. This gives

$$MP = \frac{\alpha^n NP(h) - NP(\alpha h)}{\alpha^n - 1} + \frac{\alpha^n O(h^m) - O(\alpha^m h^m)}{\alpha^n - 1}$$

By suitable argument based upon the definition of $O(h^m)$, it follows that

$$\frac{\alpha^n O(h^m) - O(\alpha^m h^m)}{\alpha^n - 1} = O(h^m)$$

Thus we have the following theorem.

THEOREM 6-7-1. With the notation previously defined, if $\alpha > 0$, $\alpha \neq 1$ and

$$MP = NP(h) + ch^n + O(h^m)$$

then

$$MP = \frac{\alpha^n NP(h) - NP(\alpha h)}{\alpha^n - 1} + O(h^m)$$

In the example for numerical differentiation, we were able to find an explicit formulation of the truncation error in the form (6-79). We had, from Eq. (6-73),

$$f'(x_0) = \frac{f_1 - f_{-1}}{2h} - \frac{h^2 f^{(3)}(x_0)}{6} + O(h^4)$$

Here $c = -f^{(3)}(x_0)/6$, $n = 2$, $m = 4$. It is usually not easy to find such a form of the truncation error. This difficulty can be overcome in the cases where Theorem 6-7-1 is applicable provided that the order n of the method under consideration is known. To illustrate, suppose that we know only that

$$f'(x_0) = \frac{f(x_0 + h) - f(x_0 - h)}{2h} + O(h^2)$$

and that Theorem 6-7-1 is applicable. Then, if $\alpha > 0$, $\alpha \neq 1$,

$$f'(x_0) = \frac{\alpha^2 \dfrac{f_{(x_0+h)} - f_{(x_0-h)}}{2h} - \dfrac{f(x_0 + \alpha h) - f(x_0 - \alpha h)}{2\alpha h}}{\alpha^2 - 1} + O(h^m)$$

where $m > 2$. Setting $\alpha = 2$ gives the formula (6-77).

Theorem 6-7-1 is applicable to all the **numerical differentiation** and **integration** formulas derived in this chapter provided that $f(x)$ is sufficiently differentiable. In addition, it is quite useful in the numerical solution of differential equations (see Exercise 47, Chap. 7).

In applying Richardson's extrapolation procedure in a particular problem there is no need to derive the higher-order formula. One merely calculates the approximate answer, using two different step sizes, and combines these answers by the rules which are used to derive the higher-order formula. The following example illustrates the basic ideas involved.

EXAMPLE 6-14. Compute $f'(2.0)$ from Table 6-8. This is a table of $f(x) = e^x$.

TABLE 6-8

x	$f(x)$
1.8	6.0496
1.9	6.6859
2.0	7.3891
2.1	8.1662
2.2	9.2050

Solution. The formula to be used is

$$f'(x_0) \approx \frac{f(x_0 + h) - f(x_0 - h)}{2h}$$

We get

$$f'(2.0) \approx \begin{cases} 7.4015 & h = .1 \\ 7.4385 & h = .2 \end{cases}$$

Next we take 4 times the first result, subtract the second, and divide by 3.

$$f'(2.0) \approx 7.3892 \quad \text{extrapolated result}$$

Notice that the result of the extrapolation process is the most accurate of the three approximations to $f'(2.0)$.

Taylor's-series methods can be used to derive formulas for higher derivatives. Suppose that the series in Eqs. (6-72) were terminated after the third-degree term.

$$f(x_0 - h) = f(x_0) - hf'(x_0) + \frac{h^2}{2} f''(x_0) - \frac{h^3}{6} f'''(x_0) + \frac{h^4}{24} f^{(4)}(\xi_{-1})$$

$$f(x_0 + h) = f(x_0) + hf'(x_0) + \frac{h^2}{2} f''(x_0) + \frac{h^3}{6} f'''(x_0) + \frac{h^4}{24} f^{(4)}(\xi_1)$$
(6-81)

The sum of these two formulas is

$$f(x_0 - h) + f(x_0 + h) = 2f(x_0) + h^2 f''(x_0) + \frac{h^4}{24}[f^{(4)}(\xi_{-1}) + f^{(4)}(\xi_1)]$$

This can be solved for $f''(x_0)$.

$$f''(x_0) = \frac{f(x_0 - h) - 2f(x_0) + f(x_0 + h)}{h^2}$$

$$-\frac{h^2}{24}[f^{(4)}(\xi_{-1}) + f^{(4)}(\xi_1)] \tag{6-82}$$

Thus, if $f \in C^4 [x_0 - h, x_0 + h]$, there exists a $\mu \in (x_0 - h, x_0 + h)$ such that

$$f''(x_0) = \frac{f(x_0 - h) - 2f(x_0) + f(x_0 + h)}{h^2} - \frac{h^2}{12} f^{(4)}(\mu) \tag{6-83}$$

The difficulty in finding higher-order derivatives numerically can be seen in this formula. The inherent error in finding $f''(x_0)$ varies as $1/h^2$, whereas the truncation error varies as h^2. It is difficult to pick h so that both these will be small unless the errors in evaluating $f(x)$ are very small.

Richardson's extrapolation process can be applied to the above formula for $f'''(x_0)$ to give a higher-order formula. The formula obtained will have error of order $O(h^4)$. This can be seen by carrying an appropriate number of additional terms in the Taylor's-series expansion of $f(x_0 - h)$ and $f(x_0 + h)$ in the derivation of Eq. (6-83). The derivation of such a formula with remainder is left as an exercise.

EXERCISE 38. Derive an expression for the inherent error in finding $f''(x_0)$, using Eq. (6-83).

EXERCISE 39. Use Richardson's extrapolation on Eq. (6-83) to derive a higher-order formula for finding $f''(x_0)$.

EXERCISE 40. Prove that the formula derived in Exercise 39 has order 4.

EXERCISE 41. Find approximations to $f'(x_0)$ at $x_0 = 1.10$, using $h = .05$, $h = .10$ and by **extrapolation**, where $f(x) = e^x$ and a table of e^x (Table 6-3) is given in Example 6-3. Compare the results with 3.0042.

6-8   ROMBERG INTEGRATION

The application of Richardson's extrapolation to integration theory gives a very useful algorithm for numerical integration. To illustrate the basic ideas involved, we shall derive **Simpson's rule** from the trapezoidal rule, by using Richardson's extrapolation.

Throughout this section we shall be trying to approximate

$$\int_a^b f(x)\, dx \tag{6-84}$$

For a given integer $m > 0$ let

$$h_0 = \frac{b-a}{m} \qquad h_1 = \frac{b-a}{2m}$$

The trapezoidal rule using step sizes h_0 and h_1 gives

$$\int_a^b f(x)\, dx \approx \frac{h_0}{2}\left[f(a) + f(b) + 2\sum_{i=1}^{m-1} f(a + ih_0) \right] \tag{6-85}$$

$$\int_a^b f(x)\, dx \approx \frac{h_1}{2}\left[f(a) + f(b) + 2\sum_{i=1}^{2m-1} f(a + ih_1) \right] \tag{6-86}$$

Recall that the remainder using the trapezoidal rule is $O(h^2)$, and notice that $h_1 = h_0/2$. Thus Richardson's extrapolation in this case is performed by subtracting (6-85) from 4 times (6-86) and dividing the result by 3. After some simplification we obtain

$$\int_a^b f(x)\, dx \approx \frac{h_1}{3}\left\{ f(a) + f(b) + 2\sum_{i=1}^{m-1} f(a + 2ih_1) \right.$$

$$\left. + 4\sum_{i=1}^{m} f[a + (2i-1)h_1] \right\} \tag{6-87}$$

Aside from a slight difference in notation, Eq. (6-87) can be seen to be identical to Eq. (6-66), which is **Simpson's rule** for $2m$ intervals of length h_1.

By applying Richardson's extrapolation to the **trapezoidal rule**, which has remainder $0(h^2)$, we obtained Simpson's rule, which has remainder $O(h^4)$. It would seem reasonable, then, to apply Richardson's extrapolation to Simpson's rule. It turns out that if this is done the result is a formula with remainder $O(h^6)$.

Further applications of the extrapolation process will yield formulas with remainder $O(h^8)$, $O(h^{10})$, etc. To implement the above idea, suppose that we let

$$h_2 = \frac{b-a}{4m}$$

Then the trapezoidal rule gives

$$\int_a^b f(x)\, dx \approx \frac{h_2}{2}\left[f(a) + f(b) + 2\sum_{i=1}^{4m-1} f(a+ih_2) \right] \tag{6-88}$$

Richardson's extrapolation applied to Eqs. (6-86) and (6-88) will give Simpson's rule for $4m$ intervals of length h_2.

$$\int_a^b f(x)\, dx \approx \frac{h_2}{3}\left\{ f(a) + f(b) + 2\sum_{i=1}^{2m-1} f(a+2ih_2) \right.$$
$$\left. + 4 \sum_{i=1}^{2m} f[a+(2i-1)h_2] \right\} \tag{6-89}$$

Now **Richardson's extrapolation** can be applied to Eq. (6-87) and (6-89). Since the remainders are $O(h^4)$ and $h_2 = h_1/2$, Richardson's extrapolation in this case is performed by subtracting Eq. (6-87) from 16 times Eq. (6-88) and dividing the result by 15. The result will be a formula with remainder $O(h^6)$.

Actually, we are not interested in deriving the higher-order formulas mentioned above. Instead, we wish to find the numerical result of applying these higher-order formulas. The process called **Romberg integration** is an algorithm for computing the successive numerical approximations to a definite integral which would result if the formulas mentioned above were applied in succession. As we shall see, there is no need to have these formulas tabulated in order to apply Romberg integration.

Our development of the Romberg algorithm will proceed in two steps. First we shall give an algorithm for the systematic computation of the trapezoidal rule approximations to Eq. (6-84) using a sequence of step sizes. The first three terms of this sequence are given by Eqs. (6-85), (6-86) and (6-88), with $m = 1$.

DEFINITION 6-2. Let $h_n = (b-a)/2^n$. Then define

$$T_n = \frac{h_n}{2}\left[f(a) + f(b) + 2\sum_{i=1}^{2n-1} f(a+ih_n) \right] \qquad n = 0,1,\ \ldots$$

The T_n defined above is the trapezoidal rule for 2^n subintervals for the integral (6-84). In particular

$$T_0 = \frac{b-a}{2}[f(a)+f(b)]$$

The following algorithm gives a systematic procedure for calculating T_n for $n = 0, 1, \ldots$. We call it the **sequential trapezoidal algorithm.**

ALGORITHM 6-1. Sequential Trapezoidal Algorithm. Let $k > 0$ be a given integer. Define

$$S_n^o = \frac{b-a}{2}[f(a)+f(b)]$$

Using the definition of h_n given above, compute

$$V_{n-1} = h_{n-1}\sum_{i=1}^{2n-1} f\left[a+\left(i-\frac{1}{2}\right)h_{n-1}\right]$$

$$S_n^o = 1/2\left(S_{n-1}^0 + V_{n-1}\right)$$

for $n = 1, 2, \ldots, k$.

THEOREM 6-8-1.

$$S_n^o = T_n$$

Proof. The proof proceeds by induction on n. The result is obvious for $n = 0$.

Suppose now that $S_{n-1}^o = T_{n-1}$. Then, using the definitions given previously,

$$T_n = \frac{h_n}{2}\left[f(a)+f(b)+2\sum_{i=1}^{2n-1} f(a+ih_n)\right]$$

$$= \frac{h_{n-1}}{4}\left[f(a)+f(b)+2\sum_{i=1}^{2n-1} f\left(a+\frac{ih_{n-1}}{2}\right)\right]$$

Now split the sum into two parts.

$$T_n = \frac{h_{n-1}}{4}\left\{f(a)+f(b)+2\sum_{j=1}^{2^{n-1}-1} f(a+jh_{n-1})\right.$$

$$\left.+2\sum_{j=1}^{2n-1} f\left[a+\left(j-\frac{1}{2}\right)h_{n-1}\right]\right\}$$

$$= \frac{1}{2}\left\{\frac{h_{n-1}}{2}\left[f(a)+f(b)+2\sum_{j=1}^{2^{n-1}-1} f(a+jh_{n-1})\right]\right.$$

$$\left.+ h_{n-1}\sum_{j=1}^{2n-1} f\left[a+\left(j-\frac{1}{2}\right)h_{n-1}\right]\right\}$$

$$= 1/2\,(T_{n-1}+\Box_{n-1})$$

$$= 1/2\left(\Box_{n-1}^0 + \Box_{n-1}\right)= \Box_n^0 \tag{6-90}$$

EXAMPLE 6-15. Use the sequential trapezoidal algorithm with $k = 2$ to approximate $\int_1^3 (dx/x)$.

Solution. Here $a = 1$, $b = 3$, and the exact solution is $\ln 3 = 1.0986$ to 4D.

$$h_0 = b - a = 2 \quad S_0^o = \frac{b-a}{2}[f(a) + f(b)] = \frac{4}{3}$$

$$V_0 = h_0 f\left(a + \frac{h_0}{2}\right) = 1$$

Thus $S_1^o = 1/2(S_0^o + V_0) = 7/6$. To continue,

$$h_1 = \frac{b-a}{2} = 1$$

$$V_1 = h_1\left[f\left(a + \frac{h_1}{2}\right) + f\left(a + \frac{3h_1}{2}\right)\right] = \frac{16}{15}$$

Thus $S_2^o = 1/2(S_1^o + V_1) = 67/60$.

We have now computed three approximations to the desired integral,

$$\begin{array}{ll} h_0 = 2 & T_0 = 1.33333 \cdots \\ h_1 = 1 & T_1 = 1.16666 \cdots \\ h_2 = 1/2 & T_2 = 1.11666 \cdots \end{array}$$

The **sequential trapezoidal algorithm** generates a sequence of approximations to the integral (6-84). The natural theoretical question to ask is: Does the sequence S_0^o, S_1^o, S_2^o... converge to the integral of $f(x)$ over $[a, b]$? We answer this by the following.

THEOREM 6-8-2. Let $f(x) \in C[a,b]$. If $\{S_n^o\}$ is geven by Algorithm 6-1, then

$$\lim_{n \to \infty} S_n^o = \int_a^b f(x)\, dx$$

This theorem follows by the use of the definition of the Riemann integral. The proof is left as Exercise 45.

EXERCISE 42. Continue Example 6.15 by finding S_3^o.

EXERCISE 43. Derive formula (6-89).

EXERCISE 44. Apply the sequential trapezoidal algorithm with $k = 3$ to $\int_0^1 x^3 dx$. Compare your results with the exact value of the integral.

EXERCISE 45. Prove Theorem 6-8-2. *Hint*: Divide $[a,b]$ into equal subintervals of length $h/2$, where $h = (b-a)/2^n$. Show that S_n^o is a Riemann sum for this partitioning of $[a,b]$.

EXERCISE 46. If $f \in C^2[a, b]$, find a bound on truncation error in the approximation S_k^o to $\int_a^b f(x)\, dx$.

EXERCISE 47. Derive a numerical-integration formula by applying Richardson's extrapolation to Eqs. (6-87) and (6-89).

The basic ideas of the repeated application of Richardson's extrapolation to the quantities $S_0^o, S_1^o, \ldots, S_k^o$ were illustrated in the first part of this section. We shall call the general algorithm the **Romberg extrapolation algorithm.**

ALGORITHM 6-2. **Romberg Extrapolation Algorithm.** Given $S_0^o, S_1^o, \ldots,$ S_k^o generated by Algorithm 6–1, let $j = 1$ and compute S_n^j by the formula

$$S_n^{\,j} = \frac{2^{2j} S_{n+1}^{j-1} - S_n^{j-1}}{2^{2j} - 1}$$

for $n = 0, 1, \ldots, k - j$. Repeat the above calculations for
$$j = 2, 3, \ldots, k$$
The quantities S_n^j may be arranged in a triangular array. Table 6-9 illustrates the case $k = 5$.

TABLE 6-9

$S_0^{\,0}$					
$S_1^{\,0}$	$S_0^{\,1}$				
$S_2^{\,0}$	$S_1^{\,1}$	$S_0^{\,2}$			
$S_3^{\,0}$	$S_2^{\,1}$	$S_1^{\,2}$	$S_0^{\,3}$		
$S_4^{\,0}$	$S_3^{\,1}$	$S_2^{\,2}$	$S_1^{\,3}$	$S_0^{\,4}$	
$S_5^{\,0}$	$S_4^{\,1}$	$S_3^{\,2}$	$S_2^{\,3}$	$S_1^{\,4}$	$S_0^{\,5}$

The first column of the array contains a sequence of trapezoidal-rule approximations to the integral (6-84). The second column contains a sequence of Simpson's-rule approximations. Each successive column consists of higher-order approximations to the integral. The ith column, i.e., the numbers S_k^{i-1}, consists of approximations of order h^{2i}. A proof of the following theorem is given in Ralston (pp. 121-128).

THEOREM 6-8-3. If $f(x) \in C[a,b]$ and $\{S_0^{\,j}\}$ is generated by Algorithm 6-2, then $\lim_{j \to \infty} S_0^{\,j} = \int_a^b f(x)\, dx$.

It is difficult to give good a priori rules telling which of the two sequences

$$(a)\ \left\{ S_n^o \right\}_{n=0,1,\ldots} \qquad (b)\ \left\{ S_0^{\,j} \right\}_{j=0,1,\ldots}$$

will converge more rapidly for a given problem. Generally speaking, if the higher derivatives of $f(x)$ do not grow too rapidly or if $f(x)$ is "polynomial like," then (b) is likely to be the more rapidly converging sequence. Conversely, if $f(x)$ is not highly differentiable or its higher derivatives grow rapidly, then (a) may possibly converge more rapidly than (b).

EXAMPLE 6-16. Apply the **Romberg-extrapolation algorithm** with $k = 2$ to Example 6-15.

Solution.

$$S_0^1 = \frac{4S_1^0 - S_0^0}{3} = \frac{10}{9}$$

$$S_1^1 = \frac{4S_2^0 - S_1^0}{3} = \frac{99}{90}$$

$$S_0^2 = \frac{16S_1^1 - S_0^1}{15} = \frac{742}{675}$$

Table 6-10 gives the result of the calculations of Example 6-15 and the above rounded to 7D. In this example S_0^2 is a better approximation to the exact answer than is S_2^0.

TABLE 6-10

S_i^0	S_i^1	S_i^2
1.3333333		
1.1666667	1.1111111	
1.1166667	1.1000000	1.0992593

EXERCISE 48. Continue Example 6-16 by finding S_0^3 (see Exercise 42).

EXERCISE 49. Apply the Romberg-extrapolation algorithm with $k = 2$ to $\int_0^1 x^3\,dx$ (see exercise 44)

In our development of **Romberg integration** a table of the form of Table 6-9 is generated one column at a time. This development emphasizes the idea that a sequence of trapezoidal approximations is used. The implementation of this algorithm in C++ is discussed below. The problem of writing a program which constructs a table like Table 6-9 one row at a time is left as Exercise 50. Such a program is very useful when it is not known in advance how many rows of the table are to be computed. The computation continues until a k is reached such that $|S_0^k - S_{k-1}^0| < \varepsilon$ or $|S_0^k - S_0^{k-1}| < \varepsilon$, where $\varepsilon > 0$ is a pre-assigned constant.

In program ROMBERG the quantities S_i^j are generated one column at a time and stored in $P(11,11)$. The part of the program down to $w = 4$ performs the **sequential trapezoidal algorithm**.

```
// Program 6 - 1 Romberg Integration
float f(float);
 int main()
 {
 float p[11][11],v,h,w,wm, a, b, x;
 int i,k,j,kc,kp;    // Beginning of sequential trapezoidal algorithm
 cout<<"Input a, b and k: ";
 cin>>a>>b>>k;
 h = b - a;
```

```
p[1][1] = .5*h*(f(a) + f(b));
kp = k + 1;
kc = 1;
for(i=1;i<=k;++i)
 {
 v = 0.;
 for(j=1;j<=kc;++j)
   {
   x = j;
   v = v + f(a + (x-.5)*h);
   }
 v = v * h;
 p[i + 1][1] = .5*(p[i][1] + v);
 kc = 2*kc;
 h = .5*h;
 }     // End of sequential trapezoidal algorithm
w = 4.;
for(i =2; i<=kp; ++i)
 {
 wm = w - 1.;
 for(j =i; j<=kp; ++j)
   p[j][i] = (w*p[j][i-1] - p[j-1][i-1])/wm;
 w = 4.*w;
   }
 for(i = 1; i<= kp; ++i)
   {
   for(j=1; j<=i; ++j)
    cout<<p[i][j]<<" ";
   cout<<endl;
   }
  return 0;
}
  float f(float x)
   {
   float ff;
   ff = 1./x;
   return(ff); }
/*
Input a, b and k: 1.  3.  2
1.33333
1.16667  1.11111
1.11667  1.1  1.09926
: */
```

The sample output above is for Example 6-15.

EXERCISE 50. Write a program to generate the Romberg-integration table one row at a time.

6-9   MULTIPLE INTEGRALS

A general treatment of the numerical evaluation of multiple integrals is a suitable topic for an advanced course in numerical analysis. Considerable current research is being directed toward this area. One approach to the numerical evaluation of multiple integrals is based upon formulas for the evaluation of one-dimensional integrals. For simplicity we shall restrict our attention to double integrals of the form

$$\int_c^d \int_a^b f(x, y) \, dx \, d \tag{6-91}$$

First we shall use the **trapezoidal rule** for one interval to get a formula for Eq. (6-91). Then we shall subdivide $[a, b]$ into n intervals and $[c, d]$ into m intervals and derive a more general integration formula. In neither case shall we derive expressions for the remainder.

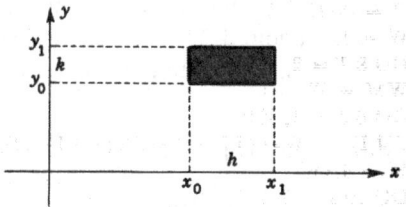

FIG. 6-1

To begin, suppose that we wish to integrate over a small rectangular region as pictured in Fig. 6-1. Using the trapezoidal rule and considering y as fixed, we can write

$$\int_{x_0}^{x_1} f(x, y) \, dx \approx \frac{h}{2} [f(x_0, y) + f(x_1, y)]$$

Now consider y as a variable, and integrate this approximation.

$$\int_{y_0}^{y_1} \int_{x_0}^{x_1} f(x, y) \, dx \, dy \approx \int_{y_0}^{y_1} \frac{h}{2} [f(x_0, y) + f(x_1, y)] \, dy$$

Finally, use the trapezoidal rule to approximate the integral on the right. The result is

$$\int_{y_0}^{y_1} \int_{x_0}^{x_1} f(x, y) \, dx \, dy \approx \frac{hk}{4} [f(x_0, y_0) + f(x_1, y_0)$$
$$+ f(x_0, y_1) + f(x_1, y_1)] \tag{6-92}$$

Next consider the more general problem (6-91). Suppose that $[a, b]$ is divided by into n integrals of length $h = (b - a)/n$ and $[c, d]$ is divided into m intervals of length $k = (d - c)/m$. Proceeding as before, we first consider y fixed. Then, using the notation $x_i = a + ih$, $y_i = c + ik$,

$$\int_a^b f(x, y) \, dx \approx \frac{h}{2} \left[f(x_0, y) + f(x_n, y) + 2 \sum_{i=1}^{n-1} f(x_i, y) \right]$$

Now integrate both sides to get

$$\int_c^d \int_a^b f(x,y)\, dx\, dy \approx \int_c^d \frac{h}{2}\left[f(x_0,y)+f(x_n,y)+2\sum_{i=1}^{n-1} f(x_i,y)\right] dy$$

Finally, use the trapezoidal rule for m intervals to approximate the integral on the right. We introduce the notation $f_{i,j}=f(x_i,y_j)$. Then, after simplification, the final result is

$$\int_c^d \int_a^b f(x,y)\, dx\, dy \approx \frac{hk}{4}\Big\{f_{0,0}+f_{0,m}+f_{n,0}+f_{n,m}$$

$$+2\left[\sum_{i=1}^{n-1}(f_{i,0}+f_{i,m})+\sum_{j=1}^{m-1}(f_{0,j}+f_{n,j})\right]+4\sum_{j=1}^{m-1}\sum_{i=1}^{n-1} f_{i,j}\Big\} \qquad (6\text{-}93)$$

The above procedure can be applied with Simpson's rule or some other integration formula in place of the trapezoidal rule. In addition, the same basic idea can be used to obtain formulas for integration in higher dimensions. In any case the formulas grow rapidly in complexity and in the number of functional evaluations required. This has served to motivate the search for more sophisticated numerical-integration formulas for higher dimensions (see Stroud).

EXERCISE 51. Derive a formula to approximate

$$\int_{y_0}^{y_2} \int_{x_0}^{x_2} f(x,y)\, dx\, dy$$

using Simpson's rule with $x_2-x_0=2h$, $y_2-y_0=2k$.

EXERCISE 52. Derive a formula to approximate

$$\int_{z_0}^{z_1} \int_{y_0}^{y_1} \int_{x_0}^{x_1} f(x,y,z)\, dx\, dy\, dz$$

Using the trapezoidal rule with $x_1-x_0=h$, $y_1-y_0=k$, $z_1-z_0=v$.

EXERCISE 53. Derive a trapezoidal-rule formula to approximate

$$\int_p^q \int_c^d \int_a^b f(x,y,z)\, dx\, dy\, dz$$

Here $[a, b]$, $[c, d]$, $[p, q]$ are to be divided into r, s, t subintervals, respectively (r, s, t are positive integers).

EXERCISE 54. Write a C++ program to implement Eq. (6-93).

Introduction to the Numerical Solution of Ordinary Differential Equations

7-1 INTRODUCTION

Many problems of science and engineering are formulated as **differential equations**. While a few of these equations are quite simple, most defy the type of analysis usually taught in an elementary course on differential equations. Indeed, a majority of differential equations arising in engineering applications cannot be solved in closed form by known methods. Thus, if a solution is needed in such cases, numerical methods must be used. In addition, it is often convenient to employ a numerical method to solve a differential equation even when a closed form of the solution is available. This might be so when the difficulty of computing values from the closed-form solution exceeds that of using a conventional numerical method for solving the equation.

Our study of the numerical solution of ordinary differential equations is divided into two parts. In this chapter we discuss the relevant theory and the basic ideas involved in the numerical solution of a differential equation. The emphasis is upon learning the meaning and usefulness of a numerical solution to a differential equation. Chapter 8 is devoted to a discussion of three of the most commonly used numerical methods. A number of examples and C++ programs are given.

In this text we shall not discuss the problem of the numerical solution of partial differential equations. We shall restrict our investigations to initial-value problems of ordinary differential equations. The abbreviation ODE is used for both the singular and plural of ordinary differential equation.

The first three sections of this chapter are concerned with certain definitions, existence and uniqueness theorems, and the conversion of a higher-order ODE into a system of first-order ODE. Much of this material is often discussed in a first course in ODE. The student may find it desirable to study these three sections concurrently with Sec. 7-4, the first section concerning the numerical solution of ODE.

Let us begin by discussing the initial-vlaue problem

$$\frac{dy}{dx} = p(x)y + g(x) \qquad y(x_0) = y_0 \qquad (7\text{-}1)$$

The ODE in Eq. (7-1) is called a **first-order linear ODE**. It is assumed that the reader is familiar with such equations from his previous studies in mathematics.

DEFINITION 7-1. If $p(x) \in C[x_0, b]$, $g(x) \in C[x_0, b]$, $x_0 \neq b$, then a solution to Eq. (7-1) on $[x_0, b]$ is a differentiable function $u(x)$ such that $u(x_0) = y_0$ and

$$\frac{d}{dx}u(x) = p(x)u(x) + g(x)$$

for all $x \in [x_0, b]$.

It is important to keep in mind that a solution to an ODE is a function. The theory of ODE has been extensively developed. Many of the most important results concern the existence and uniqueness of a solution to certain types of differential equations. Before applying a numerical method of solution to an ODE, it is essential to know that the equation has a solution and that the solution is unique. We shall discuss an example of an existence and uniqueness theorem for the linear first-order ODE.

A proof of the following theorem may be found in Birkoff-Rota.

THEOREM 7-1-1. If $p(x) \in C[x_0, b]$, $g(x) \in C[x_0, b]$ and $x_0 \neq b$, then there exists a unique solution to Eq. (7-1) on $[x_0, b]$

A theorem such as Theorem 7-1-1 tells us two things. First, it is reasonable to seek a solution to the ODE, because such a solution exists. Second, if a solution is found, it is the only possible solution.

If we suppose that the hypotheses of Theorem 7-1-1 are satisfied, then an integrating factor for the

$$\exp\left[-\int_{x_0}^{x} p(t)\, dt\right]$$

If Eq. (7-1) is multiplied by this function, then the resulting equation may be written in the form

$$\frac{d}{dx}\left\{y\exp\left[-\int_{x_0}^{x} p(t)\, dt\right]\right\} = g(x)\exp\left[-\int_{x_0}^{x} p(t)\, dt\right]$$

From this is obtained the following expression for the solution to Eq. (7-1):

$$u(x) = y_0 \exp\left[\int_{x_0}^{x} p(t)\, dt\right] + \exp\left[\int_{x_0}^{x} p(t)\, dt\right]$$
$$\int_{x_0}^{x} g(t)\exp\left[-\int_{x_0}^{t} p(s)\, ds\right] dt \qquad (7\text{-}2)$$

EXAMPLE 7-1. The ODE $y' = xy + x^2$, $y(0) = 0$, is a linear first-order ODE. By comparing it with the form (7-1) it can be seen that $p(x) = x$ and $g(x) = x^2$. Thus both $p(x)$ and $g(x)$ are continuous. If Eq. (7-2) is used, the unique solution to the ODE is found to be the function

$$e^{x^2/2} \int_0^x t^2 e^{-t^2/2} dt$$

Observe that, if the value of the solution function at a particular point x_1 is desired, a numerical integration of some sort must be used.

If $p(x)$ and $g(x)$ are continuous on $(-\infty, \infty)$ and $y(x_0) = y_0$ is given, then Eq. (7-1) will have a unique solution on any interval $[x_0, b]$ (if $b > x_0$) or $[b, x_0]$ (if $b < x_0$). In such a case the interval is often not specified, and we speak of "the" solution to Eq. (7-1).

EXERCISE 1. Verify that Eq. (7-2) is a solution to (7-1). *Hint*:

$$\frac{d}{dx} \left\{ \exp\left[\int_{x_0}^x p(t) \, dt \right] \right\} = p(x) \exp\left[\int_{x_0}^x p(t) \, dt \right]$$

EXERCISE 2. Verify that the ODE $y' = x^2 y + 3x^2$, $y(1) = 2$, satisfies the hypotheses of Theorem 7-1-1. Find the solution, using Eq. (7-2).

EXERCISE 3. Find a solution to the ODE $y' = y/(1 + x) + x$, $y(0) = 1$ on $[0, 1]$. Does a solution exist on $[-1, 0]$? Why?

EXERCISE 4. Find approximate values for the solution to the ODE of Example 7-1 at the points $x_1 = .1$, $x_2 = .2$, using the trapezoidal rule, $h = .1$.

EXERCISE 5. Derive a formula for the solution to

$$y' = y \sum_{i=0}^n a_i x^i \qquad y(0) = 1$$

where $n \geq 0$ and the a_i are given constants.

EXERCISE 6. Write a program for computing the values of the solution to Example 7-1 on $[0, 1]$, using the trapezoidal rule with $h = .1$, then with $h = .05$, then $h = .025$, then $h = .0125$.

The first-order linear ODE just discussed is an example of a more general class of ODE. If an ODE is of the form

$$g(x, y, y') = 0 \quad y(x_0) = y_0 \tag{7-3}$$

it is called a first-order ODE. We shall make an extensive study of ODE having $g(x, y, y') = y' - f(x, y)$. Then the ODE takes the form

$$y' = f(x, y)$$

An ODE written in the form $y' = f(x, y)$ is said to be in normal form.

DEFINITION 7-2. A solution to $y' = f(x, y)$, $y(x_0) = y_0$ on an interval $[x_0, b]$, where $x_0 \neq b$, is a differentiable function $u(x)$ such that $u(x_0) = y_0$ and $(d/dx) u(x) = f(x, u(x))$ for $x \in [x_0, b]$.

We have already shown that some equations of the form (7-3) are relatively easily solved. The linear first-order ODE satisfying the hypotheses of Theorem 7-1-1 have a solution on the interval $[x_0, b]$. We shall discuss two examples of nonlinear first-order ODE to show two difficulties which can easily occur. Even a relatively simple ODE may fail to have a unique solution. The following classical example is that of an ODE which has an infinite number of different solutions.

EXAMPLE 7-2. Consider the ODE $y' = y^{1/2}$, $y(0) = 0$ on an interval $[0, b]$ for any $b > 0$. We shall show that, for any c satisfying $0 < c < b$, the following function is a solution to the ODE:

$$u(x) = \begin{cases} 0 & 0 \le x < c \\ \dfrac{(x-c)^2}{4} & x \ge c \end{cases}$$

First observe that $u(0) = 0$ so that the initial value is satisfied. Now, if $0 \le x < c$, then

$$\frac{d}{dx} u(x) = \frac{d}{dx}(0) = 0$$

Because $0^{1/2} = 0$, the relation $y' = y^{1/2}$ is satisfied. If $c \le x \le b$, then $u(x) = (x-c)^2/4$ and $(d/dx)\, u(x) = (x-c)/2$. Because

$$\left[\frac{(x-c)^2}{4} \right]^{1/2} = \frac{(x-c)}{2}$$

for $x \ge c$, the relation $y' = y^{1/2}$ is satisfied.

Notice that, although the definition of $u(x)$ is a two-part definition, u(x) has a continuous first **derivative** on $[0, b]$. Because each value of c corresponds to a different function $u(x)$, there are an infinite number of different solutions to the ODE under consideration.

It may happen that an ODE has a solution on a particular interval $[x_0, b]$ but not on some longer interval $[x_0, b_1]$, where $b_1 > b$. The following example illustrates this situation (also see Exercise 3).

EXAMPLE 7-3. Consider the ODE $y' = 1 + y^2$, $y(0) = 0$ on the interval $[0, 1]$. It is easily verified that the function $y = \tan(x)$ is a solution to this ODE on $[0, 1]$. We know, however, that the function $y = \tan(x)$ is not defined (i.e., becomes infinite) at $x = \pi/2$. If the same problem is considered on the interval $[0, 2]$, there will be no function which satisfies Definition 7-2 of a solution to the ODE.

7-2   EXISTENCE THEOREMS FOR A SINGLE ODE

This section is devoted to the study of two theorems concerning the existence of a solution to a first-order ODE. Such theorems usually involve the concept of a function satisfying a **Lipschitz condition** over a particular **cartesian-product** region. The reader may need to review the definition of **cartesian product** given in Chap. 3.

DEFINITION 7-3. Let $f(x, y)$ be defined on a domain D in E^2 (the Euclidean plane). The function $f(x, y)$ satisfies a Lipschitz condition in y over D if there exists a constant K such that for all (x, y_1) and (x, y_2) in D

$$|f(x, y_1) - f(x, y_2)| \leq K|y_1 - y_2|$$

EXAMPLE 7-4. Suppose that $I = [0, 2]$, $J = [-10, 10]$, and

$$f(x, y) = |xy| + 1$$

Show that $f(x, y)$ satisfies a Lipschitz condition in y over the domain $I \times J$.

Solution.

$$|f(x, y_1) - f(x, y_2)| = |(|xy_1| + 1) - (|xy_2| + 1)|$$
$$= | |xy_1| - |xy_2| |$$
$$= |x| \cdot | |y_1| - |y_2| |$$
$$\leq |x| \, |y_1 - y_2|$$

Since $x \in [0, 2]$, $|x| \leq 2$. Thus

$$|x||y_1 - y_2| \leq 2|y_1 - y_2|$$

Hence $f(x, y)$ satisfies a Lipschitz condition in y with constant $K = 2$ on $I \times J$.

In most of the ODE that we consider the function $f(x, y)$ will be differentiable. In such a case the following theorem may be useful in calculating a Lipschitz constant for $f(x, y)$ over a domain $I \times J$.

THEOREM 7-2-1. Let I and J be intervals, and let $D = I \times J$. Suppose that $f(x, y) \in C[D]$ and that there exists a constant K such that

$$\left| \frac{\partial f(x, y)}{\partial y} \right| \leq K$$

for all $x, y \in D$. Then $f(x, y)$ satisfies a Lipschitz condition in y with constant K over the domain D.

PROOF. We shall use the **mean-value theorem** for functions of one variable in the proof. Let t be an arbitrary point in I, and suppose that y_1, y_2 are two distinct points of J. Then $f(t, y)$ is a continuous function in y for y between y_1 and y_2. Moreover,

$$\frac{\partial f(t, y)}{\partial y}$$

exists on this interval. Thus there exists a q between y_1 and y_2 such that

$$f(t, y_1) - f(t, y_2) = (y_1 - y_2) \frac{\partial f(t, q)}{\partial y}$$

Using the given bound on the partial derivative leads to

$$|f(t, y_1) - f(t, y_2)| \leq K|y_1 - y_2|$$

Because the bound does not depend on the point t in I, we have proved that $|f(x, y_1) - f(x, y_2)| \leq K|y_1 - y_2|$ for all $x \in I$.

EXAMPLE 7-5. Show that the function $f(x,y)$ in the ODE $y' = xy + x^2$, $y(0) = 0$ satisfies a Lipschitz condition in y over the region $I \times J$, where $I = [0, 1/2]$, $J = [0, 10]$.

Solution. Here $f(x,y) = xy + x^2$; so $df/dy = x$. Because it is easily seen that the maximum value of $|x|$ over the domain $I \times J$ is $1/2$, we conclude that

$$\left|\left(xy_1 + x^2\right) - \left(xy_2 + x^2\right)\right| \leq 1/2 |y_1 - y_2|$$

for all $x \in I$ and $y_1, y_2 \in J$.

EXERCISE 7. Calculate a Lipschitz constant in y for the function $f(x, y)$ in the ODE $y' = x^2 + y^2$, $y(0) = 1$, where $I = [0, 1]$, $J = [-99, 101]$.

EXERCISE 8. Let $f(x,y) = p(x)y + g(x)$. Let $I = [x_0, b]$ be a closed interval, and suppose that $p(x)$ is continuous on I. Prove that $f(x, y)$ is Lipschitz in y on $I \times J$, where $J = (-\infty, \infty)$.

We can now state an **existence theorem**. The proof of this very important theorem is beyond the scope of this text (see Henrici, 1962).

THEOREM 7-2-2. Suppose that $f(x, y)$ is defined and continuous on a domain $D = I \times J$, where $I = [x_0, b]$ and $J = (-\infty, \infty)$. Further suppose that $f(x, y)$ satisfies a Lipschitz condition in y over D. Then the ODE

$$y' = f(x, y) \qquad y(x_0) = y_0$$

has a unique solution on $[x_0, b]$.

EXAMPLE 7-6. Use Theorem 7-2-2 to prove Theorem 7-1-1.

Solution. The problem is to prove the existence of a solution to $y' = p(x)y + g(x)$, $y(x_0) = y_0$ on $[x_0, b]$ given the continuity of $p(x)$ and $g(x)$ on this interval. Because y is continuous, $p(x)y$ is continuous on $D = [x_0, b] \times (-\infty, \infty)$. Hence $p(x)y + g(x)$ is continuous on this domain.

Next, note that

$$\frac{\partial}{\partial y}\left[p(x)y + g(x)\right] = p(x)$$

Since a continuous function on a closed interval is bounded and takes on its maximum absolute value in the interval, we can let

$$K = \max_{x \in [x_0, b]} |p(x)|$$

Then $p(x)y + g(x)$ satisfies a Lipschitz condition in y with constant K on the domain D. Because all the hypotheses of Theorem 7-2-2 are satisfied, the proof is complete.

EXERCISE 9. Prove that each of the following ODE has a unique solution on the interval indicated:

(a) $y' = x^2 e^{-y^2}$ $y(0) = 1$ on $[0, 10]$

(b) $y' = x \sin y - 3x^2$ $y(-1) = -3$ on $[-1, 5]$

(c) $y' = xye^{-y^2}$ $y(-1) = 0$ on $[-1, 0]$

There are many ODE we should like to consider which do not satisfy the hypotheses of Theorem 7-2-2. The most noticeable difficulty is the requirement that a Lipschitz condition in y be satisfied over a domain $I \times J$, where J is $(-\infty, \infty)$. Such simple examples as

$$y' = x^2 + y^2 \quad y(0) = 1$$
$$y' = e^y \quad\quad\quad y(0) = 1 \quad\quad\quad (7\text{-}4)$$
$$y' = y^{3/2} \quad\quad\quad y(1) = 1$$

do not satisfy this hypothesis.

We shall state a second theorem which circumvents this difficulty. Having weaker hypotheses, however, generally means a weaker conclusion. Such is the case here. A proof of this theorem is given in Ford.

THEOREM 7-2-3. Let $y' = f(x, y)$, $y(x_0) = y_0$ be a given ODE. Suppose that there exist two positive number r, s, such that in the domain

$$D = [x_0, x_0 + r] \times [y_0 - s, y_0 + s]$$

the function $f(x, y)$ is continuous and satisfies a Lipschitz condition in y. Finally, suppose that

$$|f(x, y)| \le M$$

for all $(x, y) \in D$. Then if δ is the smaller of the two numbers r and s/M, the differential equation has a unique solution on $[x_0, x_0 + \delta]$.

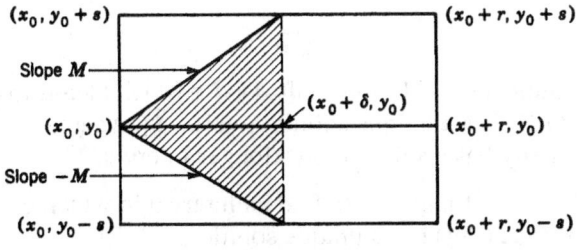

FIG. 7-1

In Theorem 7-2-3 the bound M is a bound on the slope of the solution $y(x)$ as long as $y(x)$ is in the domain D. The geometry of the situation gives some insight into the result. In Fig. 7-1, lines of slope $+M$ and $-M$ have been drawn from the initial point (x_0, y_0). The length of 5 in this case is dictated by the fact that a line of slope M escapes the domain for $x > x_0 + \delta$. The solution to the ODE $y' = f(x, y)$, $y(x_0) = y_0$ on $[x_0, x_0 + \delta]$ will satisfy

$$y_0 - s \le y(x) \le y_0 + s$$

More specifically, the solution will lie in the shaded triangular region of Fig. 7-1.

EXAMPLE 7-7. Apply Theorem 7-2-3 to the equation $y' = xy + x^2$, $y(0) = 1$ on the domain $D = [0, 1] \times [-29, 31]$.

Solution. Here $f(x, y) = xy + x^2$. We have previously established that this function satisfies a Lipschitz condition in y (see Example 7-5). On the domain D it is easy to see that

$$\left| xy + x^2 \right| \le 32$$

In the notation of Theorem 7-2-3 we have $r = 1$, $s = 30$. The smaller of r and $s/32$ is $\delta = 15/16$. Hence we conclude that the ODE has a unique solution on $[0, 15/16]$. This is clearly consistent with, but weaker than, the result we already know, namely, that the equation has a unique solution on $[0, 1]$.

EXAMPLE 7-8. Find an interval on which $y' = 1 + y^2$, $y(0) = 0$ has a unique solution.

Solution. Let us apply Theorem 7-2-3 to this ODE with $I = [0, 1]$, $J = [-50, 50]$. It is straightforward to establish that a Lipschitz condition is satisfied on $D = I \times J$. Next

$$\max_{D} \left| 1 + y^2 \right| = \max_{-50 \le y \le 50} \left| 1 + y^2 \right| = 2501$$

Thus $\delta = 50/2501$, and the ODE has a unique solution on $[0, 50/2501]$. It can be seen that the main restriction in this case is the length of the interval J. If we use $J = [-100, 100]$ then $\delta = 100/(100^2 + 1)$ results. We see that with suitable restriction

$$\delta = \frac{s}{s^2 + 1}$$

This has a maximum at $s = 1$. Using $I = [0,1]$, $J = [-1, 1]$ leads to the conclusion that the ODE has a unique solution on $[0, .5]$. Again, it is evident that the ODE actually has a solution on a longer interval.

EXERCISE 10. Using $r = 1$ and $s = 50$, find an interval in which each of the first two ODE of Eqs. (7-4) has a unique solution.

EXERCISE 11. Explain why Theorem 7-2-3 is not applicable to the example $y' = y^{1/2}$, $y(0) = 0$ given in Sec. 7-1.

EXERCISE 12. Find an interval on which the ODE $y' = y^{1/2}$, $y(0) = 1$ has a unique solution.

EXERCISE 13. Using $I = [0, 1]$, $J = [.5, 1.5]$ find an interval in which $y' = 2/y$, $y(0) = 1$ has a unique solution.

7-3   SYSTEMS OF FIRST-ORDER EQUATIONS

In this section we shall indicate how to transform higher-order equations into systems of first-order equations. We shall define what is meant by a solution to such a system. **Existence theorems** paralleling those given in Sec. 7-2 will be stated and discussed. We shall now consider the system

$$\frac{dy_1}{dx} = f_1(x, y_1, y_2, \ldots, y_n)$$

$$\frac{dy_2}{dx} = f_2(x, y_1, y_2, \ldots, y_n) \tag{7-5}$$

$$\cdots\cdots\cdots\cdots\cdots\cdots\cdots$$

$$\frac{dy_n}{dx} = f_n(x, y_1, y_2, \ldots, y_n)$$

with initial conditions $y_i(x_0) = y_{i0}$ for $i = 1, 2, \ldots, n$. The system (7-5) is called a **normal system** of n first-order differential equations for the n unknown functions $y_1, y_2, \ldots, y_n$.

DEFINITION 7-4. A solution to Eqs. (7-5) is a set of n differentiable functions $u_1(x), u_2(x), \ldots, u_n(x)$ defined on an interval $[x_0, b]$, where $x_0 \neq b$, and having the properties

$$u_1(x_0) = y_{10} \quad u_2(x_0) = y_{20} \quad \cdots \quad u_n(x_0) = y_{n0}$$

and

$$\frac{du_1(x)}{dx} = f_1(x, u_1(x), u_2(x), \ldots, u_n(x))$$

$$\cdots\cdots\cdots\cdots\cdots\cdots\cdots\cdots\cdots\cdots$$

$$\frac{du_n(x)}{dx} = f_n(x, u_1(x), u_2(x), \ldots, u_n(x))$$

for all $x \in [x_0, b]$.

EXAMPLE 7-9. Consider the system of two ODE given by

$$y_1' = y_1 + y_2 \quad y_1(0) = 0$$
$$y_2' = -y_1 + y_2 \quad y_2(0) = 1$$

A solution to this system is given by

$$y_1(x) = e^x \sin x \quad y_2(x) = e^x \cos x$$

To see this, first note that

$$y_1(0) = 0 \quad y_2(0) = 1$$

so that the initial conditions are satisfied. Next

$$y_1' = e^x \sin x + e^x \cos x = y_1 + y_2$$
$$y_2' = e^x \cos x - e^x \sin x = -y_1 + y_2$$

Hence the equation are satisfied.

As in Sec. 7-2, we shall need the fundamental concept of a function satisfying a Lipschitz condition over a **cartesian-product** region. In this case we are considering functions

$$f_i(x, y_1, y_2, \ldots, y_n)$$

for $i = 1, 2, \ldots, n$.

The domains we consider will be of the form

$$D = I \times J_1 \times J_2 \times \cdots \times J_n$$

where $I = [x_0, b]$ and the J_i's are intervals such that $y_{i,0} \in J_i$ for

$$i = 1, 2, \ldots, n$$

DEFINITION 7-5. A function $f(x, y_1, \ldots, y_n)$ defined on a domain D is said to satisfy a Lipschitz condition in $y_1, y_2, \ldots, y_n$ if there exists a constant K such that

$$\left| f(x, y_1, y_2, \ldots, y_n) - f(x, \bar{y}_1, \bar{y}_2, \ldots, \bar{y}_n) \right|$$
$$\leq K \left(|y_1 - \bar{y}_1| + |y_2 - \bar{y}_2| + \cdots + |y_n - \bar{y}_n| \right)$$

for all $y_i, \bar{y}_i, \in J_i$ $(i = 1, \ldots, n)$ and for all $x \in I$. (Here the notation $\bar{y}_i$ is meant just to indicate a point in J_i. The bar is used in place of an extra subscript or of a superscript to distinguish the point from y_i.)

EXAMPLE 7-10. Show that the function

$$f(x, y_1, y_2) = x^2 + \cos y_1 + e^{y_1} + y_1 y_2$$

satisfies a Lipschitz condition in y_1, y_2 for $x \in [0,1]$, $y_1 \in [0, 1]$, $y_2 \in [0, 1]$.

Solution. Here $D = [0, 1] \times [0, 1] \times [0, 1]$ is a unit cube in E^3.

$$\left| f(x, y_1, y_2) - f(x, \bar{y}_1, \bar{y}_2) \right|$$
$$= \left| \left(x^2 + \cos y_1 + e^{y_1} + y_1 y_2 \right) - \left(x^2 + \cos \bar{y}_1 + e^{\bar{y}_1} + \bar{y}_1 \bar{y}_2 \right) \right|$$
$$= \left| \left(\cos y_1 - \cos \bar{y}_1 \right) + \left(e^{y_1} - e^{\bar{y}_1} \right) + \left(y_1 y_2 - \bar{y}_1 \bar{y}_2 \right) \right|$$
$$\leq \left| \cos y_1 - \cos \bar{y}_1 \right| + \left| e^{y_1} - e^{\bar{y}_1} \right| + \left| y_1 y_2 - \bar{y}_1 \bar{y}_2 \right|$$

This last step is by use of the triangle inequality. Now, by use of the mean-value theorem, with $y_1, y_1 \in [0, 1]$ kept in mind, we find that there exist ξ_1, ξ_2 in $[0, 1]$ such that

$$\left| \cos y_1 - \cos \bar{y}_1 \right| = |y_1 - \bar{y}_1| \, |\sin \xi_1| \leq |y_1 - \bar{y}_1|$$
$$\left| e^{y_1} - e^{\bar{y}_1} \right| = |y_1 - \bar{y}_1| \, e^{\xi_2} \leq e |y_1 - \bar{y}_1|$$

Next observe that

$$\left|y_1y_2 - \bar{y}_1\bar{y}_2\right| = \left|(y_1y_2 - y_1\bar{y}_2) + (y_1\bar{y}_2 - \bar{y}_1\bar{y}_2)\right|$$
$$\leq |y_1|\left|y_2 - \bar{y}_2\right| + \left|\bar{y}_2\right|\left|y_1 - \bar{y}_1\right|$$
$$\leq \left|y_2 - \bar{y}_2\right| + \left|y_1 - \bar{y}_1\right|$$

for $y_1, \bar{y}_1, y_2, \bar{y}_2$ all in [0, 1].

Finally we can conclude

$$\left|f(x, y_1, y_2) - f(x, \bar{y}_1, \bar{y}_2)\right| \leq |y_1 - \bar{y}_1| + e|y_1 - \bar{y}_1| + |y_1 - \bar{y}_1| + |y_2 - \bar{y}_2|$$
$$\leq (2 + e)\left(|y_1 - \bar{y}_1| + |y_2 - \bar{y}_2|\right)$$

Thus the function satisfies a Lipschitz condition in y_1, y_2 on D, with $K = 2 + e$.

As can be seen by the example, computation of a Lipschitz constant can be quite difficult unless the problem is relatively simple. The need for an extension of Theorem 7-2-1 should be evident.

THEOREM 7-3-1. Let $D = I \times J_1 \times J_2 \times \cdots \times J_n$, where $I, J_1, \ldots, J_n$ are intervals. Suppose that $f(x, y_1, y_2, \ldots, y_n)$ is continuous on D and that $|\partial f/\partial y_i| < K_i$ in D, for $i = 1, \ldots, n$. Then Definition 7-5 is satisfied on D with

$$K = \max_{i=1,2,\ldots,n} K_i$$

Proof. To keep the notation simple, we shall write out the proof for the case $n = 2$. The more general case is proved in the same manner.

$$\left|f(x, y_1, y_2) - f(x, \bar{y}_1, \bar{y}_2)\right|$$
$$= \left|f(x, y_1, y_2) - f(x, \bar{y}_1, y_2) + f(x, \bar{y}_1, y_2) - f(x, \bar{y}_1, \bar{y}_2)\right|$$
$$\leq \left|f(x, y_1, y_2) - f(x, \bar{y}_1, y_2)\right| + \left|f(x, \bar{y}_1, y_2) - f(x, \bar{y}_1, \bar{y}_2)\right|$$

Now, if x and y_2 are considered to be fixed, then $f(x, y_1, y_2)$ satisfies the hypotheses of the mean-value theorem, in the variable y_1. Thus there exists an ξ_1 between y_1 and $\bar{y}_1$ such that

$$\left|f(x, y_1, y_2) - f(x, \bar{y}_1, y_2)\right| = |y_1 - \bar{y}_1| \left|\frac{\partial f}{\partial y_1}(x, \xi_1 y_2)\right|$$

Similarly for $f(x, \bar{y}_1, y_2)$, if $x, \bar{y}_1$ are considered as fixed and y_2 as a variable, then there exists an ξ_2 between y_2 and $\bar{y}_2$ such that

$$\left|f(x, \bar{y}_1, y_2) - f(x, \bar{y}_1, \bar{y}_2)\right| = |y_2 - \bar{y}_2| \left|\frac{\partial f}{\partial y_2}(x, \bar{y}_1, \xi_2)\right|$$

Using the definition of K_1, K_2, we have proved that

$$\left|f(x, y_1, y_2) - f(x, \bar{y}_1, \bar{y}_2)\right| \leq K_1 |y_1 - \bar{y}_1| + K_2 |y_2 - \bar{y}_2|$$

Since $K = \max(K_1, K_2)$ it follows that

$$\left| f(x, y_1, y_2) - f(x, \bar{y}_1, \bar{y}_2) \right| \le K \left(|y_1 - \bar{y}_1| + |y_2 - \bar{y}_2| \right)$$

this completes the proof.

COROLLARY 7-3-1. Let $D = I \times J_1 \times J_2 \times \cdots \times J_n$ and let $I, J_1, J_2, \ldots, J_n$ be closed finite intervals. Suppose that $f(x_1, y_1, y_2, \ldots, y_n)$ and $\partial f / \partial y_i$ for $i = 1, 2, \cdots, n$ are continuous in D. Then $f(x, y_1, y_2, \ldots, y_n)$ satisfies a Lipschitz condition in $y_1, y_2, \ldots, y_n$.

Proof. This follows from the fact that a continuous function on a closed bounded set is bounded.

EXAMPLE 7-11. Compute a Lipschitz constant in y_1, y_2 for

$$f(x, y_1, y_2) = x^2 y_1 + xy_2{}^2 + 3y_1{}^2$$

where $x \in [0, 1]$, $y_1 \in [0, 2]$, $y_2 \in [0, 2]$.

Solution. Here

$$\frac{\partial f}{\partial y_1} = x^2 + 6y_1 \qquad \frac{\partial f}{\partial y_2} = 2xy_2$$

$$D = [0,1] \times [0,2] \times [0,2]$$

$$\max_D \left| x^2 + 6y_1 \right| = 13 = K_1 \qquad \max_D \left| 2xy_2 \right| = 4 = K_2$$

Hence $K = \max(13,4) = 13$

EXAMPLE 7-12. Show that the function.

$$f(x, y_1, y_2) = x^2 y_1 = e^{-y_2{}^2} - xe^{-y_1{}^2}$$

satisfies a Lipschitz condition in y_1, y_2, for $x \in [0, 1]$, $y_1 \in (-\infty, \infty)$, $y_2 \in (-\infty, \infty)$.

Solution.

$$\frac{\partial f}{\partial y_1} = x^2 + 2y_1 xe^{-y_1{}^2} \qquad \frac{\partial f}{\partial y_2} = -2y_2 e^{-y_2{}^2}$$

$$D = [0, 1] \times (-\infty, \infty) \times (-\infty, \infty)$$

While it is not a trivial task to find

$$\max_D \left| \frac{\partial f}{\partial y_1} \right| \qquad \text{and} \qquad \max_D \left| \frac{\partial f}{\partial y_2} \right|$$

it is easy to see that each is finite. From this the desired result follows by use of Theorem 7-3-1.

EXERCISE 14. Show that $f(x, y_1, y_2) = y_1 + y_2$ satisfies a Lipschitz condition in y_1, y_2 for $-\infty < x < \infty, -\infty < y_1, y_2 < \infty$. Find a Lipschitz constant for this function over the indicated domain.

EXERCISE 15. Solve Example 7-10 by use of Theorem 7-3-1.

EXERCISE 16. Show that the function

$$f(x, y, z) = x^2 + 3xyz + z^2 - y^2 + xy$$

satisfies a Lipschitz condition in y, z for $x \in [0, 2]$, $y \in [0, 1]$, $z \in [0, 2]$. Find a Lipschitz constant for this function over the indicated domain.

EXERCISE 17. For a, b, c, d arbitrary show that

$$f(x, y_1, y_2, y_3) = ax + by_1 + cy_2 + dy_3$$

satisfies a Lipschitz condition in y_1, y_2, y_3 for $-\infty < x, y_1, y_2, y_3 < \infty$.

Both **existence theorems** of Sec. 7-2 for an ODE extend to the case of a system of ODE. Each theorem gives sufficient conditions for the existence of a unique solution to a system of first-order ODE on some interval.

THEOREM 7-3-2. Consider a system of first-order ODE as in Eqs. (7-5). Suppose for some interval $I = [x_0, b]$ that each of the functions

$$f_i(x, y_1, y_2 \ldots y_n)$$

is continuous and satisfies a Lipschitz condition in $y_1, y_2, \ldots, y_n$ for $x \in I$, and $-\infty < y_1, y_2, \ldots, y_n < \infty$. Then the system (7-5) has a unique solution on I.

EXAMPLE 7-13. Using Theorem 7-3-2, prove that the system

$$y_1' = y_1 + y_2 \qquad y_1(0) = 0$$
$$y_2' = -y_2 + y_2 \qquad y_2(0) = 1$$

of Example 7-9 has a unique solution on any interval $[0, b]$.

Solution. Here

$$f_1(x, y_1, y_2) = y_1 + y_2$$
$$f_2(x, y_1, y_2) = -y_1 + y_2$$
$$D = [0, b] \times (-\infty, \infty) \times (-\infty, \infty)$$

Clearly each of the functions is continuous. To show that each satisfies a Lipschitz condition in y_1, y_2, we examine

$$\frac{\partial f_1}{\partial y_1} = 1 \qquad \frac{\partial f_1}{\partial y_2} = 1$$

$$\frac{\partial f_2}{\partial y_1} = -1 \qquad \frac{\partial f_2}{\partial y_2} = 1$$

Then by use of Theorem 7-3-1 it follows that $f_1(x, y_1, y_2)$ satisfies a Lipschitz condition with constant $K = 1$ on D, as does $f_2(x, y_1, y_2)$. Since the hypotheses of Theorem 7-3-2 are satisfied, the proof is complete.

EXAMPLE 7-14. Investigate the existence and uniqueness of a solution to

$$y_1' = \frac{y_1 + y_2}{x - 3} \qquad y_1(0) = 1$$

$$y_2' = e^{-y_1^2} + 3e^{-y_2^2} \qquad y_2(0) = 0$$

Solution. Since the first equation is not defined for $x = 3$, we restrict our attention to an interval $I = [0, b]$, where $0 < b < 3$ is arbitrary.

$$f_1(x, y_1, y_2) = \frac{y_1 + y_2}{x - 3}$$

$$f_2(x, y_1, y_2) = e^{-y_1^2} + 3e^{-y_2^2}$$

$$D = [0, b] \times (-\infty, \infty) \times (-\infty, \infty)$$

$$\frac{\partial f_1}{\partial y_1} = \frac{1}{x - 3} \qquad \frac{\partial f_1}{\partial y_2} = \frac{1}{x - 3}$$

$$\frac{\partial f_2}{\partial y_1} = -2y_1 e^{-y_1^2} \qquad \frac{\partial f_2}{\partial y_2} = -6y_2 e^{-y_2^2}$$

On D each of these is bounded in absolute value. Thus each of the functions $f_i(x, y_1, y_2)$ is Lipschitz on D. Hence by Theorem 7-3-2 the system under consideration has a unique solution on $[0, b]$.

EXERCISE 18. Prove that the following system of ODE has a unique solution on $[0, 2]$:

$$y_1' = 3x + 4xy_1 - y_2 + y_3 \qquad y_1(0) = 1$$

$$y_2' = xe^{-y_2^2} \qquad\qquad\quad y_2(0) = -1$$

$$y_3' = x^2 + \frac{y_1 + y_2 + y_3}{x - 4} \qquad y_3(0) = 1$$

EXERCISE 19. Suppose that $a_{11}, \ldots, a_{33}$ are arbitrary constants and that $g_1(x), g_2(x), g_3(x)$ are continuous for $x \in [0, 3]$. Prove that the system below has a unique solution on $[0, 3]$.

$$y_1' = g_1(x) + a_{11} y_1 + a_{12} y_2 + a_{13} y_3 \qquad y_1(0) = y_{10}$$

$$y_2' = g_2(x) + a_{21} y_1 + a_{22} y_2 + a_{23} y_3 \qquad y_2(0) = y_{20}$$

$$y_3' = g_3(x) + a_{31} y_1 + a_{32} y_2 + a_{33} y_3 \qquad y_3(0) = y_{30}$$

Many commonly occurring systems of ODE involve functions which do not satisfy a Lipschitz condition over an unbounded domain, although a Lipschitz condition is satisfied on a bounded subdomain. Paralleling Theorem 7-2-3, we have the following result for this class of problems.

THEOREM 7-3-3. Consider a system of first-order ODE of the form (7-5). Suppose that there exist two numbers $r > 0$, $s > 0$ such that on the domain $D = I \times J_1 \times \ldots \times J_n$, where

$$I = [x_0, x_0 + r]$$
$$J_1 = [y_{10} - s, y_{10} + s]$$
$$\ldots\ldots\ldots\ldots\ldots\ldots$$
$$J_n = [y_{n0} - s, y_{n0} + s]$$

each of the functions $f_i(x, y_1, y_2, \ldots, y_n)$ is continuous and satisfies a Lipschitz condition in $y_1, y_2, \ldots, y_n$. Let M be any number such that

$$\left| f_i(x, y_1, y_2, \ldots, y_n) \right| \le M$$

for $i = 1, \ldots, n$ and for all points in D. Then the system of ODE has a unique solution on $[x, x_0 + \delta]$, where δ is the smaller of r and s/M.

EXAMPLE 7-15. Apply Theorem 7-3-3 to the system

$$y_1' = y_2 \quad y_1(0) = 1$$
$$y_2' = xy_1 \quad y_2(0) = 1$$

for $r = 2$, $s = 49$.

Solution. Here

$$f_1(x, y_1, y_2) = y_2 \quad f_2(x, y_1, y_2) = xy_1$$
$$D = [0, 2] \times [-48, 50] \times [-48, 50]$$
$$\frac{\partial f_1}{\partial y_1} = 0 \quad \frac{\partial f_1}{\partial y_2} = 1$$
$$\frac{\partial f_2}{\partial y_1} = x \quad \frac{\partial f_2}{\partial y_2} = 0$$

Thus f_1 satisfies a Lipschitz condition with constant $K = 1$, and f_2 is Lipschitz with $K = 2$. Each of the functions is clearly continuous and

$$\max_D |f_1| = 50 \quad \max_D |f_2| = 100$$

Hence the system has a unique solution on $[0, \delta]$, where $\delta = 49/100$.

If the same system is considered on the interval $I = [0, 1]$, the conclusion will be that there exists a unique solution on $[0, 49/50]$. If Theorem 7-3-2 were used, the conclusion would be that there is a unique solution on any interval $[0, b]$.

EXAMPLE 7-16. Investigate the following system of first-order ODE:

$$y_1' = 1 + y_1^2 + y_2^2 \quad y_1(0) = 1$$
$$y_2' = \frac{y_1 + y_2}{x - 4} \quad y_2(0) = 1$$

Solution. First we pick some suitable interval I. Let us consider $I = [0, 1]$.

Next, we observe that the first equation $f_1(x, y_1, y_2)$ is not Lipschitz in y_1, y_2 for $-\infty < y_1, y_2 < \infty$ so that we must use Theorem 7-3-3.

Let us arbitrarily pick $s = 2$ and determine the resulting interval $[0, \delta]$.

$$I = [0,1] \quad J_1 = [-1,3] \quad J_2 = [-2,2]$$
$$D = I \times J_1 \times J_2$$
$$f_1(x, y_1, y_2) = 1 + y_1^2 + y_2^2$$
$$f_2(x, y_1, y_2) = \frac{y_1 + y_2}{x - 4}$$

By using Theorem 7-3-1 it is easy to establish that f_1 and f_2 are Lipschitz in y_1, y_2 over D. Next

$$\max_D |f_1(x, y_1, y_2)| = 14 \quad \max_D |f_2(x, y_1, y_2)| = 5/3$$

Hence $M = 14$, and $\delta = 1/7$ results.

In this example restricting the intervals J_1, J_2 improves the result. With $s = 1$, $I = [0, 1]$,

$$D = [0,1] \times [0,2] \times [-1,1]$$
$$\max_D |f_1(x, y_1, y_2)| = 6 \quad \max_D |f_2(x, y_1, y_2)| = 1$$

Thus $M = 6$, and $\delta = 1/6$.

EXERCISE 20. Find an interval on which the following system of ODE has a unique solution:

$$y_1' = xy_1y_2 \qquad y_1(0) = 0$$
$$y_2' = x + y_1 + y_2 \quad y_2(0) = 1$$

EXERCISE 21. Find an interval on which the following system has a unique solution:

$$y_1' = x + 2y_1 + 3y_2 + 4y_3 \quad y_1(-1) = 0$$
$$y_2' = \sin y_1 y_2 y_3 \qquad \qquad y_2(-1) = 1$$
$$y_3' = \cos y_1 y_2 y_3 \qquad \qquad y_3(-1) = -1$$

EXERCISE 22. Suppose that $g_1(x)$, $g_2(x)$ are continuous on $[0, 1]$. What can you say about the existence of a unique solution to the follwing system?

$$y_1' = g_1(x) y_2 \quad y_1(0) = 1$$
$$y_2' = g_2(x) y_1 \quad y_2(0) = 1$$

We shall now show how an ordinary **differential equation** of order higher than 1 can be converted into a system of first-order equations. In this manner a higher-order equation can be studied by writing it as a system of first-order equations. Systems of equations of order higher than 1 can be treated similarly.

Definition 7-6. An nth order ODE in normal form is an equation of the form

$$y^{(n)} = f\left(x, y, y', \ldots, y^{(n-1)}\right)$$
$$y(x_0) = y_0$$
$$y'(x_0) = y'_0$$

$$\ldots\ldots\ldots\ldots\ldots$$

$$y^{(n-1)}(x_0) = y_0^{(n-1)}$$

This can be written as a system of first-order equations by the following technique: Define n new fuctions $y_1(x), y_2(x), \ldots, y_n(x)$ by

$$y_1(x) = y(x)$$
$$y_2(x) = y'(x)$$
$$y_3(x) = y''(x) \qquad\qquad (7\text{-}6)$$
$$\ldots\ldots\ldots\ldots$$
$$y_n(x) = y^{(n-1)}(x)$$

With this change of variable the nth-order ODE in Definition 7-6 becomes a system of first-order ODE. Notice for example that

$$\frac{d}{dx} y_1(x) = \frac{d}{dx} y(x) = y'(x) = y_2(x)$$

Similarly

$$\frac{d}{dx} y_2(x) = \frac{d}{dx} y'(x) = y''(x) = y_3(x)$$

This continues until

$$\frac{d}{dx} y_n(x) = \frac{d}{dx} y^{(n-1)}(x) = y^{(n)}(x) = f\left(x, y, y', \ldots, y^{(n-1)}\right)$$

The resulting system is given by

$$y'_1 = y_2 \qquad\qquad y_1(x_0) = y_0$$
$$y'_2 = y_3 \qquad\qquad y_2(x_0) = y'_0$$
$$\ldots\ldots \qquad\qquad \ldots\ldots\ldots\ldots \qquad (7\text{-}7)$$
$$y'_{n-1} = y_n \qquad\qquad y_{n-1}(x_0) = y_0^{(n-2)}$$
$$y'_n = f\left(x, y_1, y_2, \ldots, y_n\right) \quad y_n(x_0) = y_0^{(n-1)}$$

Example 7-17. Convert the following second-order ODE into a system of first-order ODE.

$$y'' = 8x^2 y' + 3e^x y - xe^{-x}$$
$$y(0) = 1 \quad y'(0) = -1$$

Solution. Let

$$y_1(x) = y(x) \qquad y_2(x) = y'(x)$$

Then

$$y'_1(x) = y_2(x) \qquad y'_2(x) = y''(x)$$

so that the system is

$$y'_1 = y_2 \qquad\qquad y_1(0) = 1$$
$$y'_2 = 8x^2 y_2 + 3e^x y_1 - xe^{-x} \ y^2(0) = -1$$

The example just given was a second-order linear ODE. The following example is of a general *nth*-order linear ODE.

EXAMPLE 7-18. Write the following *nth*-order ODE as a system of first-order

$$y^{(n)} = a_1(x) y^{(n-1)} + a_2(x) y^{(n-2)} + \cdots + a_{n-1}(x) y' + a_n(x) y + f(x)$$
$$y(x_0) = y_0 \quad y'(x_0) = y'_0 \quad y''(x_0) = y''_0 \cdots y^{(n-1)}(x_0) = y_0^{(n-1)}$$

Solution. Make the change of variable

$$y_1(x) = y(x)$$
$$y_2(x) = y'(x)$$
$$y_3(x) = y''(x)$$
$$\cdots\cdots\cdots\cdots$$
$$y_n(x) = y^{(n-1)}(x)$$

Then the equation becomes

$$y'_1 = y_2 \qquad\qquad y_1(x_0) = y_0$$
$$y'_2 = y_3 \qquad\qquad y_2(x_0) = y'_0$$
$$\cdots\cdots \qquad\qquad \cdots\cdots\cdots$$
$$y'_{n-1} = y_n \qquad\qquad y_{n-1}(x_0) = y_0^{(n-2)}$$
$$y'_n = \sum_{i=1}^{n} a_i(x) y_{n+1-i} + f(x) \qquad y_n(x_0) = y_0^{(n-1)}$$

The same technique as illustrated above can be used to write a system of higher-order ODE as a system of first-order ODE. We shall illustrate by an example.

EXAMPLE 7-19. Convert the system below to a system of first-order ODE:

$$y_1'' = g_1(x, y_1, y_2, y'_1, y'_2) \qquad y_1(x_0) = y_{10} \qquad y'_1(x_0) = y'_{10}$$
$$y_2'' = g_2(x, y_1, y_2, y'_1, y'_2) \qquad y_2(x_0) = y_{20} \qquad y'_2(x_0) = y'_{20}$$

Solution. Since subscripted *y*'s have been used in the statement of the problem, it is convenient to use a different notation. We make the following substitutions :

$$u_1(x) = y_1(x) \qquad u_3(x) = y'_1(x)$$
$$u_2(x) = y_2(x) \qquad u_4(x) = y'_2(x)$$

Then by observing $u'_3(x) = y''_1(x)$ and $u'_4(x) = y''_2(x)$ we can write the original system as

$$u'_1 = u_3(x) \qquad\qquad u_1(x_0) = y_{10}$$
$$u'_2 = u_4(x) \qquad\qquad u_2(x_0) = y_{20}$$
$$u'_3 = g_1(x, u_1, u_2, u_3, u_4) \qquad u_3(x_0) = y'_{10}$$
$$u'_4 = g_2(x, u_1, u_2, u_3, u_4) \qquad u_4(x_0) = y'_{20}$$

EXERCISE 23. Write the second-order equation below as a system of first-order equations:

$$y'' = y \qquad y(0) = 1 \qquad y'(0) = 0$$

EXERCISE 24. Convert the following equation into a system of first-order ODE:

$$xy'' - y' + 4x^3 y = 0 \qquad y(1) = 1 \qquad y'(1) = 2$$

EXERCISE 25. Convert the following equation into a system of first-order ODE:

$$y''' = 3y'' + 6xy' + 10x^2 y + \sin x$$
$$y(0) = 1 \quad y'(0) = 2 \quad y''(0) = 3$$

EXERCISE 26. Convert the following system of second-order ODE into a system of first-order ODE:

$$y'' = e^x + y' + u' + u + y \qquad\qquad y(1) = 3 \quad y'(1) = 4$$
$$u'' = e^x + 9xu + 6u' + 9y' + 10y \qquad u(1) = 1 \quad u'(1) = 2$$

EXERCISE 27. Convert the following system of two simultaneous third-order equations into a system of six simultaneous first-order equations:

$$y'''_1 = g_1(x, y_1, y_2, y'_1, y'_2, y''_1, y''_2)$$
$$y'''_2 = g_2(x, y_1, y_2, y'_1, y'_2, y''_1, y''_2)$$
$$y_1(x_0) = a_0 \quad y'_1(x_0) = a_1 \quad y''_1(x_0) = a_2$$
$$y_2(x_0) = b_0 \quad y'_2(x_0) = b_1 \quad y''_2(x_0) = b_2$$

7–4   EULER'S METHOD

Suppose that one is considering a first-order ODE $y' = f(x, y)$, $y(x_0) = y_0$ which has a unique solution $y = u(x)$ on the interval $I = [x_0, b]$. The solution $u(x)$ is a function defined at each point of I. The task of finding an approximate solution to ODE is therefore one of approximating the (usually unknown) function $y = u(x)$ on I.

In general the approximation of $u(x)$ at any one point will involve a number of arithmetic operations. Because there are an infinite number of points in I, we cannot hope to calculate an approximate value of $u(x)$ at each individual point. We therefore set as our goal the task of finding approximate values of $u(x)$ on a certain finite subset of I. The points of this set will be denoted by $x_0, x_1, \ldots, x_m$. While it is not necessary that the points be equally spaced, it is more convenient computationally to have them so. Thus, we shall assume that we wish to approximate $u(x)$ at points $x_i = x_0 + ih$ ($i = 0, 1, \ldots, m$). The quantity h is called the **step size**. The integer m is such that $x_m < b$, while $x_m + h > b$.

In the literature of **differential equations** as well as in the remainder of this text, the exact solution at a point x_i is usually denoted by $y(x_i)$, whereas an approximation to this is denoted by y_i. Thus, we seek $y_1, y_2 \ldots y_m$ which approximate $y(x_1), y(x_2), \ldots, y(x_m)$.

This section discusses **Euler's method** for solving ODE. It is perhaps the simplest of all numerical methods. In general, if the solution to an ODE is known to possess many derivatives, then the more involved methods we shall discuss later will provide more accurate approximations to the solution. However, Euler's method serves as an excellent pedagogical device and is useful in solving certain types of problems. By studying this method in some detail we shall come to understand the basic ideas involved in the numerical solution of ODE. In addition we shall come to appreciate some of the basic difficulties inherent to this subject.

ALGORITHM 7-1 (Euler). Let $f(x,y)$, x_0, y_0, h, m be given. Form the numbers $x_1, x_2, \ldots, x_m$ and $y_1, y_2, \ldots, y_m$ by the rules

$$y_{i+2} = y_i + hf(x_i, y_i)$$
$$x_{i+1} = x_i + h$$

for $i = 0, 1, \ldots, m - 1$.

EXAMPLE 7-20. Apply Algorithm 7-1 to the differential equation $y' = 2y$, $x_0 = 0$, $y_0 = 1$, with $h = .05$ and $m = 3$.

Solution.

$$y_1 = y_0 + hf(x_0, y_0) = 1 + (.05)(2) = 1.1 \qquad x_1 = .05$$
$$y_2 = y_1 + hf(x_1, y_1) = 1.1 + (.05)(2.2) = 1.21 \qquad x_2 = .10$$
$$y_3 = y_2 + hf(x_2, y_2) = 1.331 \qquad x_3 = .15$$

The exact solution to this ODE is $y = e^{2x}$. Exact values of the solution to 4D are given by

$$y(.05) = 1.1052$$
$$y(.10) = 1.2214$$
$$y(.15) = 1.3499$$

Example 7-20 suggests that the numbers generated by Algorithm 7-1 do indeed approximate the solution to an ODE. We shall delay our discussion of the underlying theory, and the statement of relevant theorems, until later.

It is instructive to have a geometric picture of **Euler's method**. We are given an initial point (x_0, y_0) on the solution curve. The slope of the solution curve at this point is given by $f(x_0, y_0)$. Thus, the tangent line to the solution curve at the initial point can be determined. **Euler's method** consists in approximating the solution function by this tangent line.

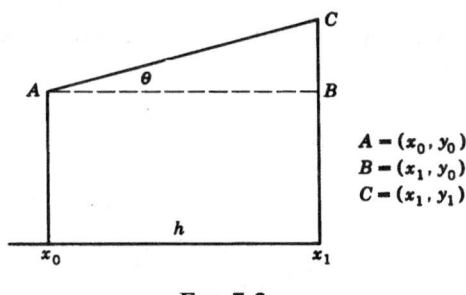

$$A = (x_0, y_0)$$
$$B = (x_1, y_0)$$
$$C = (x_1, y_1)$$

Fig. 7-2

In Fig. 7-2 the points A, B are known, as is h. The tangent of the angle θ is given by $\tan(\theta) = f(x_0, y_0)$. Hence

$$\tan \theta = f(x_0, y_0) = \frac{y_1 - y_0}{h}$$

This is easily solved for y_1, giving

$$y_1 = y_0 + hf(x_0, y_0)$$

This is one step of the Euler algorithm. If (x_1, y_1) is considered as the initial point and the entire process is repeated, the result will be the (x_2, y_2) given by Algorithm 7-1.

Even if we assume that all calculations are exact, errors still arise from several sources. Unless the solution curve $y(x)$ is itself a straight line, the value of y_1 will usually differ from $y(x_1)$. If this is so then in starting at (x_1, y_1), it will usually be the case that the slopes $f(x_1, y_1)$ and $f(x_1, y(x_1))$ will differ.

Typically, then, each subsequent step is made using an incorrect value of the slope and moving from an incorrect point under the incorrect assumption that the solution curve is a linear function. That such a process can produce anything useful is indeed remarkable. The following example indicates the nature of the underlying theory for a simple problem.

Example 7-21. Let α be a given constant. Derive an expression for the result of applying **Euler's algorithm** to

$$y' = \alpha y \qquad y(x_0) = y_0$$

and examine the behavior of the algorithm as h tends to zero.

Solution. The exact solution to this ODE is. $y(x) = y_0 e^{\alpha(x-x(0))}$. Using the Euler algorithm, we compute

$$y_1 = y_0 + h\alpha y_0 = (1+h\alpha)y_0$$
$$y_2 = y_1 + h\alpha y_i = (1+h\alpha)y_1 = (1+h\alpha)^2 y_0$$

....

$$y_n = y_{n-1} + h\alpha y_{n-1} = (1+h\alpha)y_{n-1} = (1+h\alpha)^n y_0$$

Suppose now that we pick a particular point $x > x_0$ and let the **step size** be $h = (x-x_0)/n$. Then $x_0 + h_n = x$ so that y_n is the Euler-algorithm approximation to $y(x)$.

We now examine the approximation y_n to $y(x)$ as h tends to zero. Because $h = (x - x_0)/n$, we have

$$y_n = (1+h\alpha)^n y_0 = \left[1 + \frac{\alpha(x-x_0)}{n}\right]^n y_0$$

Thus

$$\lim_{n\to\infty} y_n = \lim_{n\to\infty} \left[1 + \frac{\alpha(x-x_0)}{n}\right]^n y_0$$

Using techniques from elementary calculus, one can prove (see Exercise 28) that

$$\lim_{n\to\infty}\left[1 + \frac{\alpha(x-x_0)}{n}\right]^n = e^{\alpha(x-x_0)} \tag{7-8}$$

Thus in the limit we obtain the exact solution to the ODE.

$$\lim_{n\to\infty} y_n = y_0 e^{\alpha(x-x_0)}$$

That is, for a fixed x and a sufficiently small **step size** h, the difference between the approximation provided by Euler's algorithm and the exact solution can be made arbitrarily small.

EXAMPLE 7-22. Examine the Euler approximation to the solution to $y' = 2x$, $y(0) = 0$ on $[0,1]$, using $h = .1$.

Solution. The ODE has exact solution $y = x^2$. Euler's method gives $y_{i+1} = y_i + 2h_i$. Table 7-1 gives values of x_i, y_i and $y(x_i)$ for $h = .1$. A graph of these results is given in Fig. 7-3.

Algorithm 7-1 is easily programmed. Typically, provision is made for reading in the initial point (x_0, y_0) as well as the step size h and the end point b. A sample program is given below.

EXAMPLE 7-23. Write a program which applies Algorithm 7-1 to the problem $y' = x^2 + y^2$ for arbitrary x_0, y_0, h, b.

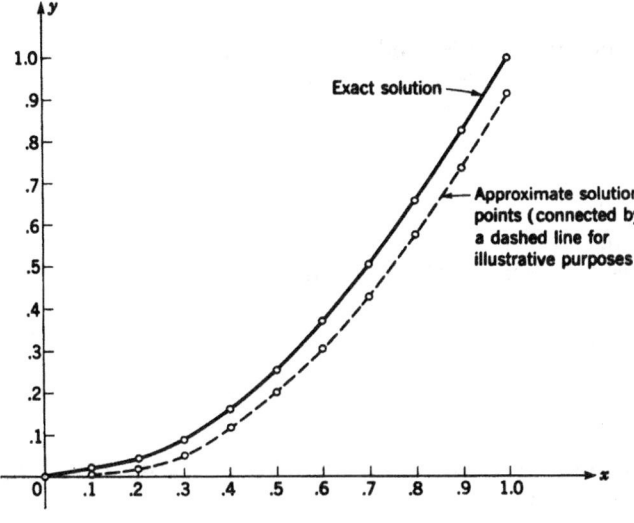

FIG. 7-3

TABLE 7-1

i	x_i	y_i	$y(x_i)$
0	0.0	.0000	.0000
1	.1	.0000	.1000
2	.2	.0200	.0400
3	.3	.0600	.0900
4	.4	.1200	.1600
5	.5	.2000	.2500
6	.6	.3000	.3600
7	.7	.4200	.4900
8	.8	.5600	.6400
9	.9	.7400	.8100
10	1.0	.9200	1.0000

Solution.

```
#include <iostream>
using namespace std;
//  PROGRAM EULER
int main(void)
  {
  int i(1);
  float x,y,h,b;
  cout<<"Input x, y, h, and b"<<endl;
```

```
cin>>x>>y>>h>>b;
cout<<"The Input Was"<<endl;
cout<<x<<' '<<y<<' '<<h<<' '<<b<<endl;
cout<<" i "<<" x "<<" y "<<endl;
while(x<b)
  {
  y = y + h*(x*x + y*y);
  x = x + h;
  cout<<i<<"  "<<x<<"   "<<y<<endl;
  i++;
  }
return 0;
}
```

Table 7-2 shows the results obtained from this program for $h = .025$, $x_0 = 0, y_0 = 1, b = .25$.

TABLE 7-2

Input x, y, h, and b
0 1 .025 .25
The Input Was
0 1 0.025 0.25

i	x	y
1	0.025	1.025
2	0.05	1.05128
3	0.075	1.07897
4	0.1	1.10822
5	0.125	1.13917
6	0.15	1.17201
7	0.175	1.20691
8	0.2	1.24409
9	0.225	1.28378
10	0.25	1.32625

Notice that, while this is an impressive list of numbers, we have no way of knowing anything about the accuracy of the results. For example, how close is $y_8 = 1.244089750$ to the value of $y(.2)$?

To give some indication of the difficulties here, the same program was run for a number of different values of the step size h. A table of the various approximations to $y(.2)$ which resulted is given in Table 7-3. Example 7-21 suggests that the smaller values of h give the more accurate approximations. From the above data we would expect that the exact value of $y(.2)$ is somewhat larger than 1.25.

EXERCISE 28. Prove the result (7-8).

EXERCISE 29. Consider the ODE $y' = x + y, y(1) = 1$. Find approximations to $y(1.2)$ using Euler's method, with $h = .1$ and $h = .05$.

EXERCISE 30. Find an expression for the exact solution to the ODE of Exercise 29. Compute $y(1.2)$, and compare with the above results.

TABLE 7-3

H	Approximation to $y(.2)$
.2	1.200000000
.1	1.222000000
.05	1.236028360
.025	1.244089750
.0125	1.248435499
.00625	1.250695414
.003125	1.251848296
.0015625	1.252430622
.00078125	1.252723277

EXERCISE 31. Apply Euler's algorithm to the equation $y' = 1 + y_2$, $y(0) = 0$ on $[0, .5]$, with $h = .1$; compare the results with the exact solution $y = \tan(x)$.

EXERCISE 32.

(a) Using Euler's method with $h = .1$, find an approximate solution to $y' = 2$, $y(0) = 1$ on $[0,1]$; compare your results with the exact solu tion.
(b) Give some additional examples for which Euler's method will give the exact solution. Can you formulate some general rule for finding such examples?

We shall now extend Euler's algorithm to a system of simultaneous first-order ODE. Suppose that the equations are given in the form

$$
\begin{aligned}
y_1' &= f_1(x, y_1, y_2, \ldots, y_n) & y_1(x_0) &= y_{10} \\
y_2' &= f_2(x, y_1, y_2, \ldots, y_n) & y_2(x_0) &= y_{20} \\
&\cdots\cdots\cdots\cdots\cdots\cdots\cdots\cdots\cdots\cdots\cdots\cdots\cdots & & \\
y_n' &= f_n(x, y_1, y_2, \ldots, y_n) & y_n(x_0) &= y_{n0}
\end{aligned}
\tag{7-9}
$$

The problem is to find approximate values for the unknown functions $y_1(x), y_2(x), \ldots, y_n(x)$ at the points $x_1, x_2, \ldots, x_m$. For a particular function, say $y_j(x)$, the exact solution is denoted by $y_j(x_0), y_j(x_1), \ldots, y_j(x_m)$ whereas an approximate solution is denoted by $y_{j0}, y_{j1}, \ldots, y_{jm}$.

We are given an initial point on each of the solution curves. The **derivative** of each curve at the initial point is easily computed. This leads to the following algorithm.

ALGORITHM 7-2 (Euler). Let $x_0, y_{10}, y_{20}, \ldots, y_{n0}, h$, and m be given. Form the numbers y_{ji} ($j = 1, \ldots, n$; $i = 1, \ldots, m$) by the rules

$$y_{1,i+1} = y_{1i} + hf_1(x_i, y_{1i}, y_{2i}, \ldots, y_{ni})$$
$$y_{2,i+1} = y_{2i} + hf_2(x_i, y_{1i}, y_{2i}, \ldots, y_{ni})$$
$$\cdots\cdots\cdots\cdots\cdots\cdots\cdots\cdots\cdots$$
$$y_{n,i+1} = y_{ni} + hf_n(x_i, y_{1i}, y_{2i}, \ldots, y_{ni})$$
$$x_{i+1} = x_i + h$$

for $i = 0, \ldots, m-1$.

EXAMPLE 7-24. Apply Algorithm 7-2 to the system below, with $h = .02$ and $m = 2$.

$$\begin{array}{ll} y_1' = x + y_1 + y_2 + y_3 & y_1(1) = 3 \\ y_2' = x + 2y_1 + 4y_2 + 6y_3 & y_2(1) = 4 \\ y_3' = x^2 + 3y_1 + 10y_3 & y_3(1) = 5 \end{array}$$

Solution.

$$y_{1,1} = 3 + .02(1 + 3 + 4 + 5) = 3.26$$
$$y_{2,1} = 4 + .02(1 + 6 + 16 + 30) = 5.06$$
$$y_{3,1} = 5 + .02(1 + 9 + 50) = 6.20$$
$$x_1 = 1.02$$
$$y_{1,2} = 3.26 + .02(1.02 + 3.26 + 5.06 + 6.20) = 3.5708$$
$$y_{2,2} = 5.06 + .02(1.02 + 6.52 + 20.24 + 37.20) = 6.3596$$
$$y_{3,2} = 6.20 + .02(1.0404 + 9.78 + 62.00) = 7.656408$$
$$x_2 = 1.04$$

EXAMPLE 7-25. Write the differential equation $y'' = y^2 + x^2$, $y(1) = 1$, $y'(1) = 2$ as a system of first-order equations. Apply Algorithm 7-2 to the system, with $h = .02$, $m = 2$.

Solution. Let $y_1(x) = y(x)$, $y_2(x) = y'(x)$. The resulting system is given by

$$\begin{array}{ll} y_1' = y_2 & y_1(1) = y_{1,0} = 1 \\ y_2' = y_1^2 + x^2 & y_2(1) = y_{2,0} = 2 \end{array}$$

The algorithm gives

$$y_{1,1} = y_{1,0} + hy_{2,0} = 1 + (.02)(2) = 1.04$$
$$y_{2,1} = y_{2,0} + h(y_{1,0}^2 + x_0^2) = 2 + .02(1^2 + 1^2) = 2.0$$
$$x_1 = 1.02$$
$$y_{1,2} = y_{1,1} + hy_{2,1} = 1.0808$$
$$y_{2,2} = y_{2,1} = h(y_{1,1}^2 + x_1^2) = 2.08244$$
$$x_2 = 1.04$$

From the above computations we conclude that

$$y(1.02) \approx 1.04 \qquad y(1.04) \approx 1.0808$$

EXAMPLE 7-26. Write a program which applies Euler's method to the system

$$y_1' = y_1 + y_2 \qquad y_1(x_0) = y_{1,0}$$
$$y_2' = -y_1 + y_2 \qquad y_2(x_0) = y_{2,0}$$

Here $x_0, y_{1,0}, y_{2,0}, h$ and m are to be read as input.

Solution. The program below illustrates how the above can be solved with a program.

Program 7-1 Euler's method for a system

```
#include <iostream>
#include <cmath>        // provide exp, sin, cos
using namespace std;
// PROGRAM EULER2
int main(void)
{
float x,y1,y2,h,f1,f2,z1,z2,y1e,y2e;
int i,m;
cin >> x>>y1>>y2>>h>>m;
y1e = exp(x)*sin(x);
y2e = exp(x)*cos(x);
cout<<" x "<<" y1 "<<" y2 "<<" y1e "<<" y2e "<<endl;
cout<<x<<" "<<y1<<" "<<y2<<" "<<y1e<<" "<<y2e<<endl;
for(i = 1; i<=m; ++i)
  {
  f1 = y1 + y2;
  f2 = y2 - y1;
  z1 = y1 + h*f1;
  z2 = y2 + h*f2;
  x = x + h;
  y1 = z1;
  y2 = z2;
  y1e = exp(x)*sin(x);
  y2e = exp(x)*cos(x);
  cout<<x<<" "<<y1<<" "<<y2<<" "<<y1e<<" "<<y2e<<endl;
  }
return 0;
}
```

Table 7-4 gives sample output from Program 7-1 for the case $x_0 = 0$, $y_1(0) = 0, y_2(0) = 1, h = .1$. The results have been rounded to 5D. The exact solution is given in Example 7-9. Values of the exact solution rounded to 5D are given in the last two columns of Table 7-4.

EXERCISE 33. Convert the ODE $y' = x^2$, $y(0) = 0$, $y'(0) = 1$ into a system of first-order ODE. Use Euler's algorithm with $h = .1$ to find approximations to $y(.1)$, $y(.2)$, $y(.3)$. Compare your results with the exact solution.

TABLE 7-4
Input: 0 0 1 .1 9 (x_0, y_1, y_2, h, m)

x_i	y_{1i}	y_{2i}	$y_1(x_i)$	$y_2(x_i)$
0	0	1	0	1
0.1	0.1	1.1	0.110333	1.09965
0.2	0.22	1.2	0.242655	1.19706
0.3	0.362	1.298	0.398911	1.28957
0.4	0.528	1.3916	0.580944	1.37406
0.5	0.71996	1.47796	0.790439	1.44689
0.6	0.939752	1.55376	1.02885	1.50386
0.7	1.1891	1.61516	1.2973	1.5402
0.8	1.46953	1.65777	1.59651	1.55055
0.9	1.78226	1.67659	1.92667	1.52891

EXERCISE 34. Use Euler's algorithm with $h = .01$ to find approximations to $y_1(.03)$, $y_2(.03)$, where

$$y_1' = y_2 \quad y_1(0) = 1$$
$$y_2' = y_1 \quad y_2(0) = 1$$

Determine the exact solution to this system, and compare your results with the exact values of $y_1(.03)$, $y_2(.03)$.

7-5   HEUN'S METHOD

In this section we shall discuss briefly a method attributed to Heun. In Sec. 7-6 we shall compare the methods of **Euler** and **Heun**. Again we caution the reader that the method under discussion is seldom used when a highly accurate approximation to the solution of an ODE is desired.

One can think of trying to solve a differential equation $y' = f(x, y)$, $y(x_0) = y_0$, by integrating each side of the equation. If $x_1 = x_0 + h$, then

$$\int_{x_0}^{x_1} y'(x) \, dx = \int_{x_0}^{x_1} f(x, y(x)) \, dx \qquad (7\text{-}10)$$

The left side may be simplified, and we obtain

$$y(x_1) = y(x_0) + \int_{x_0}^{x_1} f(x, y(x)) \, dx \qquad (7\text{-}11)$$

Now if the integral on the right side is approximated by the trapezoidal rule, we obtain the result

$$y(x_1) = y(x_0) + \frac{h}{2}\left[f(x_0, y(x_0)) + f(x_1, y(x_1))\right] + \text{remainder} \quad (7\text{-}12)$$

Finally if the quantity $y(x_1)$ in the right side of Eq. (7-12) is approximated by use of **Euler's method** and all remainder terms are ignored, the result is

$$y_1 = y_0 + \frac{h}{2}\Big[f(x_0, y_0) + f\big(x_1, y_0 + hf(x_0, y_0)\big)\Big] \qquad (7\text{-}13)$$

We state the result as an algorithm.

ALGORITHM 7-3 (Heun). Let $f(x, y)$, x_0, y_0, h, and m be given. From the numbers $y_1, y_2, \ldots, y_m$ and $x_1, x_2, \ldots, x_m$ by the rule

$$x_{i+1} = x_i + h$$

$$y_{i+1} = y_i + \frac{h}{2}\Big[f(x_i, y_i) + f\big(x_{i+1}, y_i + hf(x_i, y_i)\big)\Big]$$

for $i = 0, \ldots, m-1$.

Notice that each step of **Heun's method** involves two evaluations of the function $f(x, y)$. To get y_{i+1}, it is necessary to evaluate $f(x, y)$ at (x_i, y_i) and at $(x_{i+1}, y_i + hf(x_i, y_i))$. It is assumed that the value computed in the first evaluation will be stored so that it need not be recomputed in the second evaluation.

EXAMPLE 7-27. Apply Algorithm 7-3 to the problem $y' = 2y$, $x_0 = 0$, $y_0 = 1$, with $h = .05$, $m = 2$.

Solution.

$$x_1 = .05$$

$$y_1 = y_0 + \frac{h}{2}\ \Big[f(x_0, y_0) + f\big(x_1, y_0 + hf(x_0, y_0)\big)\Big]$$

$$= 1 + \frac{.05}{2}\ \big[2 + f(.05, 1.10)\big] = 1.105$$

$$x_2 = .10$$

$$y_2 = y_1 + \frac{h}{2}\ \Big[f(x_1, y_1) + f\big(x_2, y_1 + hf(x_1, y_1)\big)\Big]$$

$$= 1.105 + \frac{.05}{2}\ \big[2.21 + f(.10, 1.2155)\big] = 1.221025$$

Notice that these results compare more favorably with the exact results given in Example 7-20 than do the results obtained by using Euler's method.

EXAMPLE 7-28. Write a program to apply Heun's method to the ODE $y' = x^2 + y^2$. Provision should be made for reading in x_0, y_0, h, and the end point b.

Solution.

```
#include <iostream>
using namespace std;
// PROGRAM HEUN
int main(void)
 {
```

```
float x,y,h,b,w,t;
int i(1);
cin>>x>>y>>h>>b;
cout<<x<<' '<<y<<' '<<h<<' '<<b<<endl;
cout<<" i "<<" x "<<" y "<<endl;
t = .5*h;
while(x<b)
{
w = x*x + y*y;
x = x + h;
y = y + t*(w + x*x + (y+h*w)*(y+h*w) );
cout<<i<<"  "<<x<<"  "<<y<<endl;
i++;
}
return 0;
}
```

The program was run for the same initial condition and step sizes as in Example 7-23. The output for $h = .025$, $x_0 = 0$, $y_0 = 1$, $b = .25$ is given in Table 7-5. It can be seen that these figures are somewhat similar to those of Example 7-23. The general underlying theory suggests that the results are more accurate than the results obtained by Euler's method.

TABLE 7-5

0 1 .025 .25
0 1 0.025 0.25

i	x	y
1	0.025	1.02564
2	0.05	1.05266
3	0.075	1.08121
4	0.1	1.11143
5	0.125	1.14351
6	0.15	1.17764
7	0.175	1.21402
8	0.2	1.25291
9	0.225	1.29457
10	0.25	1.3393

As in Example 7-23 the above program was run for a number of step sizes. The approximate values of $y(.2)$ for various h are given in Table 7-6. An examination of the table and comparison with Table 7-3 for Euler's method suggests that **Heun's method**, for a given step size, produces a better approximation to the solution in this problem.

EXERCISE 35. Apply Heun's algorithm to the ODE $y' = x + y$, $y(1) = 1$ for $h = .1$ and $h = .05$, $b = 1.2$. Compare the results with those determined in Exercises 29 and 30.

TABLE 7-6

H	Approximation to $y(.2)$
.2	1.248000000
.1	1.251530674
.05	1.252614712
.025	1.252912559
.0125	1.252990368
.00625	1.253010234
.003125	1.253015251
.0015625	1.253016512
.00078125	1.253016828
.000390625	1.253016907

EXERCISE 36. Apply Heun's method to $y' = 1 + y^2$, $y(0) = 0$ on $[0, .5]$ with $h = .1$. Compare the results with those of Exercise 31 and the exact solution $y = \tan(x)$.

EXERCISE 37. Apply Heun's algorithm to the ODE $y' = 2x$, $y(0) = 0$ on $[0,1]$, using $h = .2$. Compare the results with the exact solution.

EXERCISE 38. Give some examples of ODE for which Heun's algorithm will give the exact solution. Can you formulate a general rule concerning such examples?

Heun's algorithm can be extended to a system of first-order ODE. The algorithm can be derived by integration and employment of the **trapezoidal rule** and Euler's method, as was done for a single first-order ODE. Because the notation becomes rather involved, we shall illustrate by using a system of two ODE.

ALGORITHM 7-4 (Heun). Let $f_1(x, y_1, y_2), f_2(x, y_1, y_2), x_0, y_{1,0}, y_{2,0}, h$, and m be given. Form the numbers $y_{1,1}, y_{1,2}, \ldots, y_{1,m}; y_{2,1}, y_{2,2}, \ldots, y_{2m};$ and $x_1, \ldots, x_m$ by the rules

$$x_{i+1} = x_i + h$$

$$e_{1,i+1} = y_{1i} + hf_1(x_i, y_{1i}, y_{2i})$$

$$e_{2,i+1} = y_{2i} + hf_2(x_i, y_{1i}, y_{2i})$$

$$y_{1,i+1} = y_{1i} + \frac{h}{2}\left[f_1(x_i, y_{1i}, y_{2i}) + f_1(x_{i+1}, e_{1,i+1}, e_{2,i+1})\right]$$

$$y_{2+i+1} = y_{2i} + \frac{h}{2}\left[f_2(x_i, y_{1i}, y_{2i}) + f_2(x_{i+1}, e_{1,i+1}, e_{2,i+1})\right]$$

for $i = 0, 1, \ldots, m - 1$.

EXAMPLE 7-29. Apply Heun's algorithm to the system

$$y_1' = y_1 + y_2 \qquad y_1(0) = 0$$
$$y_2' = -y_1 + y_2 \qquad y_2(0) = 1$$

using $h = .1$.

Solution. In this problem

$$f_1(x, y_1, y_2) = y_1 + y_2 \qquad f_2(x, y_1, y_2) = -y_1 + y_2$$

Thus

$$x_1 = .1$$
$$e_{1,1} = 0 + .1 f_1(0,0,1) = .1$$
$$e_{2,1} = 1 + .1 f_2(0,0,1) = 1.1$$
$$y_{1,1} = 0 + .05 \left[f_1(0,0,1) + f_1(.1,.1,1.1) \right] = .11$$
$$y_{2,1} = 1 + .05 \left[f_2(0,0,1) + f_2(.1,.1,1.1) \right] = 1.1$$

Table 7-7 gives the results obtained when **Heun's algorithm** was applied to the above problem, with $h = .1$ and $m = 9$. The results have been rounded to 5D; they may be compared with the Euler algorithm and exact

TABLE 7-7

x_i	y_{1i}	y_{2i}
.00	.00000	1.00000
.10	.11000	1.10000
.20	.24200	1.19790
.30	.39797	1.29107
.40	.57978	1.37640
.50	.78917	1.45026
.60	1.02761	1.50848
.70	1.29631	1.54629
.80	1.59603	1.55833
.90	1.92705	1.53860

solutions given in Table 7-4.

EXERCISE 39. Apply Heun's method to the system

$$y_1' = y_1 + y_2 + x \qquad y_1(0) = 2$$
$$y_2' = -y_1 + y_2 \qquad y_2(0) = 1$$

with $h = .05$, $m = 2$.

EXERCISE 40. Apply Heun's method to the system

$$y'_1 = 2y_1 + y_2 + x^2 \quad y_1(1) = 1$$
$$y'_2 = 3y_1 + y_2 + x \quad y_2(1) = 1$$

with $h = .05$, $m = 2$.

EXERCISE 41. Calculate approximations to $y_1(.1)$, $y_2(.1)$ for Example 7-29, using Heun's algorithm with $h = .05$.

EXERCISE 42. Repeat Exercise 41, this time using $h = .025$.

EXERCISE 43. Write a program which applies Heun's algorithm to the system

$$y'_1 = 2x + y_2 \quad y_1(0) = 0$$
$$y'_2 = x^2 - y_1 \quad y_2(0) = -2$$

for various step sizes h on $[0, 3]$. Compare your results with the exact solution, which is $y_1(x) = x^2 - 2\sin(x)$, $y_2(x) = -2\cos(x)$.

7-6   ERROR-ANALYSIS ASPECTS

The numerical solution of ODE is the most complex problem discussed in this text. The difficulties are not primarily concerned with deriving satisfactory algorithms. Instead, the difficulties lie in the error-analysis aspects of the problem.

In Chap. 5 we studied the **ill-conditioned problem** of inverting the **Hilbert matrix**. We learned that the effect due to round-off errors in the calculation can be quite large. We shall begin this section by studying an ill-conditioned system of ODE. Such a system is called **unstable**.

Suppose that we are given the system

$$y'_1 = y_2 \qquad\qquad y_1(x_0) = y_{10}$$
$$y'_2 = 10y_1 + 9y_2 \qquad y_2(x_0) = y_{20} \qquad (7\text{-}14)$$

The solution to this system is

$$y_1(x) = c_1 e^{-x} + c_2 e^{10x}$$
$$y_2(x) = -c_1 e^{-x} + 10c_2 e^{10x} \qquad (7\text{-}15)$$

where c_1, c_2 are chosen so that the initial conditions are satisfied. In particular, if the initial conditions are $y_1(0) = 1$, $y_2(0) = -1$, then the exact solution to Eqs. (7-14) is

$$y_1(x) = e^{-x}$$
$$y_2(x) = -e^{-x} \qquad (7\text{-}16)$$

Suppose that now that a small change is made in the set of initial conditions. If the initial conditions are taken as $y_1(0) = 1 + \delta$ and $y_2(0) = -1 + \varepsilon$, then the exact solution to Eqs. (7-14) is

$$\bar{y}_1(x) = \left(1 + \frac{10\delta}{11} - \frac{\varepsilon}{11}\right)e^{-x} + \frac{\delta + \varepsilon}{11}e^{10x}$$

$$\bar{y}_2(x) = -\left(1 + \frac{10\delta}{11} - \frac{\varepsilon}{11}\right)e^{-x} + \frac{10(\delta + \varepsilon)}{11}e^{10x}$$

(7-17)

Hence

$$\left|y_1(x) - \bar{y}_1(x)\right| = \left|\frac{-10\delta + \varepsilon}{11}e^{-x} - \frac{\delta + \varepsilon}{11}e^{10x}\right|$$

$$\left|y_2(x) - \bar{y}_2(x)\right| = \left|\frac{10\delta - \varepsilon}{11}e^{-x} - \frac{10(\delta + \varepsilon)}{11}e^{10x}\right|$$

(7-18)

Now no matter how small $|\varepsilon| \neq 0$ and $|\delta| \neq 0$ are, provided only that $\delta \neq -\varepsilon$, the quantities (7-18) grow quite rapidly with x. Thus, as x increases, the quantities (7-18) quickly overshadow the quantities (7-16), which are the solution to the original system. We say this system of ODE is unstable because a small change in the initial conditions produces a new problem whose solution differs greatly from the solution to the original problem.

One can think of the algorithms of Euler and Heun (and other **one-step methods** to be discussed in Chap. 8) as generating a sequence of ODE initial-value problems, along with one step of their solutions. For example, given $y' = f(x,y), y(x_0) = y_0$, one generates an approximate solution $y(x_1) \approx y_1$. Then one starts all over with the new problem $z' = f(x, z), z(x_1) = y_1$ and repeats the process. If $y(x_1) = y_1$ exactly, then the solution to $z' = f(x, z)$, $z(x_1) = y_1$ will be exactly the same as the solution to the original problem. However, if $y(x_1) \neq y_1$ and if the new ODE initial-value problem $z' = f(x, z)$, $z(x_1) = y_1$ is unstable, then its solution may bear no resemblance to the solution of the original problem.

The study of stability of **differential equations** is an important topic in applied mathematics. Often there is a close relationship between the stability of a particular physical system and the stability of the differential equations describing the system. Thus the scientist who poses a problem may have considerable insight into the feasibility of seeking a numerical solution to the problem.

EXERCISE 44. Find numerical approximations to the solution to Eqs. (7-14), with $y_1(0) = 1, y_2(0) = -1$ on $[0, 4]$, using Euler's or Heun's method for various **step sizes**. Compare your results with the exact solution.

EXERCISE 45. Discuss the stability of

$$y_1' = y_2 \qquad\qquad y_1(0) = 1$$
$$y_2' = 10y_1 + 9y_2 \qquad y_2(0) = -1$$
$$y_3' = y_2 - 2y_3 \qquad\quad y_3(0) = 1$$

for $x \geq 0$.

We continue this section by discussing a criterion standardly used to measure the level of performance of an algorithm for solving an ODE. Proofs of the results given here may be found in Henrici (1962).

The various examples and exercises in Sees. 7-4 and 7-5 suggest two things: First, in both Euler's and **Heun's method** the accuracy seems to increase as the step size decreases. Second, Heun's method appears to give greater accuracy than Euler's method, for a given step size. It is important to give a precise statement of the above observations.

Euler's method is what is known as a **first-order method**. Roughly this means that the **analytic error** (or **truncation error**) in this method varies linearly with the step size. To be more precise, suppose that an ODE $y' = f(x, y)$, $y(x_0) = y_0$ is to be solved on $[x_0, b]$. Further, suppose that a unique solution $y(x)$ exists, and $|y''(x)|$ is uniformly bounded on $[x_0, b]$. Finally, suppose that Euler's method with step size h is applied to the problem. Then, for each point x_* at which an approximate solution y_* is generated, there exists a constant L such that

$$|y(x_*) - y_*| \le Lh \tag{7-19}$$

In Henrici [1962] it is shown that the constant L depends only upon the function $f(x, y)$ and the points x_0, x_*. It does not depend upon the step size h. Thus the error in Euler''s method is $O(h)$. This of course assumes that all arithmetic is performed exactly.

The result (7-19) is quite important. To illustrate, suppose that Euler's method with step size $k = h/2$ is applied to the same problem. Although it now takes twice as many steps to reach the given point x_*, we have

$$|y(x_*) - y_*| \le Lk = L\frac{h}{2}$$

That is, the bound on the truncation error is halved by halving the step size. More generally, the bound on the truncation error at a particular point tends to zero linearly as the step size tends to zero.

As we have noted previously when discussing error bounds, the actual error may be considerably smaller than the size of the bound. In this particular problem we shall usually know that an error bound of the form (7-19) exists but shall not be able to know the size of the constant L. We can conclude, however, that the truncation error at a particular point x_* can be made arbitrarily small, provided only that h be taken small enough.

Suppose now for the problem under consideration that the hypotheses are strengthened so that $|y'''(x)|$ is assumed to be uniformly bounded. If Heun's method is applied, it can be shown that for each $x_* \in [x_0, b]$ there exists a constant L such that

$$|y(x_*) - y_*| \le Lh^n \tag{7-20}$$

The constant L depends only upon the function $f(x, y)$ and the points x_0, x_*. Again, it does not depend upon the step size h.

Heun's method is called a **second-order method**. The truncation error at a particular point (under appropriate hypotheses) is bounded by a constant times the square of the step size. For example, if the same problem is solved by using a step size of $k = h/2$, then twice as many steps are needed to reach x_* but the error bound is

$$\left| y(x_*) - y_* \right| \le Lk^2 = L\left(\frac{h}{2}\right)^2 = L\frac{h^2}{4}$$

That is, the error bound is $(1/2)^2 = 1/4$ as large.

In the next chapter we shall discuss several methods of order higher than 2. Suppose that a method with step size h is under consideration. If the truncation error at each point $x_* \in [x_0, b]$ is bounded by

$$\left| y(x_*) - y_* \right| \le Lh^n$$

where L depends only on $f(x, y)$ and x_0, x_*, then the method is said to be of nth order. The truncation error at any point x_* is said to be $0(h^n)$, and we write

$$\left| y(x_*) - y_* \right| = \phi O(h^n)$$

Higher-order methods in general require more functional evaluations or other numerical calculations per step than do lower-order methods. As we shall see, this extra computation is usually well rewarded by a considerable increase in accuracy.

In Euler's and **Heun's methods**, as well as in the Taylor and **Runge-Kutta** methods to be discussed in Chap. 8, the step size h can easily be changed in the course of a computation. The effect of a change in step size can be observed by comparing the results obtained by using two different step sizes. For example, suppose that $y_1, y_2, \ldots, y_i$ have been computed with a step size h used. Then $y(x_{i+1})$ is approximated by y_{i+1}, obtained in one additional step of the algorithm. However, $y(x_{i+1})$ can also be approximated by starting at y_i and using q steps of size h/q, where $q \ge 2$ is an integer. The number of significant digits of agreement between the two approximations to $y(x_{i+1})$ can be used as an estimate of the number of significant digits of agreement between y_{i+1} and the exact solution $y(x_{i+1})$.

Richardson's extrapolation algorithm is applicable to all the above-mentioned algorithms for the numerical solution of ODE. Thus an extrapolation can be performed on two different approximations to $y(x_{i+1})$ to provide a higher-order approximation to this quantity. This higher-order approximation can also be compared with y_{i+1} to obtain an estimate of the accuracy of the numerical computations. One can also use Richardson's extrapolation algorithm to generate higher-order algorithms for the numerical solution of ODE. An example of this is left as Exercise 47.

EXERCISE 46. In Example 7-20 calculate approximations to $y(.15)$ by starting at y_2 and using Euler's method with step sizes .05 and .025.

EXERCISE 47. Derive the algorithm

$$y_{i+1} = y_i + hf\left(x_i + \frac{h}{2}, y_i + \frac{h}{2}f(x_i, y_i)\right)$$

by performing Richardson's extrapolation on Euler's method used with step sizes h and $h/2$.

Exercise 48. Calculate y_1, y_2, y_3 for the problem $y' = 2y$, $y(0) = 1$, using $h = .05$ and the algorithm given in Exercise 47. Compare your results with those given in Examples 7-20 and 7-27.

Another possible source of major error in the numerical solution of an ODE lies in the algorithm selected. All the algorithms considered in this text are stable. To understand what this means, think of an algorithm as a set of directions which, if followed exactly, will produce a given set of numbers. Suppose that an algorithm is applied to a stable ODE and that some round-off error occurs in performing the indicated calculations. If an algorithm is stable, the round-off errors will not be magnified by subsequent calculations. Instead, the effect due to the round-off errors will remain bounded or even die off to zero. In an **unstable algorithm** the effect of a small error may be magnified and may thus produce a rapidly increasing error. Hence it is highly desirable to use only stable algorithms. An example and discussion of an unstable algorithm for the numerical solution of ODE may be found in Hildebrand (p. 207).

We conclude this section with a few words concerning the cumulative effect due to round-off error in applying an algorithm for the numerical solution of a **differential equation**. The number of steps needed to provide an approximate numerical solution on an interval $[x_0, b]$ varies inversely with the step size h. Thus, as might be expected, a bound on the cumulative effect due to round-off error will contain a factor of h in the denominator. Consequently, not only will the use of a too small step size result in a greatly increased computation time, but it will also result in a decrease in accuracy.

Numerical Solution of Ordinary Differential Equations

8-1 INTRODUCTION

A variety of numerical procedures are available to the practitioner in the area of numerical solution of **ordinary differential equations**. In this chapter three of the most commonly used classes of procedures are discussed. Each of these is suitable for use on a computer and is extensively used in actual practice. The classes of procedures to be discussed are Taylor's-series procedures (Sec. 8-2), **Runge-Kutta procedures** (Sec. 8-3), and predictor-corrector procedures (Sec. 8-4). Each class contains a number of different algorithms. We shall illustrate some of the most common algorithms by giving flow charts, computer programs and numerical examples.

8-2 TAYLOR'S-SERIES PROCEDURES

Taylor's series provide a very powerful theoretical tool for the study of ODE. In recent years there has been a significant revival of interest in the use of Taylor's-series methods for the numerical solution of ODE. This is due in part to the fact that certain error-analysis aspects of the numerical solution of ODE are more easily handled with these methods than with others. Another important factor has been the recent progress in developing computer procedures for the explicit differentiation of functions. This has allowed the development of programs which will handle automatically quite complicated systems of ODE. We shall restrict our attention to problems where the necessary differentiation does not become unwieldy.

Let us begin by considering the general first-order ODE in normal form.

$$y' = f(x, y) \quad y(x_0) = y_0 \tag{8-1}$$

We shall assume that this ODE has a unique solution $y(x)$ on $[x_0, b]$ and that $y(x) \in C^{n+1}[x_0, b]$ for some $n \geq 1$. As in Chap. 7 let $x_i = x_0 + ih$ for some $h > 0$ and $i = 0, 1, \ldots, m$. Here $x_m \leq b$ and $x_m + h > b$. Then the solution $y(x)$ can be expanded in a **Taylor's series** about any one of the points x_i.

$$y(x) = y(x_i) + (x - x_i) y'(x_i) + \frac{(x - x_i)^2}{2!} y''(x_i)$$

$$+ \cdots + \frac{(x - x_i)^n}{n!} y^{(n)}(x_i) + \frac{(x - x_i)^{n+1}}{(n+1)!} y^{(n+1)}(\xi) \tag{8-2}$$

This expansion is valid for $x \in [x_0, b]$; ξ lies between x_i and x. In particular, if $x = x_{i+1}$, then Eq. (8-2) becomes

$$y(x_{i+1}) = y(x_i) + hy'(x_i) + \frac{h^2}{2!} y''(x_i) + \cdots + \frac{h^n}{n!} y^{(n)}(x_i)$$

$$+ \frac{h^{n+1}}{(n+1)!} y^{(n+1)}(\xi_i) \tag{8-3}$$

where $x_i < \xi_i < x_{i+1}$.

The Taylor algorithm is based upon Eq. (8-3). If the remainder is truncated, the result obtained is

$$y(x_{i+1}) \approx y(x_i) + hy'(x_i) + \frac{h^2}{2} y''(x_i) + \cdots + \frac{h^n}{n!} y^{(n)}(x_i) \tag{8-4}$$

To apply Eq. (8-4), it is necessary to know $y(x_i), y'(x_i), \ldots, y^{(n)}(x_i)$. If x_i and $y(x_i)$ were known, the derivatives could be calculated as follows: First, the known values x_i and $y(x_i)$ are substituted into the differential equation to give

$$y'(x_i) = f(x_i, y(x_i))$$

Next, the differential equation (8-1) can be differentiated to give formulas for the higher derivatives of $y(x)$. Thus

$$y''(x) = \frac{d}{dx} f(x, y(x)) = f_x(x, y) + f_y(x, y) y'(x)$$

$$= f_x(x, y) + f_y(x, y) f(x, y)$$

$$\cdots\cdots\cdots\cdots\cdots\cdots\cdots\cdots\cdots\cdots$$

$$y^{(n)}(x) = \frac{d^{n-1}}{dx^{n-1}} f(x, y(x)) = \frac{d}{dx} y^{(n-1)}(x)$$

The values $y''(x_i), y'''(x_i) \ldots$ can be computed by substitution into these formulas.

$$y'(x_i) = f(x_i, y(x_i))$$

$$y''(x_i) = f_x(x_i, y(x_i)) + f_y(x_i, y(x_i)) f(x_i, y(x_i))$$

$$\cdots\cdots\cdots\cdots\cdots\cdots\cdots\cdots\cdots\cdots\cdots\cdots\cdots\cdots$$

$$y^{(n)}(x_i) = \frac{d^{n-1}}{dx^{n-1}} f(x, y(x)) \Big|_{(x_i, y(x_i))}$$

Hence, if x_i and $y(x_i)$ were known exactly, then Eq. (8-4) could be used to compute $y(x_{i+1})$ with an error

$$\frac{h^{n+1}}{(n+1)!} y^{(n+1)}(\xi_i) \qquad x_i < \xi < x_i + h$$

The initial value $y(x_0)$ is given. Therefore, with the procedure described above, $y(x_1)$ can be computed with an error which is $O(h^{n+1})$. However, a new difficulty now arises. To compute $y'(x_1), y''(x_2), \ldots, y^{(n)}(x_1)$ by the procedure given above, it is necessary to know $y(x_1)$ exactly. The use of an approximate value of $y(x_1)$ in computing these derivatives will introduce an additional error.

This is exactly what is done in the Taylor algorithm.

ALGORITHM 8-1 (Taylor). Let h, b, n, m, and the ODE

$$y' = f(x, y) \qquad y(x_0) = y_0$$

be given. Define

$$y_0 = y(x_0)$$

$$y_i^{(k)} = \frac{d^{k-1}}{dx^{k-1}} f(x, y(x)) \Big|_{(x_i, y_i)} \qquad k = 1, 2, \ldots, n$$

$$y_{i+1} = y_i + h y'_i + \frac{h^2}{2!} y''_i + \cdots + \frac{h^n}{n!} y_i^{(n)}$$

and

$$x_{i+1} = x_i + h$$

for $i = 0, 1, \ldots, m - 1$. The numbers $y_1, y_2, \ldots, y_m$ are the numerical approximations to $y(x_1), y(x_2), \ldots, y(x_m)$.

THEOREM 8-2-1. Let $f(x, y)$ have continuous partial derivatives of order n for $x_0 \le x \le b$ and for all y. With the notation given previously, if the Taylor algorithm (8-1) is applied to the ODE (8-1), then

$$|y_i - y(x_i)| = O(h^n) \qquad i = 1, 2, \ldots, m$$

A proof of this theorem may be found in Henrici (1964). As a consequence of this theorem, Algorithm 8-1 is called the **Taylor algorithm of order n**.

EXAMPLE 8-1. Apply the Taylor algorithm of order 3 to the ODE $y' = x + y$, $y(0) = 1$, using $h = .1$ to find approximate values of $y(.1)$, $y(.2)$.

Solution. Expressions for $y''(x)$ and $y'''(x)$ are easily found in this problem.

$$y' = x + y$$

$$y'' = \frac{d}{dx}(x + y(x)) = 1 + y' = 1 + x + y$$

$$y''' = \frac{d}{dx} y'' = \frac{d}{dx}(1 + y') = y'' = 1 + x + y$$

The Taylor algorithm can now be applied, with $x_0 = 0$, $y_0 = 1$ and the derivatives given above. The resulting formula is

$$y_{i+1} = y_i + h(x_i + y_i) + \frac{h^2}{2}(1 + x_i + y_i) + \frac{h^3}{6}(1 + x_i + y_i) \qquad (8\text{-}5)$$

Thus

$$y_1 = 1.0 + .1(0 + 1.0) + \frac{.1^2}{2}(1.0 + 0 + 1.0) + \frac{.1^3}{6}(1.0 + 0 + 1.0)$$

In summary, $y_1 \approx 1.11033$, $x_1 = .1$.

Next, setting $i = 1$ in Eq. (8-5) and using the values computed above we get

$$y_2 = y_1 + .1(.1 + y_1) + \frac{.1^2}{2}(1.0 + .1 + y_1) + \frac{.1^3}{6}(1.0 + 0 + y_1)$$

In summary, $y_2 \approx 1.24278$, $x_2 = .2$.

To estimate the truncation error, we can consider the effect of using a higher-order algorithm. In this case higher derivatives of the function $f(x, y) = x + y$ are easily computed. In particular $y^{(4)} = 1 + x + y$. Starting at y_1, two different approximations to $y(x_{i+1})$ are to be computed. The first is computed by using the Taylor algorithm of order 3; the second is computed by using the Taylor algorithm of order 4. Then the difference between the results will be

$$\frac{h^4}{4!}(1 + x_i + y_i)$$

With $h = .1$ this is approximately $.00000417(1 + x_i + y_i)$. Thus the additional term of the Taylor's series will make a difference of one unit in the fifth decimal place of our results.

EXERCISE 1. Apply Taylor's algorithm to $y' = x^2 + y^2$ with $y(0) = 1$. Take $h = .05$, $n = 2$, and determine approximations to $y(.05)$ and $y(.1)$. Carry the calculations to $5D$.

EXERCISE 2. Given the differential equation $y' = y$ with $y(0) = 1$, find approximate values of $y(.1)$ and $y(.2)$, using Taylor's algorithm with $n = 4$ and $h = .1$. Carry the calculations to $7D$.

EXERCISE 3. Find the exact solution to Exercise 2, and compare the true values with the approximate values for $y(.1)$ and $y(.2)$.

EXERCISE 4. Given $y' = y + \sin(x)$, find expressions for y'', y''', $y^{(4)}$ and $y^{(5)}$.

EXERCISE 5. Given $y' = x^3 y + x^2$, find expressions for y'', y''', and $y^{(4)}$.

EXERCISE 6. Given $y' = x^2 e^x y^3$ find expressions for y'', y'''.

Taylor's algorithm is relatively easy to use on a computer, provided that one can find expressions for y'', y''', $\dots$, $y^{(n)}$. This is illustrated by Example 8-2.

EXAMPLE 8-2. Write a program to apply Taylor's algorithm to $y' = y^2$, $y(0) = 1$, with $n = 5$, $h = .02$, and $m = 20$.

Solution. First we must obtain expressions for the derivatives of $y(x)$.

$$y' = y^2$$
$$y'' = 2yy'$$
$$y''' = 2(y')^2 + 2yy''$$
$$y^{(4)} = 6y'y'' + 2yy'''$$
$$y^{(5)} = 6(y'')^2 + 8y'y''' + 2yy^{(4)}$$

Program 8-1 was used to carry out the calculations. This program was written to look like the algorithm, rather than for efficiency. Notice that it is very inefficient to re-compute expressions like $H^5/120$. for each step of the procedure. It is left as an exercise to write a more efficient program for the same problem.

Program 8-1 Taylor's algorithm for an ODE

```
// Program Taylor
#include <iostream>
#include <cmath> // provides pow
using namespace std;

int main(void)
 {
 float x,y,y1,d1y,d2y,d3y,d4y,d5y,h;
 int i;
 x = 0.;
 y = 1.;
 h = .02;
 cout<<" x "<<" y "<<" y1 "<<endl;
 cout<<x<<" "<<y<<" "<<y<<endl;
  for(i = 1; i<=20; ++i)
   {
   d1y = y*y;
   d2y = 2.*y*d1y;
   d3y = 2.*(d1y*d1y + y*d2y);
   d4y = 6.*d1y*d2y + 2.*y*d3y;
   d5y = 6.*d2y*d2y + 8.*d1y*d3y + 2.*y*d4y;
   y = y + h*d1y + h*h*d2y/2. + h*h*h*d3y/6. +pow(h,4)*d4y/24.;
   y = y + pow(h,5)*d5y/120.;
   x = x + h;
   y1 = -1./(x-1);
   cout<<x<<"  "<<y<<"  "<<y1<<endl;
   }
 return 0;
 }
```

The exact solution to this ODE is given by $y(x) = -1/(x - 1)$. Table 8-1 gives the output from the above program and the exact solution at the corresponding points.

EXERCISE 7. Rewrite the program of Example 8-2 to be computationally more efficient.

EXERCISE 8. Write a program to apply Taylor's algorithm to $y' = x + y$, $y(0) = 1$ on $[0, 5]$. Use $n = 6$, $h = 0.1$.

TABLE 8-1

x	y	y^1
0	1	1
0.02	1.02041	1.02041
0.04	1.04167	1.04167
0.06	1.06383	1.06383
0.08	1.08696	1.08696
0.1	1.11111	1.11111
0.12	1.13636	1.13636
0.14	1.16279	1.16279
0.16	1.19048	1.19048
0.18	1.21951	1.21951
0.2	1.25	1.25
0.22	1.28205	1.28205
0.24	1.31579	1.31579
0.26	1.35135	1.35135
0.28	1.38889	1.38889
0.3	1.42857	1.42857
0.32	1.47059	1.47059
0.34	1.51515	1.51515
0.36	1.5625	1.5625
0.38	1.6129	1.6129
0.4	1.66667	1.66667

EXERCISE 9. Write a program to apply Heun's algorithm and Taylor's algorithm of order 2 to $y' = 1 + y^2$ with $y(0) = 0$ on $[0.0, 1.0]$ for several different sizes of h. Compare the results of these computations with the exact solution $y = \tan(x)$.

We continue this section with a discussion of the application of Taylor's series to a system of first-order simultaneous ODE. To illustrate the point, we consider a system of two ODE.

$$y'_1 = f_1(x, y_1, y_2) \quad y_1(x_0) = y_{1,0}$$
$$y'_2 = f_2(x, y_1, y_2) \quad y_2(x_0) = y_{2,0} \tag{8-6}$$

Here there are two functions $y_1(x)$, $y_2(x)$ to be determined. Following the procedure used for a single ODE, we assume that $y_1(x) \in C^{n+1}[x_0, b]$ and

$y_2(x) \in C^{n+1}[x_0, b]$. Using the same notation as before, we have

$$y_1(x) = y_1(x_i) + (x - x_i)y'_1(x_i) + \frac{(x - x_i)^2}{2!} y_1''(x_i)$$

$$+ \cdots + \frac{(x - x_i)^n}{n!} y_1^{(n)}(x_i) + \frac{(x - x_i)^{n+1}}{(n+1)!} y_2^{(n+1)}(\xi_i)$$

$$y_2(x) = y_2(x_i) + (x - x_i)y'_2(x_i) + \frac{(x - x_i)^2}{2!} y_2''(x_i)$$

$$+ \cdots + \frac{(x - x_i)^n}{n!} y_2^{(n)}(x_i) + \frac{(x - x_i)^{n+1}}{(n+1)!} y_2^{(n+1)}(\mu_i)$$

(8-7)

For $x = x_{i+1}$ we get, after truncation,

$$y_1(x_{i+1}) \approx y_1(x_i) + hy'_1(x_i) + \cdots + \frac{h^n}{n!} y_1^{(n)}(x_i)$$

$$y_2(x_{i+1}) \approx y_2(x_i) + hy'_2(x_i) + \cdots + \frac{h^n}{n!} y_2^{(n)}(x_i)$$

(8-8)

The necessary derivatives can be computed from the given functions, at a point x_i, provided that $y_1(x_i)$ and $y_2(x_i)$ are known. Thus

$$y'_1(x_i) = f_1(x_i, y_1(x_i), y_2(x_i))$$

$$y'_2(x_i) = f_2(x_i, y_1(x_i), y_2(x_i))$$

$$y_1^{(k)}(x_i) = \frac{d^{k-1}}{dx^{k-1}} f_1(x, y_1(x), y_2(x)) \Big|_{(x_i, y_1(x_i), y_2(x_i))}$$

$$y_2^{(k)}(x_i) = \frac{d^{k-1}}{dx^{k-1}} f_2(x, y_1(x), y_2(x)) \Big|_{(x_i, y_1(x_i), y_2(x_i))}$$

In general, however, $y_1(x_i)$ and $y_2(x_i)$ will be known only approximately. Thus we are led to the following algorithm.

ALGORITHM 8-2 (Taylor). Let h, b, n, m, and the system of ODE

$$y'_1 = f_1(x, y_1, y_2) \quad y_1(x_0) = y_{1,0}$$

$$y'_2 = f_2(x, y_1, y_2) \quad y_2(x_0) = y_{2,0}$$

be given. Define

$$y_{1,i}^{(k)} = \frac{d^{k-1}}{dx^{k-1}} f_1(x, y_1(x), y_2(x)) \Big|_{(x_i, y_{1,i}, y_{2,i})} \quad (k = 1, 2, \ldots, n)$$

$$y_{2,i}^{(k)} = \frac{d^{k-1}}{dx^{k-1}} f_2(x, y_1(x), y_2(x)) \Big|_{(x_i, y_{1,i}, y_{2,i})}$$

$$y_{1,i+1} = y_{1,i} + hy'_{1,i} + \frac{h^2}{2!} y''_{1,i} + \cdots + \frac{h^n}{n!} y_{1,i}^{(n)}$$

$$y_{2,i+1} = y_{2,i} + hy'_{2,i} + \frac{h^2}{2!} y''_{2,i} + \cdots + \frac{h^n}{n!} y_{2,i}^{(n)}$$

and $x_{i+1} = x_i + h$

for $i = 0, 1, \ldots, m-1$. The numbers $y_{1,i}, y_{2,i}$ are the numerical approximations to $y_1(x_i), y_2(x_i)$.

EXAMPLE 8-3. Obtain an approximate solution to the system

$$y_1' = 2y_1 + y_2 + x^2 \qquad y_1(1) = 1$$
$$y_2' = 3y_1 + 4y_2 + x \qquad y_2(1) = 2$$

at $x_1 = 1.05$, $x_2 = 1.10$, by use of Taylor's algorithm with $h = .05$, $n = 2$.

Solution. The algorithm to be used is

$$y_{1,i+1} = y_{1,i} + hy_{1,i}' + \frac{h^2}{2} y_{1,i}''$$

$$y_{2,i+1} = y_{2,i} + hy_{2,i}' + \frac{h^2}{2} y_{2,i}''$$

$$x_{i+1} = x_i + h$$

Formulas for the derivatives, and their values at $x_0 = 1$ are given by

$$y_1' = 2y_1 + y_2 + x^2 \qquad y_{1,0}' = 5$$
$$y_2' = 3y_1 + 4y_2 + x \qquad y_{2,0}' = 12$$
$$y_1'' = 2y_1' + y_2' + 2x \qquad y_{1,0}'' = 24$$
$$y_2'' = 3y_1' + 4y_2' + 1 \qquad y_{2,0}'' = 64$$

Thus

$$y_{1,1} = 1 + (.05)(5) + \frac{(.05)^2}{2}(24) = 1.28$$

$$y_{2,1} = 2 + (.05)(12) + \frac{(.05)^2}{2}(64) = 2.68$$

$$x_1 = 1.05$$

To continue,

$$y_{1,1}' = (2)(1.28) + 2.68 + (1.05)^2 = 6.3425$$
$$y_{2,1}' = (3)(1.28) + (4)2.68 + 1.05 = 15.61$$
$$y_{1,1}'' = (2)(6.3425) + 15.61 + (2)(1.05) = 30.395$$
$$y_{2,1}'' = (3)(6.3425) + (4)(15.61) + 1 = 82.4675$$

Thus

$$y_{1,2} = 1.28 + (.05)(6.3425) + \frac{(.05)^2}{2}(30.395) = 1.6351$$

$$y_{2,2} = 2.68 + (.05)(15.61) + \frac{(.05)^2}{2}(82.4675) = 3.5636$$

$$x_2 = 1.10$$

To extend the Taylor algorithm to systems of r simultaneous ODE for $r \geq 2$. It is convenient to introduce vector notation. The notation

$$Y' = F(x, Y) \qquad Y(x_0) = Y_0$$

is used to denote the system

$$
\begin{aligned}
y_1' &= f_1(x, y_1, y_2, \ldots, y_r) & y_1(x_0) &= y_{1,0} \\
y_1' &= f_2(x, y_1, y_2, \ldots, y_r) & y_2(x_0) &= y_{2,0} \\
&\cdots\cdots\cdots\cdots\cdots\cdots\cdots\cdots\cdots\cdots\cdots \\
y_r' &= f_r(x, y_1, y_2, \ldots, y_r) & y_r(x_0) &= y_{r,0}
\end{aligned}
$$

With this notation, the algorithm is given by

$$Y_{i+1} = Y_i + hY_i' + \frac{h^2}{2!}Y_i'' + \cdots + \frac{h^n}{n!}Y_i^{(n)}$$

where

$$
Y_i = \begin{bmatrix} y_{1,i} \\ y_{2,i} \\ \cdots \\ y_{r,i} \end{bmatrix}
$$

$$
Y_i^{(k)} = \begin{bmatrix} y_{1,i}^{(k)} \\ y_{2,i}^{(k)} \\ \cdots \\ y_{r,i}^{(k)} \end{bmatrix} = \begin{bmatrix} \dfrac{d^{k-1}}{dx^{k-1}} f_1(x, y_1, \ldots, y_r) \\ \dfrac{d^{k-1}}{dx^{k-1}} f_2(x, y_1, \ldots, y_r) \\ \cdots\cdots\cdots\cdots\cdots\cdots \\ \dfrac{d^{k-1}}{dx^{k-1}} f_r(x, y_1, \ldots, y_r) \end{bmatrix}_{(x_i, y_{1,i}, \ldots, y_{r,i})}
$$

An example for $r = 2$ is given below.

EXAMPLE 8.4 Obtain an approximate solution to the system

$$
\begin{aligned}
y_1' &= y_1 + y_2 & y_1(0) &= 0 \\
y_2' &= -y_1 + y_2 & y_2(0) &= 0
\end{aligned}
$$

at $x_1 = 0.1$ and $x_2 = 0.2$, using the Taylor's method just described, with $n = 3$.

Solution. We shall take $h = .1$. Then we get

$$Y_{i+1} = Y_i + .1Y_i' + \frac{0.01}{2}Y_i''' + \frac{0.001}{6}Y_i'''$$

where

$$
Y_0 = \begin{bmatrix} y_{1,0} \\ y_{2,0} \end{bmatrix} = \begin{bmatrix} 0 \\ 1 \end{bmatrix}
$$

$$
Y_i' = \begin{bmatrix} y_{1,i}' \\ y_{2,i}' \end{bmatrix} = \begin{bmatrix} y_{1,i} + y_{2,i} \\ -y_{1,i} + y_{2,i} \end{bmatrix}
$$

Thus

$$\begin{bmatrix} y_{1,i+1} \\ y_{2,i+1} \end{bmatrix} = \begin{bmatrix} y_{1,i} \\ y_{2,i} \end{bmatrix} + .1\begin{bmatrix} y'_{1,i} \\ y'_{2,i} \end{bmatrix} + \frac{.01}{2}\begin{bmatrix} y''_{1,i} \\ y''_{2,i} \end{bmatrix} + \frac{.001}{6}\begin{bmatrix} y'''_{1,i} \\ y'''_{2,i} \end{bmatrix}$$

$$= \begin{bmatrix} y_{1,i} \\ y_{2,i} \end{bmatrix} + .1\begin{bmatrix} y_{1,i} + y_{2,i} \\ -y_{1,i} + y_{2,i} \end{bmatrix} + \frac{.01}{2}\begin{bmatrix} 2y_{2,i} \\ -2y_{1,i} \end{bmatrix} + \frac{.002}{6}\begin{bmatrix} -y_{1,i} + y_{2,i} \\ -y_{1,i} - y_{2,i} \end{bmatrix}$$

Thus

$$\begin{bmatrix} y_{1,i+1} \\ y_{2,i+1} \end{bmatrix} = \begin{bmatrix} 1.099667 y_{1,i} + .110333 y_{2,i} \\ -.110333 y_{1,i} + 1.099667 y_{2,i} \end{bmatrix}$$

Using this result with $i = 0$, we get

$$\begin{bmatrix} y_{1,1} \\ y_{2,1} \end{bmatrix} = \begin{bmatrix} .110333 \\ 1.099667 \end{bmatrix}$$

Using $i = 1$, we get

$$\begin{bmatrix} y_{1,2} \\ y_{2,2} \end{bmatrix} = \begin{bmatrix} (1.099667)(0.110333) + (.110333)(1.099667) \\ (-.110333)(0.110333) + (1.099667)(1.099667) \end{bmatrix}$$

$$= \begin{bmatrix} .242660 \\ 1.197094 \end{bmatrix}$$

A computer program was written to apply the Taylor algorithm of order 3 to this problem on [0, .9], with $h = .1$. Table 8-2 gives the results obtained, along with the exact solution. This same **system of differential equations** was solved in Chap. 7 by using Euler's and also Heun's methods. The increase in accuracy with increase of the order of the method employed is quite evident.

TABLE 8-2 Taylor's Results and Exact Solutions to Example (8-4)

x_i	y_{1i}	y_{2i}	$y_1(x_i)$	$y_2(x_i)$
.0	.00000	1.00000	.00000	1.00000
.1	.11033	1.09967	.11033	1.09965
.2	.24266	1.19709	.24266	1.19706
.3	.39892	1.28963	.39891	1.28957
.4	.58097	1.37415	.58094	1.37406
.5	.79049	1.44700	.79044	1.44689
.6	1.02893	1.50401	1.02885	1.50386
.7	1.29742	1.54038	1.29730	1.54020
.8	1.59669	1.55076	1.59651	1.55055
.9	1.92692	1.52915	1.92667	1.52891

EXERCISE 10. Using Taylor's method, approximate $y(.05)$, $y(.1)$ and $y(.15)$, where $y'' = 2y' - y$, $y(0) = 1$, $y'(0) = 1$. Take $h = .05$ and $n = 3$.

EXERCISE 11. Using Taylor's method, approximate $y(.1)$ and $y(.2)$, where $y'' = y$, $y(0) = 1$, $y'(0) = 1$. Take $h = .1$ and $n = 5$. Compare your answers with $y(x) = e^x$, the exact solution.

EXERCISE 12. Find $y_1(.1)$, $y_2(.1)$, $y_3(.1)$, using the Taylor-series method, with $n = 3$, for the system

$$y'_1 = -y_1 + e^{-x} \qquad y_1(0) = 0$$
$$y'_2 = -y_2 \qquad y_2(0) = 1$$
$$y'_3 = -2y_2^2 - y_3 \qquad y_3(0) = 2$$

Take $h = .1$. Compare your answers with the exact solutions

$$y_1(x) = xe^{-x}, \qquad y_2(x) = e^{-x} \quad y_3(x) = 2e^{-2x}$$

EXERCISE 13. Write a computer program to find an approximate solution to the system of ODE given in Exercise 12. The program should have provisions for reading in the step size h and the number of steps m. Run the program for various h, and compare the results with the exact solution.

8-3 RUNGE-KUTTA METHODS

A numerical procedure for solving

$$y' = f(x, y) \qquad y(x_0) = y_0$$

for $x \in [x_0, b]$ is said to be a **one-step method** if y_{i+1}, the approximation to $y(x_{i+1})$, is obtained by use of an algorithm of the form

$$y_{i+1} = y_i + h\phi(x_i, y_i, h) \tag{8-9}$$

That is, y_{i+1} is completely determined in terms of x_i, y_i, and h. The Taylor's algorithm of order n (Algorithm 8-1) is a one-step method because

$$y_{i+1} = y_i + h\left(y'_i + \frac{h}{2!}y''_i + \cdots + \frac{h^{n-1}}{n!}y_i^{(n)} \right)$$

Here

$$\phi(x_i, y_i; h) + y'_i + \frac{h}{2!}y''_i + \cdots + \frac{h^{n-1}}{n!}y_i^{(n)}$$

where

$$y_i^{(k)} = \frac{d^{k-1}}{dx^{k-1}}f(x, y(x))\Big|_{(x_i, y_i)}$$

The biggest difficulty in applying the Taylor algorithm to a given ODE is that of finding expressions for the derivatives $y''(x), \ldots, y^{(n)}(x)$. These can easily become quite complicated and unwieldy. It was proposed by Runge that one could find one-step methods which did not involve the higher derivatives of the solution $y(x)$. Considerable work in this area was done by Kutta, and the methods which have been developed are called **Runge-Kutta methods**. We shall derive a Runge-Kutta method of order 2. The derivation is based upon the Taylor algorithm of order 2 and uses Taylor's series for a function of two variables as given in Theorem 3-4-5.

The Taylor's-series method of order 2 is

$$y_{i+1} = y_i + hy_i' + \frac{h^2}{2} y_i'' \tag{8-10}$$

where

$$y_i' = f(x_i, y_i)$$

and

$$y_i'' = f_x(x_i, y_i) + f_y(x_i, y_i) f(x_i, y_i)$$

This can be written as

$$y_{i+1} = y_i + h\varphi_t(x_i, y_i; h) \tag{8-11}$$

where

$$\varphi_t(x_i, y_i; h) = f(x_i, y_i) + \frac{h}{2}\left[f_x(x_i, y_i) + f_y(x_i, y_i) f(x_i, y_i) \right] \tag{8-12}$$

Our goal will be to construct a function $\Phi_r(x_i, y_i; h)$ which does not involve $y''(x)$ but which, if expanded in a Taylor's series, would differ from $\Phi_t(x_i, y_i; h)$ only by terms of order $O(h^2)$.

The function $f(x, y)$ can be expanded in a Taylor's series in two variables. If the expansion is performed about (x_i, y_i) and evaluated at $x = x_i + \varepsilon_i$, $y = y_i + \mu_i$, one can get

$$f(x_i + \varepsilon_i, y_i + \mu_i) = f(x_i, y_i) + \varepsilon_i f_x(x_i, y_i) + \mu_i f_y(x_i, y_i)$$
$$+ 1/2\left[\varepsilon_i^2 f_{xx}(\xi, \eta) + 2\varepsilon_i \mu_i f_{xy}(\xi, \eta) + \mu_i^2 f_{yy}(\xi, \eta) \right]$$

Consider the remainder in the special case where ε_i, μ_i are of the form

$$\varepsilon_i = ph \qquad \mu_i = qhf(x_i, y_i)$$

where p, q are constants. The remainder will contain h^2 as a factor. Thus, if f_{xx}, f_{xy}, and f_{yy} are uniformly bounded in the region under consideration, we conclude that

$$f(x_i + ph, y_i + qhf(x_i, y_i)) = f(x_i, y_i) + phf_x(x_i, y_i)$$
$$+ qhf(x_i, y_i) f_y(x_i, y_i) + O(h^2) \tag{8-13}$$

Now consider a one-step method with

$$\varphi_i(x_i, y_i; h) = a_1 f(x_i, y_i) + a_2 f(x_i + ph, y_i + qhf(x_i, y_i))$$

where a_1, a_2, p, q are constants to be specified later. By using the formula (8-13) this can be rewritten as

$$\varphi_r(x_i, y_i; h) = (a_1 + a_2) f(x_i, y_i)$$
$$+ h\left[a_2 pf_x(x_i, y_i) + a_2 qf(x_i, y_i) f_y(x_i, y_i) \right] + O(h^2) \tag{8-14}$$

Now compare this expression with Eq. (8-12). If a_1, a_2, p, q are selected so that

$$a_1 + a_2 = 1$$
$$a_2 p = 1/2$$
$$a_2 q = 1/2$$

then, except for the $O(h^2)$ remainder, φ_r will be the same as φ_t.

The second two equations imply that $p = q$. Thus, if $a_2 \neq 0$, we have the relations

$$a_1 = 1 - a_2 \quad p = q = \frac{1}{2a_2} \tag{8-15}$$

For any numbers a_1, a_2, p, and q satisfying Eqs. (8-15) we get

$$\varphi_t(x_i, y_i; h) - \varphi_r(x_i, y_i; h) = O(h^2)$$

For each $a_2 \neq 0$ a Runge-Kutta method of order 2 is

$$y_{i+1} = y_i + h\varphi_r(x_i, y_i; h) \tag{8-16}$$

where

$$\varphi_r(x_i, y_i; h) = (1 - a_2) f(x_i, y_i) + a_2 f\left(x_i + \frac{h}{2a_2}, y_i + \frac{h}{2a_2} f(x_i, y_i)\right) \tag{8-17}$$

If we take $a_2 = 1/2$ in Eq. (8-17), we get

$$\varphi_r(x_i, y_i; h) = 1/2 \left[f(x_i, y_i) + f(x_i + h, y_i + hf(x_i, y_i)) \right]$$

Then Eq. (8-16) becomes

$$y_{i+1} = y_i + \frac{h}{2}\left[f(x_i, y_i) + f(x_i + h, y_i + hf(x_i, y_i)) \right]$$

which is just Heun's method. If $a_2 = 1$ in Eq. (8-17) we get what is called the *modified Euler-Cauchy method*.

$$y_{i+1} = y_i + hf\left(x_i + \frac{h}{2}, y_i + \frac{h}{2} f(x_i, y_i)\right) \tag{8-18}$$

EXAMPLE 8-5. Apply the modified Euler-Cauchy method to the ODE $y' = x + y, y(0) = 1$, using $h = .1$ to find approximations to $y(.1)$ and $y(.2)$.
Solution. For $i = 0$ we have

$$x_0 = 0 \quad y_0 = 1 \quad f(x_0, y_0) = 1$$
$$f\left(x_0 + \frac{h}{2}, y_0 + \frac{h}{2} f(x_0, y_0)\right) = 1.1$$

Thus

$$y_1 = y_0 + hf\left(x_0 + \frac{h}{2}, y_0 + \frac{h}{2} f(x_0, y_0)\right) = 1.11$$

For $i = 1$ we have

$$x_1 = .1 \quad y_1 = 1.11 \quad f(x_1, y_1) = 1.21$$

$$f\left(x_1 + \frac{h}{2}, y_1 + \frac{h}{2} f(x_1, y_1)\right) = 1.3205$$

Thus

$$y_2 = y_1 + hf\left(x_1 + \frac{h}{2}, y_1 + \frac{h}{2} f(x_1, y_1)\right) = 1.24205$$

The exact solution to this ODE is given by

$$y(x) = 2e^x - x - 1$$

Thus, to 6D, $y(.1) = 1.110342$, $y(.2) = 1.242806$.

EXERCISE 14. Apply the modified Euler-Cauchy method to $y' = x^2 + y^2$ with $y(0) = 1$. Take $h = .05$, and determine approximations to $y(.05)$ and $y(.1)$. Carry the calculations to 5D. Compare your results with those found in Exercise 1.

EXERCISE 15. Write a program which applies the modified Euler-Cauchy method to $y' = x^2 + y^2$ for arbitrary x_0, y_0, h, and b. Run the program with $x_0 = 0$, $y_0 = 1$, $b = .25$, and $h = .025$, $h = .0125$, $h = .00625$. Compare your results with those obtained in Example 7-23.

EXERCISE 16. Show that the parameters a_1, a_2, p and q in Eq. (8-14) cannot be chosen so that $\Phi_r(x_i, y_i; h)$ agrees with the Taylor's method of order 3. [Hint: Carry out the two-dimensional Taylor's-series expansion for $\Phi_r(x_i, y_i; h)$ through the term involving h^3. Then compare this with the Taylor's method of order 3.]

The most commonly used Runge-Kutta algorithm is the fourth-order Runge-Kutta method given by

$$y_{i+1} = y_1 + h\varphi(x_i, y_i; h) \tag{8-19}$$

where

$$\varphi(x_i, y_i; h) = 1/6(k_1 + 2k_2 + 2k_3 + k_4) \tag{8-20}$$

and

$$k_1 = f(x_i, y_i)$$

$$k_2 = f\left(x_i + \frac{h}{2}, y_i + \frac{h}{2} k_1\right)$$

$$k_3 = f\left(x_i + \frac{h}{2}, y_i + \frac{h}{2} k_2\right) \tag{8-21}$$

$$k_4 = f(x_i + h, y_i + hk_3)$$

As might be expected, the derivation of a **Runge-Kutta** fourth-order method is rather complicated. The basic idea is to find a $\Phi(x_i, y_i; h)$ which, when expanded in a Taylor's series and used in Eq. (8-19), gives a formula which differs from the Taylor algorithm of order 4 by terms which are of order $O(h^5)$. A detailed derivation of some fourth-order Runge-Kutta methods is given by Romanelli in Ralston and Wilf.

Example 8-6. Apply the fourth-order Runge-Kutta method [Eqs. (8-19) to (8-21)] to $y' = x + y$, $y(0) = 1$, using $h = .1$ to find approximations to $y(.1)$ and $y(.2)$.

Solution. For $i = 0$

$$x_0 = 0 \qquad y_0 = 1$$

$$k_1 = f(x_0, y_0) = 1$$

$$k_2 = f\left(x_0 + \frac{h}{2}, y_0 + \frac{h}{2}k_1\right) = \left(0 + \frac{.1}{2}\right) + \left[1 + \frac{.1}{2}(1)\right] = 1.1$$

$$k_3 = f\left(x_0 + \frac{h}{2}, y_0 + \frac{h}{2}k_2\right) = \left(0 + \frac{.1}{2}\right) + \left[1 + \frac{.1}{2}(1.1)\right] = 1.105$$

$$k_4 = f(x_0 + h, y_0 + hk_3) = (0 + .1) + \left[1 + (.1)(1.105)\right] = 1.2105$$

Thus

$$\varphi(x_0, y_0; h) = 1/6\left[1.0 + (2)(1.1) + (2)(1.105) + 1.2105\right] = 1.1034167$$

Hence

$$y_1 = 1.0 + (.1)(1.1034167) = 1.11034167$$

For $i = 1$

$$x_1 = .1 \qquad y_1 = 1.11034167$$

$$k_1 = f(x_1, y_1) = .1 + 1.11034167 = 1.21034167$$

$$k_2 = f\left(x_1 + \frac{h}{2}, y_1 + \frac{h}{2}k_1\right) = \left(.1 + \frac{.1}{2}\right) + \left[1.11034167 + \frac{.1}{2}(1.21034167)\right]$$

$$= 1.32085875$$

$$k_3 = f\left(x_1 + \frac{h}{2}, y_1 + \frac{h}{2}k_2\right) = \left(.1 + \frac{.1}{2}\right) + \left[1.11034167 + \frac{.1}{2}(1.32085875)\right]$$

$$= 1.32638461$$

$$k_4 = f(x_1 + h, y_1 + hk_3) = (.1 + .1) + \left[1.11034167 + (.1)(1.32688461)\right]$$

$$= 1.44303013$$

Thus

$$\varphi(x_1, y_1; h) = 1/6\left[1.21034167 + (2)(1.32085875)\right.$$

$$+ (2)(1.32638461) + 1.44303013\right]$$

$$= 1.32472642$$

Hence

$$y_2 = 1.11034167 + (.1)(1.32472642) = 1.24281431$$

Exact values for the solution of this ODE are given in Example 8-5.

Figure 8-1 gives a flow chart of an algorithm for the fourth-order **Runge-Kutta** method.

EXERCISE 17. Apply the fourth-order Runge-Kutta method to $y' = x^2 + y^2$, with $y(0) = 1$. Take $h = .1$, and determine approximations to $y(.1)$ and $y(.2)$. Carry the results to 6D (see Exercise 14).

EXERCISE 18.
(a) Write a program for the flow chart given in Fig. 8-1, the Runge-Kutta method. Write the program so that $f(x, y)$ is given by a function statement or subprogram.
(b) Run the program in (a) to solve $y' = x^2 + y^2$, with $y(0) = 1$. Take $h = .1$ and $b = 2$.

EXERCISE 19. Apply the fourth-order Runge-Kutta method to $y' = (1 - x)y$, with $y(1) = 1$. Take $h = .2$, and determine approximations to $y(1.2)$ and $y(1.4)$. Carry the results to 6D.

We now state the extension of the fourth-order Runge-Kutta method to solve the first-order system of two equations

$$\begin{aligned} y'_1 &= f_1(x, y_1, y_2) \quad y_1(x_0) = y_{10} \\ y'_2 &= f_2(x, y_1, y_2) \quad y_2(x_0) = y_{20} \end{aligned} \tag{8-22}$$

for $x \in [x_0, b]$. Using the notation $x_i = x_0 + ih$, and $y_{1,i}, y_{2,i}$ as the numerical approximations to $y_1(x_i), y_2(x_i)$ we have

$$\begin{aligned} y_{1,i+1} &= y_{1,i} + h\varphi_1(x_i, y_{1,i}, y_{2,i}; h) \\ y_{2,i+1} &= y_{2,i} + h\varphi_2(x_i, y_{1,i}, y_{2,i}; h) \end{aligned} \tag{8-23}$$

where

$$\begin{aligned} \varphi_1(x_i, y_{1,i}, y_{2,i}; h) &= 1/6(k_{11} + 2k_{12} + 2k_{13} + k_{14}) \\ \varphi_2(x_i, y_{1,i}, y_{2,i}; h) &= 1/6(k_{21} + 2k_{22} + 2k_{23} + k_{24}) \end{aligned} \tag{8-24}$$

and

$$\begin{aligned} k_{11} &= f_1(x_i, y_{1,i}, y_{2,i}) \\ k_{21} &= f_2(x_i, y_{1,i}, y_{2,i}) \\ k_{12} &= f_1\left(x_i + \frac{h}{2}, y_{1,i} + \frac{h}{2}k_{11}, y_{2,i} + \frac{h}{2}k_{21}\right) \\ k_{22} &= f_2\left(x_i + \frac{h}{2}, y_{1,i} + \frac{h}{2}k_{11}, y_{2,i} + \frac{h}{2}k_{21}\right) \end{aligned} \tag{8-25}$$

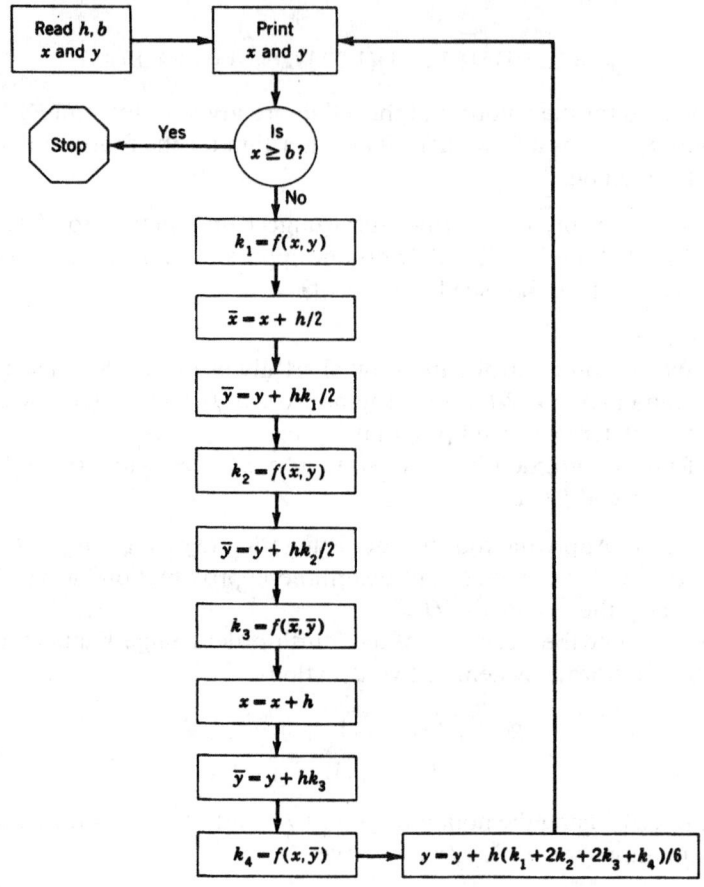

FIG. 8-1   Flow chart for Runge-Kutta method.

$$k_{13} = f_1\left(x_i + \frac{h}{2}, y_{1,i} + \frac{h}{2}k_{12}, y_{2,i} + \frac{h}{2}k_{22}\right)$$

$$k_{23} = f_2\left(x_i + \frac{h}{2}, y_{1,i} + \frac{h}{2}k_{12}, y_{2,i} + \frac{h}{2}k_{22}\right)$$

$$k_{14} = f_1\left(x_i + h, y_{1,i} + hk_{13}, y_{2,i} + hk_{23}\right)$$

$$k_{24} = f_2\left(x_i + h, y_{1,i} + hk_{13}, y_{2,i} + hk_{23}\right)$$

EXAMPLE 8-7. Obtain an approximate solution to the system

$$y'_1 = y_1 + y_2 \qquad y_1(0) = 0$$
$$y'_2 = -y_1 + y_2 \qquad y_2(0) = 1$$

at $x = .1$. Take $h = .1$, and use the fourth-order Runge-Kutta method.

Solution. For $i = 0$ we have

$$x_0 = 0 \qquad y_{10} = 0 \qquad y_{20} = 1$$

$$k_{11} = f_1(x_0, y_{10}, y_{20}) = 1$$

$$k_{21} = f_2(x_0, y_{10}, y_{20}) = 1$$

$$k_{12} = f_1\left(x_0 + \frac{h}{2}, y_{10} + \frac{h}{2}k_{11}, y_{20} + \frac{h}{2}k_{21}\right) = 1.1$$

$$k_{22} = f_2\left(x_0 + \frac{h}{2}, y_{10} + \frac{h}{2}k_{11}, y_{20} + \frac{h}{2}k_{21}\right) = 1.0$$

$$k_{13} = f_1\left(x_0 + \frac{h}{2}, y_{10} + \frac{h}{2}k_{12}, y_{20} + \frac{h}{2}k_{22}\right) = 1.105$$

$$k_{23} = f_2\left(x_0 + \frac{h}{2}, y_{10} + \frac{h}{2}k_{12}, y_{20} + \frac{h}{2}k_{22}\right) = .995$$

$$k_{14} = f_1(x_0 + h, y_{10} + hk_{13}, y_{20} + hk_{23}) = 1.2100$$

$$k_{24} = f_2(x_0 + h, y_{10} + hk_{13}, y_{20} + hk_{23}) = .9890$$

Hence

$$\varphi_1(x_0, y_{10}, y_{20}; h) = 1.103333$$

$$\varphi_2(x_0, y_{10}, y_{20}; h) = .996500$$

$$y_{11} = .110333$$

$$y_{21} = 1.099650$$

The **Runge-Kutta fourth-order** method for a system of first-order ODE is relatively easily programmed for use on a computer. In the sample program (Program 8-2) the array r[3][5] is used to contain the elements k_{ij}, $i = 1, 2; j = 1, 2, 3, 4$. The functions $f_1(x, y_1, y_2)$ and $f_2(x, y_1, y_2)$ are given in a FUNCTION. In this particular program $f_1(x, y_1, y_2) = y_1 + y_2$, and $f_2(x, y_1, y_2) = -y_1 + y_2$

PROGRAM 8-2   RUNGE-KUTTA FOR A SYSTEM

```
/*
The system y1(prime) = y1 + y2      y1(0) = 0
           y2(prime) = -y1 + y2     y2(0) = 1
  h = .1, b = 1, x = 0
was solved using the following Runge-Kutta program for a system.
*/
#include <iostream>
using namespace std;

float f(int,float,float []);

int main(void)

{
```

```
float y[3],r[3][5],yb[3],yb1[3];
float x,h,b,h2,xb;
int i;
cout<<"Input h, b and x:";
cin>>h>>b>>x;
h2 = h/2.0;

cout<<"Input y[1] and y[2]:";
cin>>y[1]>>y[2];

cout<<" x      y1      y2 "<<endl;

while(x < b)
 {
 cout<<x<<" "<<y[1]<<" "<<y[2]<<endl;
  for(i = 1; i<=2; ++i)
   {
   r[i][1] = f(i,x,y);
   yb[i] = y[i] + h2*r[i][1];
   }
   xb = x + h2;
   x = x + h;
  for(i=1; i<=2; ++i)
   {
   r[i][2] = f(i,xb,yb);
   yb1[i] = y[i] + h2*r[i][2];
   }
  for(i = 1; i<=2; ++i)
   {
   r[i][3] = f(i,xb,yb1);
   yb[i] = y[i] + h*r[i][3];
   }
  for(i = 1; i<=2; ++i)
   r[i][4] = f(i,x,yb);
  for(i = 1; i<=2; ++i)
   y[i] = y[i] + (h/6.0)*(r[i][1]+2.0*(r[i][2]+r[i][3]) + r[i][4]);
 } //end of while
return 0;
}

float f(int i, float x, float y[])
 {
 if(i == 1)
  return(y[1] + y[2]);
 if(i == 2)
  return(y[2] - y[1]);
 }
```

```
/*   THE OUTPUT FOR THE SYSTEM FOLLOWS
Input h, b and x:.1 1 0
Input y[1] and y[2]:0 1


  x      y1       y2
  0      0        1
  0.1  0.110333 1.09965
  0.2  0.242656 1.19706
  0.3  0.398912 1.28957
  0.4  0.580946 1.37406
  0.5  0.790443 1.44689
  0.6  1.02885  1.50386
  0.7  1.2973   1.5402
  0.8  1.59651  1.55055
  0.9  1.92668  1.52891
*/
```

This program was run for the case $y_1(0) = 0$, $y_2(0) = 1$, with $h = .1$. This is the same problem as was considered in Example 8-4 and previously in Chap. 7. The results obtained on [0, .9] differed from the exact solution by at most .00001.

EXAMPLE 8-8. Using Program 8-2, solve the second-order differential equation

$$\frac{d^2 y}{dx^2} + \frac{1}{x}\frac{dy}{dx} + \left(1 - \frac{1}{4x^2}\right)y = 0 \qquad (8\text{-}26)$$

with $y(\pi/2) = 1$, $y'(\pi/2) = -1/\pi$ for $x \in [\pi/2, \pi]$. Take $h = .1$.

Solution. Reducing Eq. (8-26) to a system of two first-order equations, we get

$$y'_1 = y_2 \qquad\qquad y_1\left(\frac{\pi}{2}\right) = 1$$

$$y'_2 = -\frac{y_2}{x} - \left(1 - \frac{1}{4x^2}\right)y_1 \qquad y_2\left(\frac{\pi}{2}\right) = \frac{-1}{\pi}$$

where $y_1 = y$ and $y_2 = y'$.

The C++ FUNCTION to evaluate the above expressions is given below.

```
      float f(int i, float x, float y[ ])
      {
      If(i == 1)
      {
      f = y[2];
      return(f);
      }
      else
      {
```

```
    f = -y[2] /x – (1.0 – 1.0/(4.0*x*x))*y[1] ;
    return(f);
    }
}
```

The problem can now be solved by use of Program 8-2. The exact solution is $y(x) = (\pi/2x)^{1/2} \sin(x)$. Table 8-3 gives sample results.

TABLE 8.3 Runge-Kutta and Exact Solutions to Example (8-8)

x_i	$y_{1,i}$	$y_{2,i}$	$y(x_i)$
$.15708E + 01$	$.1000000000E + 01$	$-.3183098861E + 00$	$.1000000000E + 01$
$.16708E + 01$	$.9647684691E + 00$	$-.3855148725E + 00$	$.9647684249E + 00$
$.17708E + 01$	$.9230628092E + 00$	$-.4477490020E + 00$	$.9230627202E + 00$
$.18708E + 01$	$.8753930949E + 00$	$-.5047534506E + 00$	$.8753929611E + 00$
$.19708E + 01$	$.8222948134E + 00$	$-.5562805256E + 00$	$.8222946358E + 00$
$.20708E + 01$	$.7643272707E + 00$	$-.6021028996E + 00$	$.7643270512E + 00$
$.21708E + 01$	$.7020711804E + 00$	$-.6420208072E + 00$	$.7020709214E + 00$
$.22708E + 01$	$.6361256140E + 00$	$-.6758676733E + 00$	$.6361253192E + 00$
$.23708E + 01$	$.5671044543E + 00$	$-.7035144901E + 00$	$.5671041279E + 00$
$.24708E + 01$	$.4956324635E + 00$	$-.7248731566E + 00$	$.4956321106E + 00$
$.25708E + 01$	$.4223410628E + 00$	$-.7398989310E + 00$	$.4223406891E + 00$
$.26708E + 01$	$.3478639069E + 00$	$-.7485921052E + 00$	$.3478635189E + 00$
$.27708E + 01$	$.2728323263E + 00$	$-.7509989796E + 00$	$.2728319309E + 00$
$.28708E + 01$	$.1978707044E + 00$	$-.7472122022E + 00$	$.1978703092E + 00$
$.29708E + 01$	$.1235918502E + 00$	$-.7373705204E + 00$	$.1235914632E + 00$
$.30708E + 01$	$.5059242422E - 01$	$-.7216579893E + 00$	$.5059205368E - 01$
$.31708E + 01$	$-.2055153220E - 01$	$-.7003026742E + 00$	$-.2055187775E - 01$

EXERCISE 20. Use Program 8-2 to solve system

$$y'_1 = x + y_2^2 \quad y_1(0) = 0$$
$$y'_2 = y_1 - x \quad y_2(0) = 1$$

for $x \in [0,1]$. Take $h = 0.1$.

EXERCISE 21. Use Program 8-2 to solve the second-order equation $y'' = x + y^2 - 2y'$, with $y(0) = 1$ and $y'(0) = 1$, and $x \in [0,1]$. Take $h = .1$.

The extension of Eqs. (8-19) to (8-21) to a system

$$Y' = F(x,Y) \qquad Y(x_0) = Y_0$$

of r simultaneous first-order ODE is given by

$$Y_{i+1} = Y_i + h\Phi(x_i, Y_i; h)$$

where

$$\varphi(x_i, Y_i; h) = 1/6\left(K_{i,1} + 2K_{i,2} + 2K_{i,3} + K_{i,4}\right)$$

and

$$K_{i,1} = F\left(x_i, Y_i\right)$$

$$K_{i,2} = F\left(x_i + \frac{h}{2}, Y_i + \frac{h}{2}K_{i,1}\right)$$

$$K_{i,3} = F\left(x_i + \frac{h}{2}, Y_i + \frac{h}{2}K_{i,2}\right)$$

$$K_{i,4} = F\left(x_i + h, Y_i + hK_{i,3}\right)$$

Here each capital letter denotes a vector. Thus, for example,

$$Y_i = \begin{bmatrix} y_{1i} \\ y_{2i} \\ \cdots \\ y_{ri} \end{bmatrix}$$

$$\Phi\left(x_i, Y_i; h\right) = \begin{bmatrix} \varphi_1\left(x_i, y_{1i}, y_{2i}, \ldots, y_{ri}; h\right) \\ \varphi_2\left(x_i, y_{1i}, y_{2i}, \ldots, y_{ri}; h\right) \\ \cdots \\ \kappa_r\left(x_i, y_{1i}, y_{2i}, \ldots, y_{ri}; h\right) \end{bmatrix}$$

$$K_{i,1} = \begin{bmatrix} f_1\left(x_i, y_{1i}, y_{2i}, \ldots, y_{ri}\right) \\ f_2\left(x_i, y_{1i}, y_{2i}, \ldots, y_{ri}\right) \\ \cdots \\ f_r\left(x_i, y_{1i}, y_{2i}, \ldots, y_{ri}\right) \end{bmatrix}$$

Program 8-2 is easily modified to handle a system of r simultaneous first-order ODE. This is left as Exercise 22.

EXERCISE 22.

  (a) Modify Program 8-2 so that it is capable of solving a system of r simultaneous first-order differential equations.
  (b) Use the program in part (a) to solve the system of ODE given in Exercise 12.

8-4   PREDICTOR-CORRECTOR METHODS

Both the **Taylor algorithms** and the **Runge-Kutta algorithms** are based upon **Taylor's series**. The procedures to be presented in this section are based upon the quadrature formulas of Chapter 6. We shall see that **predictor-corrector formulas** are used in conjunction with a Taylor or Runge-Kutta algorithm (to provide starting values). Under appropriate conditions predictor-corrector formulas will require fewer computations per step than Runge-Kutta formulas.

In addition, predictor-corrector formulas of orders considerably higher than 4 can easily be given and used. In this section, however, our ultimate goal is a fourth-order algorithm.

We begin by considering once again an ODE of the form

$$y' = f(x, y) \qquad y(x_0) = y_0 \qquad (8\text{-}27)$$

We assume that the unique solution $y(x)$ exists on $[x_0, b]$ and is sufficiently differentiable to allow the development of the underlying theory. Let $x_i = x_0 + ih$. Then from Eqs. (8-27) we get

$$\int_{x_i}^{x_{i+1}} y'(x)\,dx = \int_{x_i}^{x_{i+2}} f(x, y(x))\,dx$$

The left side can be simplified, leading to

$$y(x_{i+1}) = y(x_i) + \int_{x_i}^{x_{i+1}} f(x, y(x))\,dx \qquad (8\text{-}28)$$

The basic idea of this section is to replace the integral on the right side of Eq. (8-28) by a quadrature formula. Suppose that

$$\int_{x_i}^{x_{i+1}} f(x, y(x))\,dx = Q(f; x_i, x_{i+1}) + E(f; x_i, x_{i+1})$$

where Q and E denote a quadrature formula and associated error term, respectively, on the interval $[x_i, x_{i+1}]$. Then Eq. (8-28) can be written as

$$y(x_{i+1}) = y(x_i) + Q(f; x_i, x_{i+1}) + E(f; x_i, x_{i+1}) \qquad (8\text{-}29)$$

To illustrate, suppose that we use formula (6-54),

$$\int_{x_i}^{x_{i+1}} g(x)\,dx = hg(x_i) + \frac{h^2}{2} g'(\xi_i)$$

where $x_1 < \xi_i < x_{i+1}$. The result is

$$y(x_{i+1}) = y(x_i) + hf(x_i, y(x_i)) + \frac{h^2}{2} f'(\xi_i, y(\xi_i))$$

Since $f'(x,y) = y''(x)$, this can also be written as

$$y(x_{i+1}) = y(x_i) + hf(x_i, y(x_i)) + \frac{h^2}{2} y''(\xi_i)$$

The result, upon truncating the error term $(h^2/2)y''(\xi_i)$, is just Euler's method.

$$y_{i+1} = y_i + hf(x_i, y_i) \qquad (8\text{-}30)$$

Next let us try a more complicated quadrature formula, namely, (6-55),

$$\int_{x_i}^{x_{i+1}} g(x)\,dx = \frac{h}{2}\left[-g(x_{i-1}) + 3g(x_i)\right] + \frac{5h^3}{12} g''(\xi_i)$$

Here $x_{i-1} < \xi_i < x_{i+1}$. The use of this formula in Eq. (8.29) yields

$$y(x_{i+1}) = y(x_i) + \frac{h}{2}\left[-f(x_{i-1}, y(x_{i-1})) + 3f(x_i, y(x_i))\right] + \frac{5h^3}{12} y'''(\xi_i) \qquad (8\text{-}31)$$

If we truncate the remainder, we are led to the algorithm

$$y_{i+1} = y_i + \frac{h}{2}\left[-f\left(x_{i-1}, y_{i-1}\right) + 3f\left(x_i, y_i\right)\right] \tag{8-32}$$

The algorithm of Eq. (8-32) is of a fundamentally different nature from the algorithms considered previously for the solution of an ODE. As can be seen, to compute y_{i+1}, it is necessary to know both y_i and y_{i-1}. Since only $y(x_0) = y_0$ is given, we would need to apply some other numerical procedure to calculate y_1 before Eq. (8-32) could be used to compute $y_2, y_3, \ldots$. That is, Eq. (8-32) is not **self-starting.**

It is common to use a **Runge-Kutta algorithm** or **Taylor algorithm** to provide the starting values needed in an algorithm of the type of Eq. (8-32). It is proved in Henrici (1962) that if the error in computing y_1 is $O(h^2)$ then the $y_2, y_3, \ldots$, generated by Eq. (8-32) will satisfy

$$\left|y_i - y\left(x_i\right)\right| = O\left(h^2\right)$$

EXAMPLE 8-9. Apply the algorithm of Eq. (8-32) to solve $y' = x + y$, $y(0) = 1$. Take $h = .1$, and use the modified Euler-Cauchy method [see Eq. (8-18)] to determine y_1. Compute approximations to $y(.2)$ and $y(.3)$.

Solution. From Example 8-5 we have $y_1 = 1.11$. For $i = 1$, $x_0 = 0$, $y_0 = 1$, and $x_1 = .1$, $y_1 = 1.11$.

$$f\left(x_0, y_0\right) = 1$$
$$f\left(x_1, y_1\right) = .1 + 1.11 = 1.21$$

Thus by Eq. (8-32) we get

$$y_2 = y_1 + \frac{h}{2}\left[-f\left(x_0, y_0\right) + 3f\left(x_1, y_1\right)\right] = 1.2415$$

For $i = 2$, $x_2 = 0.2$, $y_2 = 1.2415$,

$$f\left(x_2, y_2\right) = 1.4415$$

$$y_3 = y_2 + \frac{h}{2}\left[-f\left(x_1, y_1\right) + 3f\left(x_2, y_2\right)\right] = 1.397225$$

It is important to note that the calculation of y_3 required but one evaluation of $f(x, y)$ plus the use of quantities previously calculated. The calculations of $y_4, y_5, \ldots$, would similarly require only one evaluation of $f(x, y)$ per step. The two other second-order methods we have studied (Heun's and the modified Euler-Cauchy) each requires two functional evaluations per step.

EXERCISE 23. Write a program which solves $y' = 1 + y^2$, $y(0) = 0$ on $[0.0, 1.0]$ for various step sizes h, using:

(a) Taylor algorithm of order 2
(b) Heun's method
(c) Euler-Cauchy method
(d) Algorithm of Eq. (8-32), using (c) to give the starting value. Compare the results with the exact solution $y = \tan(x)$ (see Exercise 9).

To continue, let us try still another quadrature formula in Eq. (8-29). This times we try the trapezoidal rule

$$\int_{x_i}^{x_{i+1}} g(x)\,dx = \frac{h}{2}\left[g(x_i)+g(x_{i+1})\right]-\frac{h^3}{12}g''(\xi_i)$$

Used in Eq. (8-29), this gives

$$y(x_{i+1}) = y(x_i)+\frac{h}{2}\left[f(x_i,y(x_i))+f(x_{i+1},y(x_{i+1}))\right]-\frac{h^3}{12}y'''(\xi_i) \qquad (8\text{-}33)$$

Proceeding as usual, we drop the remainder to get

$$y_{i+1} = y_i +\frac{h}{2}\left[f(x_i,y_i)+f(x_{i+1},y_{i+1})\right] \qquad (8\text{-}34)$$

Formula (8-34) introduces an entirely new difficulty. Notice that y_{i+1} appears on both sides of the expression. For a given function $f(x, y)$, Eq. (8-34) will give an equation involving x_i, y_i, x_{i+1}, y_{i+1} and h. Under appropriate conditions that equation can be solved for y_{i+1} in terms of x_i, y_i, x_{i+1} and h. In the light of Chap. 1 it seems natural to consider solving Eq. (8-34) for y_{i+1} by the method of successive substitution.

To illustrate, consider the case $i = 0$. The equation to be solved is

$$y_1 = y_0 +\frac{h}{2}\left[f(x_0,y_0)+f(x_1,y_1)\right] \qquad (8\text{-}35)$$

An initial guess $y_1(0)$ is made to y_1. Here, and subsequently in this section, it will be necessary to use superscripts to denote the various elements in a sequence. Thus

$$y_1^{(1)} = y_0 +\frac{h}{2}\left[f(x_0,y_0)+f(x_1,y_1^{(0)})\right]$$

and in general

$$y_1^{(j+1)} = y_0 +\frac{h}{2}\left[f(x_0,y_0)+f(x_1,y_1^{j})\right] \qquad (8\text{-}36)$$

To continue, we need some hypotheses concerning $f(x, y)$ and Eq. (8-35). We shall assume that $f(x, y)$ satisfies a **Lipschitz condition** in y, with constant K, in the region under consideration. This is the same hypothesis used in the existence and uniqueness theorems for the solution of an ODE given in Chapter 7. With this hypothesis it follows that Eq. (8-35) has a solution $y_1 = p$ provided that h is sufficiently small. A proof of a special case of this result is outlined in Exercise 26.

Now from Eqs. (8-35) and (8-36) we get

$$y_1^{(j+1)} - p =\frac{h}{2}\left[f(x_1,y_1^{(j)})-f(x_1,p)\right]$$

where p is a solution to Eq. (8-35). Then, using the Lipschitz condition, we have

$$\left| y_1^{(j+1)} - p \right| \leq \frac{Kh}{2} \left| y_1^{(j)} - p \right| \tag{8-37}$$

If Eq. (8-37) is used recursively, we obtain

$$\left[y_1^{(j+1)} - p \right] \leq \left(\frac{Kh}{2} \right)^{j+1} \left| y_1^{(0)} - p \right|$$

Then, if h is small enough so that $Kh/2 < 1$, it follows that

$$\lim_{j \to \infty} y_1^{(j)} = p$$

To summarize, we have shown that under suitable conditions y_{i+1} can be computed by fixed-point iteration on Eq. (8-34). To specify a particular method for solving an ODE, we specify how to make the initial guess and how many fixed-point iterations to perform. In the special case where we use **Euler's method** to make the initial guess each time and perform just one iteration, we get

$$y_1^{(0)} = y_0 + hf(x_0, y_0)$$

$$y_1^{(1)} = y_0 + \frac{h}{2} \left[f(x_0, y_0) + f(x_1, y_1^{(0)}) \right]$$

$$= y_0 + \frac{h}{2} \left[f(x_0, y_0) + f(x_1, y_0 + hf(x_0, y_0)) \right]$$

Once again we have derived Heun's method.

Heun's method is our first example of a **predictor-corrector method**. To advance the solution from x_i to x_{i+1} we use a predictor (the **Euler method**) to approximate the solution and a corrector (the **trapezoidal rule**) to perform one or more fixed-point iterations to improve the approximation. The number of fixed-point iterations to be performed can be specified in advance or modified as the need arises in order to ensure good accuracy in solving Eq. (8-34) for y_{i+1}. The latter idea is illustrated in the following example.

EXAMPLE 8-10. Find approximations to $y(.1)$, $y(.2)$ in the ODE $y' = x + y$, $y(0) = 1$, using the ideas of the predictor-corrector scheme just discussed. Perform the calculations correct to $6D$. The exact solution is given in Example 8-5.

Solution. The Euler formula to be used as predictor is

$$y_1^{(0)} = y_0 + hf(x_0, y_0)$$

Hence

$$y_1^{(0)} = 1 + (.1)(0 + 1) = 1.1$$

We correct, using the trapezoidal rule iteratively.

$$y_1^{(j+1)} = y_0 + \frac{h}{2} \left[f(x_0, y_0) + f(x_1, y_1^{(j)}) \right]$$

Hence

$$y_1^{(1)} = 1 + .05\left(1 + (.1 + 1.1)\right) = 1.11$$

$$y_1^{(2)} = 1 + .05\left(1 + (.1 + 1.11)\right) = 1.1105$$

$$y_1^{(3)} = 1 + .05(1 + 1.2105) = 1.110525$$

$$y_1^{(4)} = 1 + .05(1 + 1.210525) = 1.110526$$

$$y_1^{(5)} = 1.110526$$

To $6D$ we have the result 1.110526; so we use

$$y_1 = 1.110526$$

To continue, we predict $y_2^{(0)}$, using Euler's method,

$$y_2^{(0)} = y_1 + hf(x_1, y_1)$$
$$= 1.110526 + .1(.1 + 1.110526) = 1.231579$$

We correct, using the trapezoidal rule iteratively.

$$y_2^{(j+1)} = y_1 + \frac{h}{2}\left[f(x_1, y_1) + f(x_2, y_2^{(j)})\right]$$

Hence

$$y_2^{(1)} = 1.110526 + .05(1.210526 + 1.431579) = 1.242631$$

$$y_2^{(2)} = 1.110526 + .05(1.210526 + 1.442631) = 1.243184$$

$$y_2^{(3)} = 1.110526 + .05(1.210526 + 1.443184) = 1.243212$$

$$y_2^{(4)} = 1.110526 + .05(1.210526 + 1.443212) = 1.243213$$

Hence $y_2 = 1.243213$.

The algorithm just illustrated is a second-order method for solving an ODE. In Example 8-10 the results obtained were just slightly better than those obtained by using Heun's method. The significant advantages of **predictor-corrector methods** do not appear until we consider higher-order algorithms.

EXERCISE 24. Approximate $y(.05)$ and $y(.10)$ in Example 8-10 by using $h = .05$ in the predictor-corrector scheme illustrated in that example.

EXERCISE 25. Write a program for the algorithm of Example 8-10. Run the program for $y' = x + y$, $y(0) = 1$, using various h. In your program use the convergence criterion that $|y_i^{(j+1)} - y_i^{(j)}| < \varepsilon$, where ε is inputted. Have the program print out the number of iterations used to satisfy this convergence criterion.

EXERCISE 26.
   (a) Let $I = [x_0, b]$ and $J = (-\infty, \infty)$. Suppose that $f(x, y) \varepsilon C[I \times J]$ and that there exists an $M < \infty$ such that $|f(x, y)| \leq M$ for $(x, y) \varepsilon I \times J$. Define

$$g(\alpha) = y_0 + \frac{h}{2}\left[f(x_0, y_0) + f(x_1, \alpha)\right]$$

where $x_1 \in I$. Prove that $g(\alpha)$ is continuous and carries the interval $[y_0 - hM, y_0 + hM]$ into itself. (*Hint*: Find a bound on $|g(\alpha) - y_0|$.)

(b) Suppose that $|f_y(x_1, y)| \leq K$ for $y \in [y_0 - hM, y_0 + hM]$ and that h is such that $hK/2 < 1$. If $y_1^{(0)} \in [y_0 - hM, y_0 + hM]$ and

$$y_1^{(j+1)} = y_0 + \frac{h}{2}\left[f(x_0, y_0) + f\left(x_1, y_1^{(j)}\right)\right] \quad j = 0, 1, \ldots$$

use Theorem 1-5-3 to prove that $y_1 = \lim_{j \to \infty} y_1^{(j)}$ exist and

$$y_1 = y_0 + \frac{h}{2}\left[f(x_0, y_0) + f(x_1, y_1)\right]$$

EXERCISE 27.

(a) Consider an ODE of the form

$$y' = cy + dx + e \quad y(x_0) = y_0$$

where c, b, e are given constants. Show that for this problem Eq. (8-34) can be solved directly for y_{i+1} to give

$$y_{i+1} = \frac{2}{2 - hc}\left[y_i + \frac{h}{2}(cy_i + dx_i + dx_{x+1} + 2e)\right]$$

(b) Use the formula derived above to compute y_1 and y_2 in Example 8-10. Carry the arithmetic to 6D.

We shall now discuss a general class of predictor-corrector methods attributed to Adams, Bashforth, and Moulton. As in Eq. (8-28) we integrate each side of the differential equation to obtain

$$y(x_{i+1}) = y(x_i) + \int_{x_i}^{x_{i+1}} f(x, y(x))dx \tag{8-38}$$

Suppose now that the integral in Eq. (8-38) is approximated by a **quadrature formula** which involves values of the integrand at $x_{i+1}, x_i, x_{i-1}, \ldots, x_{i-p}$ for some positive integer p. Such formulas are of the form

$$\int_{x_i}^{x_{i+1}} g(x)dx \approx A_{i+1}g(x_{i+1}) + A_i g(x_i) + \cdots + A_{i-p}g(x_{i-p})$$

The result is the following equation:

$$y_{i+1} = y_i + A_{i+1}f(x_{i+1}, y_{i+1}) + A_i f(x_i, y_i) + \cdots + A_{i-p}f(x_{i-p}, y_{i-p}) \tag{8-39}$$

In this equation everything but y_{i+1} is considered to be known.

Equation (8-39) is to be solved by fixed-point iteration. An initial guess is obtained by approximating the integral in Eq. (8-38) by a quadrature formula involving values of the integrand at $x_i, x_{i-1}, \ldots, x_{i-q}$ for some positive integer q. Such formulas are of the form

$$\int_{xi}^{x_{i+1}} g(x)dx \approx B_i g(x_i) + B_{i-1}g(x_{i-1}) + \cdots + B_{i-q}g(x_{i-q})$$

The result is a formula for y_{i+1} which involves only known quantities.

$$y_{i+1} = y_i + B_i f(x_i, y_i) + B_{i-1}f(x_{i-1}, y_{i-1}) + \cdots + B_{i-q}f(x_{i-q}, y_{i-q}) \tag{8-40}$$

It may be used as a predictor to generate a starting value for fixed-point iteration in Eq. (8-39).

If the quadrature formulas (6-57) and (6-56) are used as predictor and corrector, respectively, the result is a fourth-order method for the numerical solution of ODE. For simplicity we shall call it **Adams's method**. The algorithm is not self-starting; commonly one approximates $y(x_1)$, $y(x_2)$, $y(x_3)$ using the fourth-order **Runge-Kutta algorithm**.

ALGORITHM 8-3 (Adams). Let h, m, and the ODE.

$$y' = f(x,y) \qquad y(x_0) = y_0$$

be given. Compute y_1, y_2, y_3 by a one-step method of at least fourth order. To compute y_{i+1} for $3 \le i \le m-1$, use

$$y_{i+1}^{(0)} = y_i + \frac{h}{24}[-9f(x_{i-3},y_{i-3})+37f(x_{i-2},y_{i-2}) \\ -59f(x_{i-1},y_{i-1})+55f(x_i,y_i)] \tag{8-41}$$

as the predictor and

$$y_{i+1}^{(j+1)} = y_i + \frac{h}{24}[f(x_{i-2},y_{i-2})-5f(x_{i-1},y_{i-1}) \\ +19f(x_i,y_i)+9f(x_{i+1},y_{i+1}^{(j)})] \tag{8-42}$$

$(j = 0, 1, \ldots)$ iteratively as the corrector. The value of y_{i+1} is taken as $y_{i+1}^{(k)}$ where k is specified in advance or where $\{y_i^{(j)}\}_{j=0\ldots k}$ satisfies a specified convergence criterion.

EXAMPLE 8-11. Apply Adams's method to $y' = x + y$, $y(0) = 1$ to compute approximations to $y(.4)$, $y(.5)$.

Solution. The standard approach to this problem would be to compute starting values y_1, y_2, y_3 by a fourth-order Runge-Kutta or Taylor algorithm. Since the main purpose of this example is to illustrate how the algorithm functions after y_1, y_2, y_3 are found, we shall use the exact values of the solution (to 6D) for starting values. The exact solution is $y = 2e^x - x - 1$.

$$y(0) = 1.0$$
$$y(.1) = 1.110342$$
$$y(.2) = 1.242806$$
$$y(.3) = 1.399718$$

To apply Eq.(8-41)for $i = 3$, we compute

$$y(x_0,y_0) = 1.000000$$
$$y(x_1,y_1) = 1.210342$$
$$y(x_2,y_2) = 1.442806$$
$$y(x_3,y_3) = 1.699718$$

Then

$$y_4^{(0)} = y_3 + \frac{h}{24}(-9f_0 + 37f_1 - 59f_2 + 55f_3)$$

where $f_i = f(x_i, y_i)$. Thus

$$y_4^{(0)} = 1.399718 + \frac{.1}{24}[-9 + 37(1.210342) - 59(1.442806) + 55(1.699718)]$$

$$= 1.583641$$

Now using Eq. (8-42), we get

$$y_4^{(1)} = y_3 + \frac{h}{24}(f_1 - 5f_2 + 19f_3 + 9f_4^{(0)})$$

where

$$f_4^{(0)} = x_4 + y_4^{(0)} = .4 + 1.583641 = 1.983641$$

Thus

$$y_4^{(1)} = 1.399718 + \frac{.1}{24}[1.210342 - 5(1.442806) + 19(1.699718) + 9(1.983641)]$$

$$= 1.583650$$

Because this is only slightly different from the predictor value, we take $y_4 = y_4^{(1)}$. To continue,

$$f_4 = x_4 + y_4 = 1.983650$$

By Eq. (8-41) we

$$y_5^{(0)} = y_4 + \frac{h}{24}(-9f_1 + 37f_2 - 59f_3 + 55f_4)$$

$$= 1.583650 + \frac{.1}{24}[-9(1.210342) + 37(1.442806) - 59(1.699718)$$

$$+ 55(1.983650)]$$

Thus

$$y_5^{(0)} = 1.797434$$

Now, using Eq. (8-42), we get

$$y_5^{(0)} = y_4 + \frac{h}{24}(f_2 - 5f_3 + 19f_4 + 9f_5^{(0)})$$

where

$$f_5^{(0)} = x_5 + y_5^{(0)} = .5 + 1.797434 = 2.297434$$

Thus

$$y_5^{(1)} = 1.583650 + \frac{.1}{24}[1.442806 - (5)(1.699718)$$

$$+ (19)(1.983650) + (9)(2.297434)]$$

or
$$y_5^{(1)} = 1.797444$$

Again nothing the small difference between $y_5^{(0)}$ and $y_5^{(1)}$, we set $y_5 = y_5^{(1)}$.
Thus our approximations are

$$y(.4) = 1.583650 \quad \text{and} \quad y(.5) = 1.797444$$

The example just given illustrates one of the most important points about **Adams's algorithm**. On the assumption that previously computed values are stored, only one evaluation of $f(x, y)$ is needed in the computation of $y_{i+1}^{(0)}$, the predicted value. Then only one additional evaluation of $f(x, y)$ is needed for each iteration of the corrector. If, as in the example just given, only one iteration is used, then y_{i+1} is computed from the previous values by just two evaluations of $f(x, y)$. This compares favorably with the four functional evaluations per step of the Runge-Kutta fourth-order method. In general one commonly adjusts the step size h so that not more than two iterations of the corrector formula are needed per step of the algorithm (see Hull and Creemer). In such a case Adams's method is computationally more efficient than the fourth-order **Runge-Kutta method.**

EXERCISE 28. Continue Example 8-11 by calculating y_6.

EXERCISE 29. Write a FUNCTION for Algorithm 8-3. Let y_0, y_1, y_2, y_3, $f(x_0, y_0)$, $f(x_1, y_1)$, $f(x_2, y_2)$, $f(x_3, y_3)$, x_0, h, b, and k be inputs to the function. Here k is the number of iterations of the corrector to be performed at each step.

EXERCISE 30. Using the Runge-Kutta fourth-order method, calculate starting values y_1, y_2, y_3 for the ODE $y' = x^2 + y^2$, $y(0) = 1.0$, using $h = .1$. Then use the function of Exercise 29 to find an approximate solution to the ODE, with $b = 2.0$, $k = 1$.

We conclude this section with an extension of Adams's algorithm to a system of n simultaneous first-order ODE. We shall use the vector notation developed earlier in this chapter. Let

$$Y' = F(x,Y) \qquad Y(x_0) = Y_0 \qquad (8\text{-}43)$$

denote the system

$$y_1' = f_1(x, y_1, y_2, \ldots, y_n) \quad y_1(x_0) = y_{1,0}$$
$$y_2' = f_2(x, y_1, y_2, \ldots, y_n) \quad y_2(x_0) = y_{2,0}$$
$$\cdots\cdots\cdots\cdots\cdots\cdots\cdots\cdots\cdots\cdots\cdots\cdots\cdots$$
$$y_n' = f_n(x, y_1, y_2, \ldots, y_n) \quad y_n(x_0) = y_{n,0}$$

Then the predictor is given by

$$Y_{i+1}^{(0)} = Y_i + \frac{h}{24}[-9F(x_{i-3}, Y_{i-3}) + 37F(x_{i-2}, Y_{i-2})$$
$$- 59F(x_{i-1}, Y_{i-1}) + 55F(x_i, Y_i)] \qquad (8\text{-}44)$$

The corrector is given by

$$Y_{i+1}^{(j+1)} = Y_i + \frac{h}{24}[F(x_{i-2},Y_{i-2})-5F(x_{i-1},Y_{i-1})$$
$$+19F(x_i,Y_i)+9F(x_{i+1},Y_{i+1}^{(j)})] \tag{8-45}$$

for $j = 0,1,\ldots$.

A function for Adams's method for a system of ODE is given below. It is assumed that the system of ODE is represented by the function subroutine $F(L, X, Y)$ as follows:

$$F(1,X,Y) = f_1(x,y_1,y_2,\ldots,y_n)$$
$$F(2,X,Y) = f_2(x,y_1,y_2,\ldots,y_n)$$
$$\ldots\ldots\ldots\ldots\ldots\ldots\ldots\ldots\ldots\ldots$$
$$F(N,X,Y) = f_n(x,y_1,y_2,\ldots,y_n)$$

It is assumed that all starting values have been computed previously and are available as arguments for the function ADAMS. The arguments are N (the number of equations), XO (the initial point), H (the step size), B (the end point), K (the number of iterations of the corrector to be used), Y [containing $y_{1,3}, y_{2,3}, y_{3,3}, \ldots, y_{n,3}$, the numerical approximations to $y_1(x_3)$, $y_2(x_3); \ldots, y_n(x_3)$], and F_1 [containing in its four columns the vectors $F(x_0,Y_0), F(x_1, Y_1), F(x_2,Y_2), F(x_3, Y_3)$].

EXERCISE 31.

(a) Write a program to solve a system of ODE by using the fourth-order Runge-Kutta method to compute starting values and Adams's method to continue the solution.

(b) Run the program on the problem of Exercise 12, and compare the results with the exact solution.

Appendix A: C++ Programming

Lesson 1

Getting Started

1.1 PREPROCESSOR DIRECTIVES

Before beginning our study of **C++**, we will examine two short programs. Don't worry about understanding the details of these programs yet; it will all be explained as we go along.

```
#include <iostream> /*This is for proper C++ input and output*/
/*All programs will need this statement*/
int main()              // Sample 1
  {
    cout << "My first C++ program" << endl;
    return 0;
  }
```

The first line of this example is called the **PRE-PROCESSOR DIRECTIVE**; it tells the compiler to include routines necessary for input and output. The /* ... */ is used to comment or explain to the user what is happening. More detail on this line will be given later. The next line **INT MAIN()** is necessary for all C++ programs and is where execution starts. The **left {and right} braces** will enclose the body of the program.

1.2 PROGRAM BODY

The statements starting with the { brace and ending with the } brace are called the **PROGRAM BODY** and contain the **STATEMENTS** which will be executed by the computer. The program body consists of the symbol {, followed by some executable statements (statements telling the computer to perform certain tasks), followed by the symbol }. There is only one executable statement in this example–the **cout** statement. When the program is executed the **cout** (pronounced c out) statement displays the string "My first C++ program" on the screen. The **"endl"** on the end of the line, *will not be printed. Even though it will not be printed it instructs* the computer to start a new line for the next output item.

Use the program menu to execute the above program. The **C++ editor** can be used to build the file. After being successful with this program, save the program, and add another line to print **"this is fun"**. If all did not go well, **REVISE** the file and make any necessary changes. Save the file and continue.

Look at the following program:

```
int main()        // Sample 2
{
 int N1, N2, SUM;
cout << "Enter two numbers:";
cin >> N1 >> N2;
SUM = N1 + N2;
cout << N1 << "plus" << N2 << "equals" << SUM;

return 0;

}
```

1.3   PROGRAM VARIABLES

C++ and most other common programming languages perform their calculations by manipulating items called **VARIABLES**. In a C++ program a variable is written as a string of letters and digits that begins with a letter. Some sample names for C++ variables are:

X Y Z SUM N1 N2 KELLY rate distance Time

The memory associated with C++ variables can hold numbers or other types of data. For now we will confine our attention to variables that hold only numbers. In C++ all **variables** must be **declared** before they can be used in the body of a program. The variables are declared at the beginning of the program. Each declaration **(int, float, char)** contains the type of the variable, followed by the **variable name**. All three variables in this sample 2 program are integers. A variable of type **int** can store whole numbers only. If we wanted to store numbers with decimals, we would declare the variable to be of type **float**. Single characters such as "a", "b", "c", "1", "2" are declared as **char**.

The lines between { and } contain the statements to be carried out by the computer. The semicolons at the ends of the statements are used to separate the statements. The semicolon is not necessary after the last statement since there is no subsequent statement that it needs to be separated from. Since { is not a program statement, there is no semicolon after it. Semicolons before} are optional.

The first executable statement in the body of the program -**Sample 2**-causes the following phrase to appear on the screen:

Enter two numbers

Suppose that in response to this, an obedient user enters two numbers on the keyboard, say 4, followed by a space, followed by a 5, followed by pressing the RETURN or ENTER key (**data placed on the input line will be separated by one or more blank spaces**) .

The next statement, called an **INPUT** statement,

cin >>...;

tells the computer what to do with these numbers. It tells the computer to input the two numbers called $N1$ and $N2$ and store them in the memory space allocated to variables $N1$ and $N2$. Four will be stored in the memory position allocated for $N1$ and 5 will be stored in the memory position allocated for $N2$.

1.4   ASSIGNMENT STATEMENT

The next statement in our sample 2 program is:

SUM = N 1 + N 2;

The symbol = is called the **ASSIGNMENT OPERATOR**, and this statement is called an **ASSIGNMENT STATEMENT.** There can be spaces between the equal sign and $N1$. The assignment operator has no standard pronunciation but most people pronounce it "*gets the value*" or "*is assigned*". Whatever is on the left hand side of the operator gets the value of whatever is on the right hand side. The above assignment statement changes the value of what is stored in memory for SUM to the value of $N1$ plus the value of $N2$. Finally, the **cout** statement, which is the OUTPUT statement, produces the output.

Enter and run the above program. Use the same procedure as before. Do not forget to include the pre-processor directive(s) which is needed for every program.

PROBLEM: What is the output produced by the following four lines (when correctly embedded in a complete program)? The variables are of type *integer:*

```
X = 2;
Y = 3;
Y = X * 7;

cout << X << Y;
```

ANSWER:

PROGRAM #1: Write a C++ program that will print your name on one line and your hometown on the next line. Use the C++ editor to create the program. Use the name prog1.cpp, or a similar name, when you create the file so it will be saved under that name. Be sure to Save and Execute it.

The following example is a program to print the letter *L* in magnified form:

int main()

/* prints a letter in magnified form */
```
   {
       cout << "       **          " << endl;
       cout << "       **          " << endl;
       cout << "       **          " << endl;
       cout << "       ******      " << endl;
       cout << "       ******      " << endl;
       return 0;
   }
```

The /* print a letter in magnified form */ is called a comment line. The /* and */ at the beginning and end of a comment are necessary. Comments of this type can be extended over several lines. Another form of comment is the // comment. Comments of this type cannot be extended over lines. The // comment can be placed at the end of any C++ statement or can be on a line by itself. The /* .... */ type of comment can be included within a statement.

Examples of comments are:

cin>>a>>b; // this is an input statement

cout<<a<<b; /* this is an output statement */

PROGRAM #2: Write a program that will print your initials in magnified form. Call this prog2.cpp.

1.5 DATA TYPES

A **DATA TYPE** is a description of a category of data. Each variable can hold only one type of data. In our sample program all variables were of type *integer*. That means that their values must be a whole number such as:

38 0 1 89 3987 −12 −5

Numbers that include a fractional part, such as the ones below, are of type float or real:

2.7192 0.098 −15.8 1000053.98

To declare the variables *x* and *y* as float you will use the following declaration:

float x, y;

The type used for letters, or, more generally, any single symbol is **char**, which is short for character. Values of this type are declared as follows:

char x, y;

A variable of type char can hold any **(one)** character available on the input keyboard. Therefore, x could hold any of the characters A, a, +, or 6.

It is perfectly acceptable to have variables of more than one type in a program. In such cases they are all declared at once following the format illustrated below:

```
{
int n1;
float time;
char initial;
}
```

1.6 MATHEMATICAL OPERATORS

The mathematical operations that are available in C++ and their symbol is given below:

OPERATION	SYMBOL	EXAMPLE
addition	+	A + B
subtraction	–	C – 14
multiplication	*	5 * B
division	/	A / B
integer portion	%	A % B

To add two numbers stored using the variable names x and y and store their sum in z the following assignment statement would be used:

```
Z = x + y;
```

More complicated assignment statements would look as follows:

```
Z = 3+8/2 - (5+7)/6;

Z = x+y/3. - (p+q)/(a+b);
```

Mathematical operations are performed from left to right with division (/), multiplication (*) and integer portion (%) taking precedence over addition (+) and subtraction (-). In the first assignment statement above the operations would be performed as follows:

```
Z = 3+8/2 - (5+7)/6;
        4
    7   -   12/6
    7   -    2
        5
```

z would be stored as 5.

PROBLEMS:

1. What is the output produced by the following three lines?
(when correctly embedded in a program)? The variables are of type integer.

X = 2;
X = X + 1;
cout << X;

2. What is the output produced by the following? The variables are of type char: (Be careful on this one - it is tricky!)

A = 'B';
B = 'C';
C = A;
cout << A << B << C;

PROGRAM #3: Write a C++ program that will read two integers and will then output their sum, difference, and product.

Lesson 2

Strings and Formatted Output

2.1 STRING DATA TYPE

Often we would like to store a whole word or combination of words in a variable rather than just one character. To do this in C++ it is necessary to define a new data type: **STRING**. We do it in either of the following ways: (here string is just a set of characters).

> **string name**;
>
> **char name[60]**;

If we declare this type before we declare our variables we can use "name" as a set of characters just like integer, real, and char. C++ will let us do it as follows.

```
int main()
  {
char name[60];
int test1, test2;
float average;
cout << "What is your name?:";
cin >> name;
cout << endl << "What are your test scores?:";
cin >> test1 >> test2;
average = (test1 + test2)/2;
cout << endl << name <<" "<< average;
return 0;
  }
```

In the above example we declared name to be 60 characters in length (a maximum of 59 characters as one character is used as an end of string marker). However, we could declare it to be of almost any length, for example, 25. The string type or array of characters will make more sense when we study arrays later on. When you try the above program you will notice that the average was printed in a rather strange way. The reason it looks like this is because we have not specified how we would like decimal numbers to be printed. The next section deals with that topic.

PROGRAM #4: Write a C++ program that will ask for a name and 3 test scores and will print the person's name and average.

PROGRAM #5: Write a program that will ask for a person's name and age and then respond with the statement:

"Hello, _____. You do not look _____ year old."

2.2   FORMATTED OUTPUT

The output for integer, float, and character data can be controlled by formatting. You can specify the field width through the setw() function as well as the number of places to the right of the decimal point through the setprecision function. To use the following formatting function, you must include the library file (pre-processor directive):

#include <iomanip.h>

The output statement:
cout << setw(10) << 100 << setw(10) << 50 << setw(10) << 25;
Would produce:
_ _ _ _ _ _ _ 100 _ _ _ _ _ _ _ _ 50 _ _ _ _ _ _ _ _ 25

cout << setw(3) << 78634;
OUTPUT: 78634

cout << setprecision(3) << setw(10) << 765.432;
OUTPUT: _ _ _ 765.432

cout << setprecision(4) << setw(10) <<.00456;
OUTPUT: _ _ _ _ _ _ .0046

Check out the following example:

```
   int main()
 /* a program to convert feet to inches */
  {
    float feet, inches;
    cout << "Enter number of feet:";
    cin >> feet;
    inches = feet * 12.0;
    cout << setw(8) << feet << "feet equal"<< setw(8) << inches
    <<"inches" <<
    endl;
    return 0;
  }
```

PROGRAM #6: Write a program to convert hours and minutes to minutes only. Ask the user for the number of hours and the number of minutes, and then print the answer. For example, suppose the user's response for the number of hours is 3 and for the number of minutes is 10. The following message should be printed:

3 hours and 10 minutes equals 190 minutes

The next example computes the amount of interest that you would receive if you saved $100.00 in the bank for one year at an interest rate of 12 percent per year.

```
int main()

    /* computes interest on $100 for one year at 12% */
    {
    float principal ;
    float interest ;
    principal = 100;
    interest = principal * 0.12;
    cout << principal << "  "<<interest << endl;
    return 0;
    }
```

Rather than using the setw for formatting the above output the quotes were used to separate principal and interest. The number of white spaces placed between the two quotes will be the separation between principal and interest.

PROGRAM #7: Write a program that asks the user for the amount of money he/she wants to invest at a rate of 7% per year and then prints a message proclaiming the amount of interest earned in one year. For example, if the user responded that he has 200 dollars to invest, the computer will respond:

"You will earn $14 interest at the end of one year if you invest $200 with us."

Lesson 3

Loops

3.1 GENERAL INFORMATION ON LOOPS

In order to convert temperature from degrees Centigrade to degrees Fahrenheit, we just multiply the Centigrade temperatures by 9/5 and then add 32. The following is an example of a simple program that converts 100 degrees Centigrade to degrees Fahrenheit.

```
int main()
  {
  int C;
  float F;
  C = 100;
  F = (C*9)/5. + 32.;     /* formula for Fahrenheit given Centigrade */
  cout << C << " "<<F ;
  return 0;
  }
```

Notice that this program only converts one value, 100, from Centigrade to Fahrenheit. It would be more useful and interesting to calculate the Fahrenheit temperature for lots of different values for Centigrade; for example 100, 110, 120, . . . etc. This is accomplished by using a loop. There are three types of loops that we will consider: the **WHILE loop**, the **DO/WHILE loop**, and the **FOR loop**. Before studying the loop structure we will first learn about relational operators.

3.2 RELATIONAL OPERATORS

A relational operator is used to compare two numeric values and give a result of true or false. The relational operators available in C++ that can be used in a relational expression such as a WHILE loop, a DO/WHILE loop, and a FOR loop are:

== equal
< less than
> greater than
!= not equal
<= less than or equal
>= greater than or equal

For example, the way to use one of these in a WHILE loop is:

while(num1 < num2)

Don't worry too much about understanding the **WHILE** statement now, that will be explained in the next section. For now just understand that the loop will only work when num1 is less than num2.

Another use of the Relational Operators is in a comparison or control statement called the **IF statement**. The IF statement will compare two items using the Relational Operators, and if the comparison is true, the commands following the IF statement will be executed; otherwise, those commands will be skipped. Below is the basic usage of the *IF* statement.

if(num1 < num2) /*No Semicolon*/

cout << "num1 is less than num2";

The IF statement will be further explained in Chapter 4.

3.3 THE WHILE LOOP

The next example uses a **WHILE** loop to continuously calculate degrees Fahrenheit for new values of *C*. Each time through the loop, C is being incremented by 10. Thus we will be finding degrees Fahrenheit (F) for C = 100, 110, 120, 130. When will we stop? We will arbitrarily use 200 as our stopping criterion. Every loop must have a stopping criterion; otherwise you will be stuck in something known as an infinite loop. It is like being stuck in a revolving door with no way out. The following program demonstrates the use of the **WHILE** loop.

```
int main()
  {
  int C ;
  float F ;
  C = 100;
  while (C <=200) /* notice no semicolon after statement */
    {
    F = (C*9)/5. + 32.;
    cout << C << F;
    C = C + 10;   /* increment C by 10 each time through */
    }
  return 0;
  }
```

In the line containing the WHILE statement, the value of C is compared to 200. If C is less than or equal to 200 the loop is executed. If C is greater than 200 the entire WHILE loop (the three statements between the {and} braces) is skipped. Notice here we have used the braces { } to denote a group of statements that we want to execute as long and the relational expression C <=200 is true.

PROBLEM: Write a program that reads a list of student names and prints the name that comes first alphabetically.

Let's solve the problem together. If you were trying to find the alphabetically first name in a long list of names *without* using the computer, how would you do it? You would probably read down the list of names, one at a time, and remember only the name that was alphabetically first so far. If you read a name that was alphabetically before the one you were remembering, then you would remember the new name instead. We can use this as a model for telling the computer how to solve the problem. This method of problem solving is called finding an ALGORITHM for the problem. An **algorithm** is simply a recipe for finding the solution. We will tell the computer to read a list of names one at a time and to save the *smallest* (the one that comes first in alphabetical order) name it has processed so far in the variable **SMALLNAME**. Since we don't know how many names will be in the list, we will use the *sentinel value* "DONE" (a value that tells the computer that no more values will be read), since that should not be anyone's name. A sentinel value is often used to terminate a loop, especially if the number of input values is not known in advance. An outline or algorithm for our program will look like this:

Read the first name and store it in NAME
Initialize, or define, SMALLNAME to be NAME
WHILE NAME is not equal to "DONE"
{
If NAME is less than SMALLNAME then store it in SMALLNAME
Read the next name into NAME
}

PROGRAM #8: Write a program that continually asks the user for a name until the user says "DONE" and then prints out the smallest (alphabetically first) name. For example, if the user responded with the following names:

SALLY
JOE
KENNETH
ANNE
MABEL
ZYGMUND
BOBBY
DONE
the computer would respond with

The smallest name is ANNE.

If we know the number of data items to be processed beforehand, a counter can be used to control the **WHILE** loop. For example, we could change our algorithm to:

Ask the user how many names he wants to give and store this number in NUMNAMES

> Read the first name into NAME
> Initialize SMALLNAME to NAME
> Initialize COUNT to 1
> WHILE COUNT is less than NUMNAMES
> > {
> Read the next name into NAME
> Store the smaller of NAME and SMALLNAME in SMALLNAME
> Increase COUNT by 1 (COUNT = COUNT +1)
> > }

PROGRAM #9: Modify PROG8 so that it asks the user beforehand for the number of names to be processed. This time, do **not** use "DONE" to terminate the loop.

3.4   THE DO/WHILE LOOP

The **DO/WHILE** loop works the same way as the WHILE loop except that the comparison (to determine whether or not to quit the loop) is not made until the end of the loop. (We say it is a **post-test** loop and the While Loop is a **pre-test** loop.) For example:

```
int main()

{
float C, F;
C = 100;
do      /* Remember, do not use upper case letters in C++ keywords
           such as float, cin, cout, etc.! */
  {
    F = (C*9)/5. + 32.;
    cout << "\n"<< C<<" "<< F ;
    C = C + 10;
  }
while (C <= 200);
return 0;
  }
```

Notice that a semicolon follows the WHILE statement in a DO/WHILE but does not follow the WHILE in just the WHILE statement.

   The following program inputs a student's name and two scores and then prints the name and average score. The process is continued until the average of the scores is 0.0. The 0.0 average is used as the *sentinel* value - a condition to terminate the loop.

int main()

```
  {
  char  name[60];
  int  score1, score2;
  float average;
  do
    {
    cout << "What is your name?";
    cin >> name;        // cannot have spaces in name
    cout << "Input your two scores, separated by a space." << endl;
    cin >> score1 >> score2;
    average = (score1+score2)/2.0;
    cout << name <<" "<< average;
    }
  while (average != 0);
  return 0;
    }
```

Run the program with the following data:

    Bonnie, 90,85
    Elizabeth, 80, 90
    Beverly, 85, 92
    Dawn, 95, 98
    THATSALL, 0, 0

3.5   THE **FOR** STATEMENT

The **FOR** loop is handy to use when we know exactly how many times we need to execute a loop. Suppose we know that we want to go through a loop 5 times. Using a FOR loop we can do that:

int main()

```
  {
  int count;
  float number;
  for (count=1; count <= 5; count = count + 1)
      {
      cout << "Give me a number, please.";
      cin >> number;
      cout << "Your number is " << number << endl;
      }
    cout << "Thank you for the numbers." << endl;
    return 0;
  }
```

The variable COUNT is called the running variable and takes the initial value of 1. The block of statements is then executed and COUNT takes on the value of 2 and is then checked against 5. If COUNT is less than or equal to 5 the block is executed again. This process continues until COUNT is greater than 5.

The language C++ provides **short cut methods** to add 1 or to subtract one from a memory position. Consider: count = count + 1 can be replaced by count++ or ++count; count = count - 1 can be replaced by count-- or --count.

PROGRAM #10: Write a program that prints your name 10 times. Use a FOR loop with a counter to keep track of the number of times the loop has executed.

PROGRAM #11: Write a program to input a student's name and three test scores. Print the name and average score. Continue this process until the name "END" is encountered. Hint: use the program that you wrote for prog10 and just make a few changes to it.

PROGRAM #12: Write a program to input a student's name and score that they received on their recent test, then calculate and print the class average for the test. Do not use a trip value to quit the loop, but assume there are 5 students in the class.

The next program finds and prints the sum of the integers 1 through 100:

int main()

```
  {
int sum, num ;
sum = 0;
num = 1;
  while (num <= 100)
    {
    sum = sum + num;
    num = num + 1;
    }
cout << "The sum of first 100 integers is  " << sum;
return 0;
  }
```

PROGRAM #13: Change the above program so that it finds and prints the sum of the odd integers 1 through 99. (Form the sum $1 + 3 + 5 + \ldots +99$).

Pretend that you are looking for a job for the month of July. Suppose that someone offers you a job that pays one cent for the first day, two cents for the second day, four cents for the third day, eight cents for the fourth day, etc. The amount of your pay is doubling each day. Would you take this job or one that offered you $1000 per day for the entire month? Before you answer, you may want to run the following program. It determines the day in which the day's pay for the first job first exceeds $10,000.

```
int main()
  {
  int day;
  float daypay;
    day = 1;
  daypay = 0.01;
  while (daypay < 10000.)
    {
    day = day + 1;
    daypay = daypay * 2;
    }
  cout << "The day in which pay first exceeds $10,000 is " << day;
  return 0;
  }
```

PROGRAM #14: Using the above program, change it so that it prints the total salary you will receive for the month of July. Hint: there are 31 days in July.

Lesson 4

Decision Statements

4.1 IF STATEMENTS

One of the most used CONTROL STATEMENTS in the C++ language is the **IF** statement. In its simplest form, it evaluates an expression and then proceeds on the basis of the result of that comparison. For example, the following program **reads in a set of scores** and counts the number of students that made an A. An A is considered to be a score of 90 or above. A sentinel value of 0 is used. This signals that no more data will follow and normally this is not a good data value – here we would not expect 0 to be one of the scores.

```
int main()
{
int acount, score; /* acount is a variable for the number of A's */
acount = 0;
do
    {
    cin >> score;
    if (score >= 90)
       acount = acount + 1;
    }
while (score != 0);     /* trip on score of 0 */

cout << "There were  " << acount << "  A scores.";

return 0;
}
```

Use the following as input:

98 45 17 90 91 87 34 99 100 0

How many times will the loop be executed?

What will be printed out?

Again, if you have more than one statement to be executed when the IF expression evaluates to true, you can use the braces { } to designate the block to be executed.

4.2   IF – ELSE Statements

Whenever you want one thing done when the **IF** expression evaluates to true and another thing done when it evaluates to false, it is handy to use the **IF - ELSE statement**. Here is a short program that asks the user to guess the digit the computer is thinking. If the digit agrees with the one the program is trying to match (which is 6 in our case), the program prints "Right. Good job." If the digit is not 6 the program prints "Sorry, wrong number."

```
int main()
  {
  int x ;
  cout << "I am thinking of a digit.  What is it?";
  cin >> x;
  if (x == 6)
    cout << "Right. Good job.";
  else
    cout << "Sorry, wrong number.";
  return 0;
  }
```

PROGRAM #15: Put the above program in a loop so that it will continue to play the game until the person guesses the digit. Have it count the number of guesses it takes. It should print a message to the user telling how many guesses it took him/her to get the correct digit.

Suppose we need a program to count the number of students who are freshmen, sophomores, juniors, and seniors in a class of 10 students. This program accomplishes that by combining IF and ELSE statements.

```
int main()
  {
  int  nfr,nso,njr,nsr,grade,count;

  nfr = 0;
  nso = 0;
  njr = 0;
  nsr = 0;
  count = 1;
```

```
cout << "Enter your grade level as either 9,10,11,or 12.";
cin >> grade;
while (count <= 10)
  {
  if (grade == 9)
    nfr = nfr + 1;
  else if (grade == 10)
    nso = nso + 1;
  else if (grade == 11)
    njr = njr + 1;
  else if (grade == 12)
    nsr = nsr + 1;
  cout << "Please enter a 9,10,11, or 12.  Try again.";
  count = count + 1;
  }
cout << "There were  " << nfr << "  freshmen."<<endl;
cout << "There were  " << nso << " sophomores."<<endl;
cout << "There were  " << njr << " juniors."<<endl;
cout << "There were  " << nsr << "  seniors";
return 0;
}
```

Run the program and test it out.

In the following program a company pays every employee $5.00 per hour. This program asks for name and number of hours worked. It then calculates and prints the total wages for the week.

```
int main()
  {
float hours, paydue ;
char name[30];
cout << "Enter your name: ";
cin >> name;
cout << "hours worked:";
cin >> hours;
paydue = hours * 5.00;
cout << name<<" "  << hours <<" "<< paydue;
return 0;
  }
```

PROGRAM #16: Write a program that pays an employee $5.00 per hour for all hours less than or equal to 40 and pays twice that amount for all hours over 40. For example, if you worked for this company last month for 52 hours you would receive $5.00 per hour for 40 hours (200 dollars) plus $10.00 per hour for 12 hours (120 dollars) for a total of 320 dollars.

PROGRAM #17: Write a program to determine how much fine you should pay for a speeding ticket. Assume the fine is computed as follows:

```
Amount over limit (miles/hour)    Fine
 1 - 10                           $10
11 - 20                           $20
21 - 30                           $25
31 - 40                           $40
41 or more                        $60
```

Your program should ask for both the speed limit and the driver's speed. It should then print

THE FINE IS $_____

PROGRAM #18: Using a FOR loop, write a program that prints the numbers 1 - 10, along with their square and cube. An example of the output: (Format the output to get a chart like this.)

NUMBER	SQUARE	CUBE
1	1	1
2	4	8
.	.	.
.	.	.
.	.	.
10	100	1000

Lesson 5

Arrays

5.1 ARRAY DEFINITION

Suppose we wish to process grades for a class of several students. The number of students in each class will vary, but we wish to write a program that will calculate the average grade and then find the number of grades less than the average grade in the class. We cannot calculate the average until all the grades for the class have been entered. So we must save the grades and count the ones below average after the average is calculated.

Should we save each grade in a variable of different name? This would lead to a long repetitious program. Fortunately, C++ lets us solve the problem in a very nice way using **ARRAYS**. An **ARRAY** is a data structure used for storing a collection of data items that are all the same type (an array stores items in adjacent memory positions). Usually, we first describe the structure of an array in an *array type declaration*. The array GRADE of 20 elements is declared as

float grade[20];

C++ reserves 20 memory cells for the GRADEs; these memory cells will be adjacent to each other in memory. Each cell of GRADE will contain a single student's grade. We can use each one, just as we did variables, by calling it by its name. The elements are stored in the array GRADE as GRADE[0], GRADE[1], . . . , GRADE[19]. GRADE[1] means the grade of the second student in the class.

To read in the scores of students into the array GRADE, we would do this:

int main()
```
{
float grade[20];
int i, sum ;
float average;
 for (i=0; i< 20; ++i)
```

```
{
cout << "Please enter your grade:";
cin >> grade[i];
}
return 0;
}
```

To find the average of the class, we would change the main part of the program to this:

```
sum = 0;
for (i=0; i<20; ++i)
   {
     cout << "Please enter your grade:";
     cin >> grade[i];
     sum = sum + grade[i] ;
   }
average = sum / 20.0;
   }
```

5.2   EXAMPLES

PROGRAM #19: Change the above program so that it finds the average grade for a class of 10 students; then it prints each grade and whether it is "below average", "average", or "above average".

The following program finds the largest element in a list of numbers, and prints the number of the cell where it was found (its position within the array).

```
int main()

{
int n[10];
int i, largest, position(0);

largest = 0;
for (i=0; i<10; ++i)
  {
    cin >> n[i];
    if ( n[i] > largest )
    {
      largest = n[i];
      position = i;
    }
  }
cout << "largest is   " << largest<<endl;
cout << "position is  " << position;
return 0;
}
```

PROGRAM #20: Write a program to print the smallest element in a list of numbers and its position(s) within the list. (Be careful with this one, test all possible cases).

PROGRAM #21: Write an interactive program that plays the game of HANGMAN. Hint: Let WORD be an array of characters. Read the word to be guessed into successive elements of the array WORD. The player must guess the letters that belong to WORD. The game ends when either all letters have been guessed correctly (player wins) or a specified number of incorrect guesses have been made (computer wins.) Use an array SOLUTION to keep track of the solution so far. Initialize SOLUTION to a string of symbols "*". Each time a letter in WORD is guessed, replace the corresponding "*" in SOLUTION with that letter and show it to the player before the next guess is made. Play several games to try it out.

Searching an array of elements for a particular element or sorting a list to place in order lexicographically (alphabetic or numeric order) are large topics in computer science. Program #20 is an example of searching. To sort a list of elements (numbers) in order from largest number to smallest number consider the following algorithm:

1) Define or input the list of N numbers and store in an array
2) Search the list starting at the first element for the largest element and swap places with the largest element and the element in the first position
3) Search the list starting with the second element for the largest element and swap places with the new largest element and the element in the second position
4) Continue this process until you have considered the n-1st and n^{th} element
5) Output the list – it should now be in order from largest element to smallest element

The above algorithm is spoken of as the selection algorithm or sort.

PROGRAM #22: Write a program, using the selection sort, to arrange a list of number in order from high number to low number.

Lesson 6

Functions

6.1 BASICS OF FUNCTIONS

C++ has some built-in **functions** (special purpose programs) that come in quite handy. For example, the function **SQRT** finds the square root of a number. The following part of a program shows how the function is called:

```
{
cout << sqrt(4)<<endl;
cout << sqrt(25)<<endl;
cout << sqrt(81)<<endl;
cout << sqrt(100)<<endl;
}
```

This would result in the following being printed:

2
5
9
10

Other similar functions are **SIN, COS, EXP, LN, LN10, POW, FABS, ABS** and others. Each of the pre-defined function is a **special purpose** program that performs a given task. The functions listed above are used often in science, math and engineering and alleviates the task of the programmer having to write the code when one of these functions is needed. Also, if one is needed several times with a program, only one statement is needed, rather than several, each time the function is used.

The following illustrates how to call some of the pre-defined functions:

```
z = 1.57;
```

x = sin(z); /*the number stored for *z* is passed to the function sin and the value for sin(1.57) is calculated and returned to the calling statement and stored in x. */

x = pow(5,2); /*the function raises 5 to the power of 2 and the result is stored in x /*

x = exp(x); /*the number e(2.71828) is raised to the power of 1.57 and the result is stored in x */

You will want to try each of these in a manner similar to the test cases above for the SQRT function. The POW function has two arguments and all others listed have one argument. To use these functions you will need to include the **pre-processor directive <cmath>**.

Often, we need some functions other than the ones C++ provides. In that case, we get to make our own. For example, suppose we wanted a function that would return the larger of two integers we gave it. We have to give the function a **name** and then define it for C++. Let's call it **LARGER**. We would define it like this:

```
int larger(int first, int second)
{
int la;
if (first > second)
la = first;
else
la = second;
return(la);
}
```

The first line may be a little confusing. We are telling C++ that we are defining what is meant by the function LARGER. We are saying that we will give it two integers and it should give us an integer in return. If we said in our program that **x = larger(23,67)**, C++ should assign x to be 67. Here is how it would look in the entire program.

```
int larger(int, int);
int main()
    {
int num1,num2,num3,num4,temp1,temp2,big ;
cout << "Enter 4 numbers.";
cin  >> num1 >> num2 >> num3 >> num4;
temp1 = larger(num1,num2);
temp2 = larger(num3,num4);
big = larger(temp1,temp2);
cout << "The largest number is  " << big;
return 0;
    }
```

```
int larger (int first,int second)

  {
int la;
if (first > second)
  la = first;

else

  la = second;

  return(la);

  }
```

Notice that we had to define the function before we reached the main body of the program. The method of defining a function in this manner is spoken of as defining the **function prototype**. This would not be necessary if we had placed the actual function before the main program.

PROGRAM #23: Write a function CUBE that finds the cube of a number. For example, the cube of 5 is 125 (5*5*5) and the cube of 2 is 8 (2*2*2). Test your function by writing a program that makes a table of the numbers 1 through 10 with their cubes. It should look something like this: (Remember to format the output to get the chart.)

NUMBER	CUBE
1	1
2	8
3	27
.	.
.	.
.	.

PROGRAM #24: Write a function POWER that finds the power of an integer. For example POWER(2,5) should return 32, since 2 to the fifth power = 2*2*2*2*2 = 32. Test it out in a program.

6.2 RANDOM NUMBER GENERATORS

Functions to generate random numbers are very common. Random number generators normally produce a sequence of real numbers between 0 and 1 that for all practical purposes appear to be random. Most random number generators start with an initial real value called a **seed**. Other random numbers are determined once the seed is specified. Here is a function which is a random number generator:

randomize();
n = random(100); /* include the pre-processor <stdlib.h> */

The numbers generated are all between 0 and 99. The next function uses **random** but finds random numbers in the range you specify.

```
int n, range;
randomize();
cout << "Enter the range.";
cin >> range;
n = random(range);
```

The above sequence will define random numbers in the range between 0 and (range −1). Each time random is called **one** of the numbers in the specified range will be returned.

Now that we have a random number generator, we can write some more interesting games. The number guessing game that we wrote back in program 15 can now be made more interesting.

PROGRAM #25: Write a guessing game program that will ask the user to guess a number. If the user is correct, the computer will respond with a nice message and tell how many guesses it took. If the user is wrong, the computer will tell the user HIGHER or LOWER and ask for another guess. This will continue until the user gets it right. Change the program so that the user can select the range of the integer. Also, you might want to give the user a maximum number of guesses, in case the user becomes confused and never gets the right answer.

Lesson 7
Additional Problems

PROGRAM #26: If you have watched Wheel of Fortune you have noticed that the most popular letters chosen are *R, S, T, L, N,* and *E*. Pretend that you are the producer and you want to judge how hard the puzzle will be to solve, given only those letters. Write a program that will allow you to input a phrase, a person, a place, or something in some other category, and then display only those six letters and blanks everywhere else. Have the category printed above it. Test its difficulty by having your friends solve it.

Example: category: place

         L E _ _ N _ T _ N
         _ E N T _ _ _

PROGRAM #27: The square root of a number *N* can be approximated by repeated calculations using the formula

$NG = 0.5(LG + N/LG)$

where NG stands for next guess and LG stands for last guess. Write a program which implements this process. The initial guess will be the starting value for LG. The procedure will compute a value for NG using the formula above. The difference between NG and LG is checked to see whether these two guesses are almost identical. If so, NG is the square root; otherwise, the last guess LG becomes NG and the process is repeated (another value is computed for NG, the difference is checked, etc.). For this program the loop should be repeated until the difference between NG and LG is less than 0.005. Use an initial guess of 1.0 and test the program for the numbers: 4, 120.5, 88, 36.01 and 10000. Print your results.

PROGRAM #28: Write a program that will provide **change for a dollar** for any item purchased that costs less than one dollar. Print out each unit of change (quarters, dimes, nickels, or pennies) provided. Always give the biggest denomination coin possible. For example, if there are 37 cents left in change, give a quarter, a dime, and two pennies. Note that we are not using half-dollars. The input should be the amount of purchase and the output should be the number of each coin type given in change.

EXAMPLE:

purchase price: 18 cents
change: 3 quarters, 1 nickel, 2 pennies.

PROGRAM #29: Write a program that accepts two integers, then finds and prints the largest positive integer that divides both of them evenly. This number is called the greatest common divisor. Have the program continue to run until 0, 0 is entered.

EXAMPLE:

integers? 24, 36
The greatest common divisor of 24 and 36 is 12.

PROGRAM #30: An integer is said to be perfect if it is equal to the sum of all of its factors except itself. For example, 28 is perfect since 28 = 1 + 2 + 4 + 7 + 14. If the sum of the factors of a number is greater than the number, it is called an abundant number; if less, deficient.

Write a program that inputs a number and tells if the number is perfect, abundant, or deficient. Allow the program to continue to input integers until 0 is entered.

PROGRAM #31: The number 55 has the interesting property that its square can be "sliced" into two two-digit integers whose sum is 55. That is, 55*55=3025, and 30+25=55. Write a program that outputs all such two-digit numbers.

PROGRAM #32: Write a program that will input numbers and keep a rolling sum. If the number is less than 50, double it and add to the sum; if it is greater than 50, add half of the number to the sum. If the number is 50, stop inputting numbers, and don't add the 50 to the sum. The program should output the "sum" and the "average" of the altered numbers.

PROGRAM #33: Write a program that inputs the current month and year numerically. After inputting the person's name, the person's birth month and year should be entered. Finally, the program should output the age of the person.

EXAMPLE:

Current date? 6,1991
　Name? John
　John's birth date? 3,1970
　John is 21 years old.

PROGRAM #34: Write a program that will input a string of **six different characters**, then print all possible permutations of the string of characters. Hint: there should be 720 different items printed.

PROGRAM #35: Write a program to determine the frequency of each vowel in some English language text. The input will consist of sentences running over a number of lines. The end of the text is indicated by the special symbol '*'. Don't forget to consider both upper and lower case vowels. Remember: vowels are *a, e, i, o,* and *u* (and never *y* in our case.)

PROGRAM #36: The Sieve of Eratosthenes is a method of finding prime numbers. The method starts with a list of consecutive integers and crosses out all multiples of two. Moves to the next number still in the list and crosses out all multiples of that number. Continue the process until the list is exhausted. The remaining list is the list of prime numbers. Using an array, construct a program that uses the method of the sieve to output all prime numbers less than 300. Note: 1 is not a prime.

PROGRAM #37: Write a program that inputs a numerator and a denominator of a fraction, then outputs the number as a mixed numeral in lowest terms. For example, input of 45 and 30 would produce 1½. Terminate the program with 0,0, and make sure that the program can handle bad data (like 5,0).

PROGRAM #38: In an experiment, a ball is dropped from a certain distance above the ground. When it hits the ground, it bounces back up, but only goes half as high as its previous bounce. That is, if a ball was dropped from 8 feet, it would bounce 4 feet high, then 2 feet, the 1 foot, and so forth.

Write a program that inputs the height from which the ball was dropped and the number of bounces desired. The program should print out how far the ball traveled in bouncing the number of times specified by the user. Don't forget that in that previous example, the ball actually travels 8 feet between the first and second bounces, 4 feet up, then 4 feet back down!

PROGRAM #39: Write the following "wicked game". Have the computer tell the user that it has selected a number between 1 and 10, and that the person had 9 guesses to find out that number. After every guess, the computer should tell the user that he/she is wrong. After the last guess the computer should then say what the number was. Of course, the number that the computer says or displays should not be one of the numbers that the person guessed.

PROGRAM #40: Three sailors, shipwrecked on a desert island, have gathered a pile of coconuts that are to be divided early the next morning. During the night one sailor arises, divides the pile into three equal parts, and finds one left over, which he gives to the monkey. He then hides his share and replaces the other two shares into a single pile. Later during the night, each of the other two sailors arises separately and repeats the performance of the first sailor. In the morning all three sailors arise, divide the pile into three equal shares, and find one left over, which they give to the monkey.

Your program needs to find the smallest positive integer that could represent the original number of coconuts possible such that the monkey will get one and only one coconut during each division of the pile.

Appendix B: Answers to Selected Exercises

1.

i	p_i	e_i
0	3	$(2)^2$
1	7/3	$(2/3)^2$
2	47/21	$(2/21)^2$
3	2,207/987	$(2/987)^2$

2. Given the system

$$a_{11}x_1 + a_{12}x_2 = b_1$$
$$a_{21}x_1 + a_{22}x_2 = b_2$$

let $D = a_{11}a_{22} - a_{12}a_{21}$. Then, if $D \neq 0$, the solution is

$$x_1 = \frac{1}{D}\begin{vmatrix} b_1 & a_{12} \\ b_2 & a_{22} \end{vmatrix} \qquad x_2 = \frac{1}{D}\begin{vmatrix} a_{11} & b_1 \\ a_{21} & b_2 \end{vmatrix}$$

3. (*a*) Begin by setting $B = x_1$. For each $i = 2, 3, \ldots, n$ compare B with x_i. If $B \geq x_i$, go on to the next element of the set. If $B < x_i$, set $B = x_i$ and go on to the next element of the set.

4. Set $i = 1$. Using results of Exercise 3, find a largest element from the set $x_i, x_{i+1}, \ldots, x_n$. Interchange this element with x_i, increase i by 1, and repeat the process.

5. If $n \neq 2$ and n is even, then n is not a prime. If $n \geq 3$ is odd, then determine whether or not n is divisible by m, where m takes the values 3, 5, 7, 9, ... subject to $m^2 \leq n$.

9. (a) Take $N \geq 1/\varepsilon$.

   (b) Take $N \geq 2/\varepsilon$.

10. (a) Take $N \geq 5/8\varepsilon$.

   (b) Take $N \geq 1/\varepsilon$.

11. (a) Take $\delta = \varepsilon/3$.

14. (a) Use the theorem that differentiability implies continuity.

   (b) Differentiability implies continuity on [0,1]. Use an ε, δ argument at $x = 0$.

16. Use the rule for differentiation of a product.

17. (a) Problem has a solution in [0,1].

   (b) No solution.

   (c) Problem has infinitely many solutions. One solution lies in $[0, \pi/2]$.

   (d) Problem has a solution in [0,1].

18. Observe that if $|x|$ is large enough then the sign of $P(x)$ is the same as the sign of $a_0 x^3$.

20. Use the mean-value theorem. Note that usually a direct proof of Rolle's theorem is first given; Rolle's theorem is then used to prove the mean-vlaue theorem.

21. Apply Rolle's theorem $n - 1$ times to successive pairs of roots of f.

22. [3,4] is a solution.

23. [2,3] is a solution.

24. $[\frac{1}{4},1]$ is a solution.

25. Observe that $p_3 \in [2,3]$.

26. If p_0 is near but not at $5^{\frac{1}{2}}$, then p_1 will be a worse approximation to $5^{\frac{1}{2}}$ then p_0.

27. $\left\{ \left(\frac{5}{2}\right)^{\frac{1}{2}}, \quad (2)(5)^{\frac{1}{2}} - \left(\frac{5}{2}\right)^{\frac{1}{2}} \right\}$.

30. The use of $k = \frac{1}{2}$ gives the interval $[3/4A, 5/4A]$.

31. $x = (1/3)(2x + A/x^2)$. If $p_0 = 1$, then $p_2 = 2 + 208/450$. If $p_0 = 3$, then $p_2 = 2 + 2,848/77,841$.

32. If $p_0 = 1.1$, then $p_1 = .99$ and $p_2 = 1.00$ (rounded to two decimal places).

33. $P_{n+1} = p_n - \dfrac{(p_n - p_{n-1})f(p_n)}{f(p_n) - f(p_{n-1})}$

40. If $p_0 > 0$, the iteration converges. There are infinitely many negative starting values which lead to a p_n such that $p_n = 0$. For example, $p_0 = -(A/2)^{1/3}$ gives $p_1 = 0$, p_2 undefined.

42. The iteration converges for all $p_0 > 0$, $n \geq 2$.

44. $|p_{n+1} - p_n| > |p_{n+1} - \text{solution}|$.

45. Suppose that $p_0 > 0$ and $p_{n+1} = (\frac{1}{3})(2p_n + A/p_n^2)$. If $n \geq 1$, then $p_n > A^{\frac{1}{3}}$ and $A/p_n^2 < A^{\frac{1}{3}}$. Hence $|p_{n+1} - A^{\frac{1}{3}}| < (\frac{2}{3})|p_n - A/p_n^2|$.

CHAPTER 2

1. $A+B = B+A = \begin{bmatrix} 5 & 10 \\ -1 & 1 \\ 1 & 4 \end{bmatrix}$ $A+C$ *and* $C-B$ undefined

$B-A = \begin{bmatrix} 1 & 2 \\ 1 & -5 \\ 1 & 8 \end{bmatrix}$ $C-C = \begin{bmatrix} 0 & 0 & 0 \\ 0 & 0 & 0 \end{bmatrix}$

$3A+2B = \begin{bmatrix} 12 & 24 \\ -3 & 5 \\ 2 & 6 \end{bmatrix}$ $3A-2B = \begin{bmatrix} 0 & 0 \\ -3 & 13 \\ -2 & -18 \end{bmatrix}$

2. $A = \begin{bmatrix} \frac{2}{5} & \frac{4}{5} \\ \frac{2}{5} & \frac{6}{5} \end{bmatrix}$ $B = \begin{bmatrix} -\frac{3}{5} & -\frac{6}{5} \\ -\frac{3}{5} & -\frac{9}{5} \end{bmatrix}$

3. $AB = \begin{bmatrix} 4 & -2 \\ -1 & -6 \end{bmatrix}$ $BA = \begin{bmatrix} 1 & 8 & 1 \\ 0 & -6 & 2 \\ 1 & 2 & 3 \end{bmatrix}$

4. $1x_1 + 2x_2 + 1.5x_3 + 3x_4 = 0$

$2x_1 + 4x_2 + 1.5x_3 + 2x_4 = 2$

$8x_1 + 2.5x_2 + 0x_3 + x_4 = 1.5$

$-3x_1 - 3.5x_2 + 0x3 + 3x_4 = -2.5$

5. $AB = B$ and $BA = B$.

6. If A is $n \times m$ and B is $m \times k$, then (i,j) components of $a(AB)$, $(aA)B$, and $A(aB)$ are

$$a\sum_{s=1}^{m} a_{is}b_{sj}, \quad \sum_{s=1}^{m}(aa_{is})b_{sj}, \quad \text{and} \quad \sum_{s=1}^{m}a_{is}\left(ab_{sj}\right)$$

respectively. It is evident that these are all equal.

7. The matrices $(A + B)C$ and $AC + BC$ have the same dimension. The (i,j)th components are, respectively,

$$\sum_{s=1}^{k}(a_{is} + b_{is})c_{sj} \quad \sum_{s=1}^{k}a_{is}c_{sj} + \sum_{s=1}^{k}b_{is}c_{sj}$$

These are easily seen to be equal.

9. (a) $D(A+B+C) = D((A+B)+C)$

$= D(A+B)+DC$

$= DA+DB+DC$

(b) $C(dA+eB) = C(dA)+C(eB)$

$= dCA+eCB$

(c) $(A+B)(C+D) = A(C+D)+B(C+D)$

$= AC+AD+BC+BD$

10. $\left| \dfrac{h}{3}\left(\varepsilon_0 + \varepsilon_{2m} + 2\sum_{i=1}^{m-1} \varepsilon_{2i} + 4\sum_{i=1}^{m} \varepsilon_{2i-1}\right) \right| \le (b-a)\varepsilon$

12. $x_1 = -14/3,\ x_2 = -91/6,\ x_3 = -17/3,\ x_4 = 17,\ x_5 = 7.$

13. 24.

14. $x_1 = -1,\ x_2 = 0,\ x_3 = 1,\ x_4 = 5.$

16. Yes. The only nonzero coefficients lie on the diagonal.

17. $X = B.$

18. The (i,j) th element of the product can be written in the form

$$\sum_{s=1}^{n} a_{is} b_{sj} = \sum_{s=1}^{j} a_{is} b_{sj} + \sum_{s=j+1}^{n} a_{is} b_{sj}$$

For a below-the-diagonal element $i > j$. If $i > j$ in the right side above the first sum is zero, because these $a_{is} = 0$, and the second sum is zero, because these $b_{sj} = 0$.

23. To interchange equation i and j, where $i \ne j$:
1. Add row i to row j.
2. Add -1 times the new jth row to the ith row.
3. Multiply the new ith row by -1.
4. Add -1 times the new ith row to jth row.

24. If an equation is multiplied by zero, the effect is identical to deleting the equation from the system.

25. $x_1 = 2,\ x_2 = 2,\ x_3 = 0.$

26. $x_1 = 1,\ x_2 = 2,\ x_3 = 0,\ x_4 = -4.$

28. If at any stage a nonzero pivot element connot be found, then the column under consideration has zeros on and below the diagonal. Thus, proceed to the next column, and continue as before.

29. $(a)\quad A^{-1} = \dfrac{1}{186}\begin{pmatrix} 39 & -15 & -21 \\ 70 & 16 & 10 \\ -7 & 17 & -1 \end{pmatrix}$

$(b)\quad ((b-a)h_k^2/12)\max\limits_{a\le x\le b}|f''(x)|,\ \text{where } h =$

30. $(a)\quad A^{-1} = \dfrac{1}{73}\begin{bmatrix} 73 & 0 & 0 & -73 \\ -29 & -8 & 7 & 62 \\ -31 & -1 & 10 & 26 \\ 27 & 15 & -4 & -25 \end{bmatrix}$

$(b)\ x_1 = -2,\ x_2 = 1,\ x_3 = -1,\ x_4 = 0.$

35. $10^{-8}/1.362.$

36. $(a)\ |c_1| \le 3.432 \times 10^{-2}\qquad |c_2| \le 6.348 \times 10^{-1}$
$\quad |c_3| \le 2.7363 \qquad\qquad\quad |c_4| \le 4.1328$
$\quad |c_5| \le 2.0223$
$(b)\ \max|e_i| \le 10^{-9}/4.1328.$

CHAPTER 3

3. $x_1 = x_2 = 0$ and $x_1 = x_2 = \pi/2 + 2k\pi$ for $k = 0, \pm1, \pm2, \dots$.
4. $P_3 = (511/256, 523/512)$.
5. $P_3 = (2,1)$.
7. $P_3 = (-1/432, 1/108)$.
8. $P_2 = (497/256, 2,079/2,048)$.
9. $P_3 = (-5/5,832, 341/34,992)$.
11. $6^{1/2}, 69^{1/2}, 95^{1/2}, 55^{1/2}$.
15. $(1,0)$
18. Prove that $|p_{1n}| \le |p_{2n}| \le 2^{n-1}/3^n$.
26. $\|B\|_r = .9, \|B\|_c = .8, \|B^2\|_r = .54, \|B^2\|_c = .48$.
32. $P_2 = (-401/400, 401/400)$.
35. $P_1 = (-12/5, -151/7, 271/70)$.

37. $\begin{pmatrix} \pi/2 & 0 \\ 0 & 0 \end{pmatrix}$.

38. $J(P)$ is independent of P.
41. $P_2 = (0, 1/128)$.
44. The column norm of the Jacobian matrix does not exceed 11/24 on R.
45. The column norm of the Jacobian matrix does not exceed 17/24 on R.
49. $P_1 = (.70714, .70714)$.
50. $P_1 = (-.0235, -.0059)$, $P = (0,0)$.

CHAPTER 4

3. $P(2) = 37, P(-3) = -1,828$.
4. $P(3) = 304, P'(3) = 625$.
5. (a) No.
    (b) 113, 236, 362.
7. (a) $P(x) = (x^2 - 3x + 1)(2x^2 - 6x + 7) + 24(x - 3) + 63$.
    (b) $P(x) = (x^2 + 2x - 4)(2x^2 - 4x + 7) - 21(x + 2) + 68$.
13. $x - x^2/2$, In $1.2 \approx .18$, with error bound .0027.
14. $f(x) = 1 - x^2 + (x^4/24)[384t^4/(1 + t^2)^5 - 288/(1+ t^2)^4$
    $+24/(1 + t^2)^3]$, where t is between 0 and x.
15. 2/3.
16. Solve $x = 1 - x^2$ to get $(5\frac{1}{2} - 1)/2$.
17. 2.15, 2.155, 2.155083.
18. The terms of the series expansion alternate in sign and decrease in magnitude. The error is bounded by $f_0'(t^5dc)/6!$
19. Use the sign alternation in sin $x = x - x^3/3! + x^5/5! \dots$ to conclude that the error does not exceed $|x^3/6|$ for small $|x|$. An interval is $|x| \le (.1)6^{1/6}$.
20. $\phi(x) \approx x - x^3/6 + x^5/40)/\sqrt{2\pi}$.
22. .9980, error bound .0012.
23. .7126, error bound .0003.
24. Error bound .0017.
25. Error bound $.00625e^3$.

26. 1.02261, error bound .04.
27. $h \le .02/e^{1/2}$.
29. .2066, error bound .00006.
30. Error bound .00012.
31. $n = 2$. Hint: use the results in Example 4–20.
34. 1.2428, error bound 4.2×10^{-7}.
36. .415.
42. .62840 (calculations rounded to 5D).
43. .62840 (calculations rounded to 5D).
44. .56754.
45. .56714.
46. 2.4048.

CHAPTER 5

1. 85.75, 86.61, 87.49.
2. 83.145, 86.610, 90.135.
3. 86.03, 86.61, 87.19.
4. 81.81, 91.70.
5. .000404.
6. 63.33, 70.00.
7. If $2^{1/2} \approx 1.414$ is correctly rounded to three decimal places, then the absolute error does not exceed .0005/1.4135.
9. The absolute value of the error may be arbitrarily large.
10. The distance is 6.65 centimeters, with the absolute error bounded by .2 centimeter. The assumption of a linear variation probably contains analytic error.
11. (a) $x = R \cos \alpha \cos \beta$
$y = R \sin \alpha \cos \beta$
$z = R \sin \beta$
(b) $\Delta x = (R + \Delta R) \cos (\alpha + \Delta \alpha) \cos (\beta + \Delta \beta) - R \cos \alpha \cos \beta$
$\Delta y = (R + \Delta R) \sin (\alpha + \Delta \alpha) \cos (\beta + \Delta \beta) - R \sin \alpha \cos \beta$
$\Delta y = (R + \Delta R) \sin (\beta + \Delta \beta) - R \sin \beta$
Here $|\Delta R| \le 50$, $|\Delta \alpha| \le .001$, $|\Delta \beta| \le .001$. By the use of appropriate trigonometric identities and by neglecting terms involving $(\Delta \alpha)^2$, $(\Delta \beta)^2$, etc., one gets, when $\alpha = \beta = \pi/4$, the result

$$\Delta x \approx \frac{\Delta R - R(\Delta \alpha + \Delta \beta)}{2}$$
$$\Delta y \approx \frac{\Delta R + R(\Delta \alpha - \Delta \beta)}{2}$$
$$\Delta z \approx \frac{\Delta R + R \Delta \beta}{2^{1/2}}$$

Thus $\max|\Delta x| = \max |\Delta y| \approx 45$ meters and $\max |\Delta z| \approx 49.5$ meters.
12. $L_0(x)\varepsilon_0 + L_1(x)\varepsilon_1 + L_2(x)\varepsilon_2$.

13. If the interpolation is performed on 1.0, 1.1, 1.2, then the inherent error does not exceed .00625 and the analytic error does not exceed .0000625$e^{1.2}$.

14. $\pi = C^2h/4V$. Errors may arise from incorrect measurements on C, h, V.

15. $-h(2C\delta + \delta^2)/4V$.

16. .0264.

17. $t = 9.833$ seconds. A lower bound on t is found by solving the same problem, with velocities 84.84 and 90.9 used. An upper bound is found by using velocities 83.16, 89.1.

18. (a) .125 × 10^6 (b) .351 × 10^{-4}
   (c) .100 × 10^{-1} (d) .250 × 10^2
   (e) −.100 × 10^{-3} (f) −.824 × 10^2

19. (a) .31416 × 10^1 (b) .27183 × 10^1
   (c) .6023 × 10^{23} (d) −.2588 × 10^2
   (e) .6387 × 10^2

20. (a) .173 × 10^4 (b) .986 × 10^1
   (c) .556 × 10^{-2} (d) .100 × 10^{11}

23. (a) −.121 × 10^2 (b) .185 × 10^3
   (c) .185 × 10^3 (d) Zero
   (e) −.176 × 10^2

27. The distributive law need not hold. An example is given by $A = .30 \times 10^1$, $B = .27 \times 10^0$, $C = .77 \times 10^0$, with $t = 2$ arithmetic used.

28. .111 × 10^1, .333 × 10^0, .556 × 10^0.

29. .1800 × 10^1, .5556 × 10^0, .1000 × 10^1.

30. Result in all cases is .14 × 10^1.

31. .1414 × 10^1.

34. .14 × 10^{-1}.

36. The result is exact.

37. Using upper triangulation and back substitution gives $x = .70$, $y = 2.5$ in two-digit arithmetic.

38. $x = .970$, $y = 2.05$.

41. 3.8.

42. 200.

43. 100.

44. 5.

45. .77.

CHAPTER 6

1. .41503, with error bound .00193.
   .41763, with error bound .00125.
   .41430, with error bound .00333.

2. .49730, with error bound .00333.
   .49101, with error bound .01333.

5. First show that

$$\frac{d}{dx}\prod_{j=0}^{n}(x-x_j) = \sum_{i=0}^{n}\prod_{\substack{j=0\\j\neq i}}^{n}(x-x_j)$$

6. $f'(x_1) = \left(\frac{1}{12}h\right)(-9f_0 + 8f_1 + f_2) - h^2 f'''(\xi)/2$.

7. .45790, with error bound .005.

8. $f'(x_1) = \left(\frac{1}{2}h\right)[-f(x_0) + f(x_2)] - \left(h^2/2\right)(d/dx)f''(\xi(x))\big|_{x=x_1}$.

10. $f'(x) = 1/(1 + x^2)$, $f'(.5) = .8$.

h	$f'(.5)$	Bound on inherent error	Bound on truncation error
.005	.800000	10^{-5}	10^{-6}
.010	.799995	5×10^{-6}	3.9×10^{-6}
.020	.799998	2.5×10^{-8}	1.6×10^{-5}
.025	.799974	2×10^{-8}	2.6×10^{-5}

11. $(\frac{1}{2}h)(-3\varepsilon_0 + 4\varepsilon_1 - \varepsilon_2)$.
    $(\frac{1}{2}h)(\varepsilon_0 - 4\varepsilon_1 + 3\varepsilon_2)$.

12. .95375, .95533.

13. If all $|\varepsilon_i| \le \varepsilon$, then $|(1/12h)(\varepsilon_0 - 8\varepsilon_1 + 8\varepsilon_3 - \varepsilon_4)| \le 3\varepsilon/2h$. Higher-order formulas tend to have smaller truncation error and larger inherent error.

14. The middle formula of the third group has an error bound of .51 when $h = .07421$.

15. $\frac{1}{2}$, error $-1/6$.

16. $.5[f(0) + f(1)]$.

17. $(3/8)[f(0) + 3f(1) + 3f(2) + f(3)]$.

18. $(1/12)[5f(0) - 16f(1) + 23f(2)]$.

19. $(1/24)[f(0) - 5f(1) + 19f(2) + 9f(3)]$.

20. $C_0 = C_2 = 1/3$, $C_1 = 4/3$. The result is Simpson's rule.

21. $(3h/8)[f(x_0) + 3f(x_1) + 3f(x_2) + f(x_3)]$.

22. $(h/12)[5f(x_0) - 16f(x_1) + 23f(x_2)]$.

23. $(h/24)[f(x_0) - 5f(x_1) + 19f(x_2) + 9f(x_3)]$.

24. Multiply each C_i by h, and replace $f(i)$ by $f(x_i)$.

25. (a) .8596 (b) .8588
    (c) Exact (to 4D), .8588

26. $(h/2)[-f(x_0) + 3f(x_1)] + (5h^3/12)f''(\mu)$, where $\mu \in [x_0, x_2]$.

27. $(-3/80)f^{(4)}(\mu)$, where $\mu \in [0, 3]$.

28. $(-3h^5/80)f^{(4)}(\mu)$, where $\mu \in [x_0, x_3]$

30. $(h/12)[5f(x_0) - 16f(x_1) + 23f(x_2)] + [3h^4f'''(\mu)]/8$, where $\mu \in [x_0, x_3]$.

31. 25/36, error bound 1/120.

32. 111/160, error bound 1/270.

34. $m \ge 51$.

35. $m \ge 164$.

37. If all $|\varepsilon_i| \le \varepsilon$, then

$$\left| \frac{h}{3}\left(\varepsilon_0 + \varepsilon_{2m} + 2\sum_{i=1}^{m-1} \varepsilon_{2i} + 4\sum_{i=1}^{m} \varepsilon_{2i-1} \right) \right| \le (b-a)\varepsilon$$

38. If all $|\varepsilon_i| \le \varepsilon$, then $|(|\varepsilon_{-1} - 2\varepsilon_0 + \varepsilon_1)/h^2| \le 4\varepsilon/h^2$.

39. $(1/12h^2)(-f_{-2} + 16f_{-1} - 30f_0 + 16f_1 - f_2)$.

40. Carry two more terms in the Taylor's-series expansions used in the derivation of the second-order formula.

41. 3.000, 3.000, 3.000.

42. $V_2 = 3,776/3,465$
$S_3^0 \approx 1.1032$.

44. .5, .3125, .265625, .253906.

46. $((b-a)h_k^2/12) \max\limits_{a \le x \le b} |f''(x)|$, where $h_k = (b-a)/2^k$.

48. $S_2^1 = 1.0987334$, $S_1^2 = 1.0986983$, $S_0^2 = 1.0986802$.

49.

.500000		
.312500	.250000	
.265625	.250000	.250000

51. $(hk/9)\{[f(x_0,y_0) + 4f(x_0,y_1) + f(x_0,y_2)] + 4[f(x_1, y_0) + 4f(x_1,y_1) + f(x_1,y_2)] + [f(x_2, y_0) + 4f(x_2,y_1) + f(x_2,y_2)]\}$.

52. $(hkv/8)\sum\limits_{k=0}^{1}\sum\limits_{j=0}^{1}\sum\limits_{i=0}^{1} f(x_i,y_j,z_k)$.

CHAPTER 7

2. $u(x) = 5e^{(x^3 - 1)/8} - 3$.

3. $u(x) = (1+x)\left[1 + \int_0^x (t\,dt)/(1+t)\right]$.

4. $u(.1) = .000500$.
$u(.2) = .003015$.

5. $y = \exp\left[\sum\limits_{i=0}^{n} a_i x^{i+1}/(i+1)\right]$.

7. 202.

8. Use the fact that a continuous function on a closed interval achieves its maximum and minimum values.

10. $\delta = 25/1,301$, $\delta = 50/e^{51}$.

11. The function $y^{1/2}$ does not satisfy a Lipschitz condition on any domain in which y can be zero.

12. $\delta = t/\sqrt{1+t}$ for any $t \in (0,1)$.

13. $[0,.125]$.

14. $K = 1$.

15. $K = 2 + e$.

16. $K = 14$.

17. $K = \max(|b|, |c|, |d|)$.

20. The use of $r = 1$, $s = 2$ gives $[0, 1/3]$.

22. There exists a unique solution on $[0,1]$.

23. $y_1' = y_2$ $y_1(0) = 1$
$y_2' = y_1$ $y_2(0) = 2$

24. $y_1' = y_2$ $y_1(1) = 1.$
$xy_2' - y_2 + 4x^3y_1 = 0$ $y_2(1) = 2.$

25. $y_1' = y_2$ $y_1(0) = 1$
$y_2' = y_3$ $y_2(0) = 2$
$y_3' = 3y_3 + 6xy_2 + 10x^2y_1 + \sin x$ $y_3(0) = 3$

26. Let $w_1 = y, w_2 = y', w_3 = u, w_4 = u'$.

$w_1' = w_2$ $w_1(1) = 3$
$w_2' = e^x + w_2 + w_4 + w_3 + w_1$ $w_2(1) = 4$
$w_3' = w_4$ $w_3(1) = 1$
$w_4' = e^x + 9xw_3 + 6w_4 + 9w_2 + 10w_1$ $w_4(1) = 2$

28. The result $\lim_{n \to \infty}(1 + v/n)^n = e^v$ can be proved by taking the ln of $(1 + v/n)^n$ and then applying l'Hospital's rule.

29. 1.43, 1.4465 (to 4D)

30. $y(x) = 3e^{(x-1)} - x - 1, y(1.2) = 1.4642.$

31. $y_1 = .1, y_2 = .201, y_3 = .30504, y_4 = .41435, y_5 = .53152.$

32. (b) $y' = c$, where c is a constant.

33. $y_{1,1} = .1, y_{1,2} = .2, y_{1,3} = .3001.$ Exact solution $y(x) = x^2 + x.$

34. $y_{1,3} = 1.0303, y_{2,3} = 1.0303.$ Exact solution $y_1(x) = y_2(x) = e^x.$

35. $h = .1$ gives $y_2 = 1.4631. h = .05$ gives $y_4 = 1.4639.$

36. $y_1 = .10000, y_2 = .20560, y_3 = .31315, y_4 = .42408, y_5 = .54003.$

38. $y' = ax + b$, where a and b are constants.

39. $y_{1,1} = 2.15375$ $y_{2,1} = .94500$
$y_{1,2} = 2.31486$ $y_{2,2} = .87912$

40. $y_{1,1} = 1.16256$ $y_{2,1} = 1.21750$
$y_{1,2} = 1.35481$ $y_{2,2} = 1.47575$

41. $y_{1,2} = .11018$ $y_{2,2} = 1.09692$

42. $y_{1,2} = .11026$ $y_{2,2} = 1.09825$

45. Observe that the first two equations do not involve y_3 and are the same system studied in Exercise 44.

46. $h = .05$ gives 1.331. $h = .025$ gives 1.334025.

48. $y_1 = 1.105, y_2 = 1.221025, y_3 = 1.34923.$

CHAPTER 8

1. $y_1 = 1.05250, y_2 = 1.11106$

2. $y_1 = 1.1051709, y_2 = 1.2214027.$

3. $y(x) = e^x.$

4. $y'' = y' + \cos x, y''' = y'' - \sin x, y^{(4)} = y''' - \cos x, y^{(5)} = y^{(4)} + \sin x.$

5. $y^{(4)} = 6y + 18xy' + 9x^2y'' + x^3y'''.$

10. 1.05127, 1.10517, 1.16183.

11. Your results should agree with the exact solution to at least 6*D*.

12. $y_{1,1} = .090500, y_{2,1} = .904834, y_{3,1} = 1.637334$.

14. $y_1 = 1.05256, y_2 = 1.11406$.

17. $y_1 = 1.111463, y_2 = 1.247724$.

19. $y_1 = .980198, y_2 = .923116$.

24. $y_1^{(3)} \equiv y_1 = 1.052564, \quad y_2^{(3)} \equiv y_2 = 1.110387$.

27. (*b*) $y_1 = 1.110526, y_2 = 1.243213$.

28. $y_6^{(1)} \equiv y_6 = 2.044240$.

Bibliography

Aitken, A.C., On Interpolation by Iteration of Proportional Parts, without the Use of Differences, Proc. Edinburgh Math. Soc., vol. 3, ser. 2, pp. 56–76, 1932.

Atkinson, K., An Introduction to Numerical Analysis, Wiley, (2nd ed.), 1989.

Acton, Forman S., Numerical Methods That Work, Harper Collins, 1970.

Arden, Bruce W. & Astill, Kenneth N., Numerical Algorithms: Origins and Applications, Addison-Wesley, 1970.

Birkhoff, G., & Rota, G., Ordinary Differential Equations, Ginn and Company, Boston, 1962.

Carnahan, Brice, Luther, H.A. & Wilkes, James O., Applied Numerical Methods, John Wiley & Sons, 1969.

Carnahan, Brice & Wilkes, J.C., Digital Computing and Numerical Methods (with FORTRAN IV, WATFOR, and WATFIV PROGRAMMING), John Wiley & Sons, 1973.

Cohen, A.M., Cutts, J.F, Feilder, R., Jones, D.E., Ribbans, J. & Stuart, E., Numerical Analysis, Halsted Publishing, 1973.

Conte, S.D., Elementary Numerical Analysis, 2nd Ed. McGraw Hill Co., 1974.

Conte, S & de Boor, C, Elementary Numerical Analysis, McGraw-Hill, 1980.

Dahlquist, Germund, Ake Bjorck & Anderson, Ned, Numerical Methods, Prentice Hall, 1974.

Daniel, James W. & Moore, Ramon E., Computation and Theory in Ordinary Differential Equations, W.H. Freeman, 1970.

Davis, Phillip J. & Rabinowitz, Philip, Methods of Numerical Integration, Academic Publishing, 1975.

Deuflhard, P. & Hohmann, A., Numerical Analysis in Modern Scientific Computing, 2nd ed., Springer, 2003.

Dorn, W.S. & Greenberg, H.J., Mathematics and Computing with FORTRAN Programming, John Wiley & Sons, 1967.

Dorn, W.S. & McCracken, D., Numerical Methods with FORTRAN IV Case Studies, John Wiley & Sons, 1972.

Evans, G., Practical Numerical Analysis, John Wiley, 1995.

Faddeev, D. K. & Faddeeva, V. N., Computational Methods of Linear Algebra, translated by Robert C. Williams, W. H. Freeman and Company, San Francisco, 1963.

Ford, L. R., "Differential Equations," 1st ed., McGraw-Hill Book Company, New York, 1933.

Forsythe, G.E, Malcolm, M.P. & Moler, C.B., Computer Methods for Mathematical Computations, Prentice Hall, 1977.

Fulks, W., Advanced Calculus: An Introduction to Analysis, John Wiley & Sons, Inc., New York, 1961.

Gear, C. William, Numerical Initial Value Problems in Ordinary Differential Equations, Prentice Hall, 1971.

Gerald, Curtis F., Applied Numerical Analysis, Addison-Wesley, 1970.

Greenspan, Donald, Introduction to Numerical Analysis and Applications, Rand Publishing, 1971.

Hamming, Richard W., Introduction to Applied Numerical Analysis, McGraw Hill, 1971.

Hamming, Richard W., Numerical Methods for Scientist and Engineers, 2nd Ed., McGraw Hill, 1973.

Heath, M. T., Scientific Computing: An Introductory Survey, 2nd ed., McGraw-Hill, 2002.

Henrici, P., Elements of Numerical Analysis, John Wiley & Sons, Inc. New York, 1964.

Hildebrand, F.G, Introduction to Numerical Analysis, 2nd Ed., McGraw Hill, 1974.

Hull, T. E., and A. L. Creemer, The Efficiency of Predictor-Corrector Procedures, J. Assoc. Comput. Mach., vol. 10, pp.291–301, 1963.

Isaacson, E. & Keller, H., Analysis of Numerical Methods, Wiley, 1966 (or Dover 1994).

Kahaner, D. & Moler, C. & Nash, S., Numerical Methods and Software, Prentice-Hall, 1989.

Kincaid, D. & Cheney, W., Numerical Analysis: Mathematics of Scientific Computing, Brooks/Cole, 1996.

Neville, E. H., Iterative Interpolation, J. Indian Math. Soc., vol. 20, pp. 87–120.

Ralston A., A First Course in Numerical Analysis, McGraw-Hill Book Company, New York, 1965.

Ralston, A. & Rabinowitz, P., A First Course in Numerical Analysis, McGraw-Hill, 1978.

Stiefel, E. L., "An Introduction to Numerical Analysis," translated by Werner C. Rheinboldt and Cornelie J. Rheinboldt, Academic Press Inc., New York, 1963.

Todd, J., Survey of Numerical Analysis, McGraw-Hill Book Company, New York, 1962.

Wilkinson, J. H., Rounding Errors in Algebraic Processes, Prentice-Hall, Inc. Englewood Cliffs, N.J., 1963.

Index

A CATALOG OF SELECTED
DOVER BOOKS
IN SCIENCE AND MATHEMATICS

Mathematics–Bestsellers

HANDBOOK OF MATHEMATICAL FUNCTIONS: with Formulas, Graphs, and Mathematical Tables, Edited by Milton Abramowitz and Irene A. Stegun. A classic resource for working with special functions, standard trig, and exponential logarithmic definitions and extensions, it features 29 sets of tables, some to as high as 20 places. 1046pp. 8 x 10 1/2. 0-486-61272-4

ABSTRACT AND CONCRETE CATEGORIES: The Joy of Cats, Jiri Adamek, Horst Herrlich, and George E. Strecker. This up-to-date introductory treatment employs category theory to explore the theory of structures. Its unique approach stresses concrete categories and presents a systematic view of factorization structures. Numerous examples. 1990 edition, updated 2004. 528pp. 6 1/8 x 9 1/4. 0-486-46934-4

MATHEMATICS: Its Content, Methods and Meaning, A. D. Aleksandrov, A. N. Kolmogorov, and M. A. Lavrent'ev. Major survey offers comprehensive, coherent discussions of analytic geometry, algebra, differential equations, calculus of variations, functions of a complex variable, prime numbers, linear and non-Euclidean geometry, topology, functional analysis, more. 1963 edition. 1120pp. 5 3/8 x 8 1/2. 0-486-40916-3

INTRODUCTION TO VECTORS AND TENSORS: Second Edition--Two Volumes Bound as One, Ray M. Bowen and C.-C. Wang. Convenient single-volume compilation of two texts offers both introduction and in-depth survey. Geared toward engineering and science students rather than mathematicians, it focuses on physics and engineering applications. 1976 edition. 560pp. 6 1/2 x 9 1/4. 0-486-46914-X

AN INTRODUCTION TO ORTHOGONAL POLYNOMIALS, Theodore S. Chihara. Concise introduction covers general elementary theory, including the representation theorem and distribution functions, continued fractions and chain sequences, the recurrence formula, special functions, and some specific systems. 1978 edition. 272pp. 5 3/8 x 8 1/2. 0-486-47929-3

ADVANCED MATHEMATICS FOR ENGINEERS AND SCIENTISTS, Paul DuChateau. This primary text and supplemental reference focuses on linear algebra, calculus, and ordinary differential equations. Additional topics include partial differential equations and approximation methods. Includes solved problems. 1992 edition. 400pp. 7 1/2 x 9 1/4. 0-486-47930-7

PARTIAL DIFFERENTIAL EQUATIONS FOR SCIENTISTS AND ENGINEERS, Stanley J. Farlow. Practical text shows how to formulate and solve partial differential equations. Coverage of diffusion-type problems, hyperbolic-type problems, elliptic-type problems, numerical and approximate methods. Solution guide available upon request. 1982 edition. 414pp. 6 1/8 x 9 1/4. 0-486-67620-X

VARIATIONAL PRINCIPLES AND FREE-BOUNDARY PROBLEMS, Avner Friedman. Advanced graduate-level text examines variational methods in partial differential equations and illustrates their applications to free-boundary problems. Features detailed statements of standard theory of elliptic and parabolic operators. 1982 edition. 720pp. 6 1/8 x 9 1/4. 0-486-47853-X

LINEAR ANALYSIS AND REPRESENTATION THEORY, Steven A. Gaal. Unified treatment covers topics from the theory of operators and operator algebras on Hilbert spaces; integration and representation theory for topological groups; and the theory of Lie algebras, Lie groups, and transform groups. 1973 edition. 704pp. 6 1/8 x 9 1/4. 0-486-47851-3

A SURVEY OF INDUSTRIAL MATHEMATICS, Charles R. MacCluer. Students learn how to solve problems they'll encounter in their professional lives with this concise single-volume treatment. It employs MATLAB and other strategies to explore typical industrial problems. 2000 edition. 384pp. 5 3/8 x 8 1/2. 0-486-47702-9

NUMBER SYSTEMS AND THE FOUNDATIONS OF ANALYSIS, Elliott Mendelson. Geared toward undergraduate and beginning graduate students, this study explores natural numbers, integers, rational numbers, real numbers, and complex numbers. Numerous exercises and appendixes supplement the text. 1973 edition. 368pp. 5 3/8 x 8 1/2. 0-486-45792-3

A FIRST LOOK AT NUMERICAL FUNCTIONAL ANALYSIS, W. W. Sawyer. Text by renowned educator shows how problems in numerical analysis lead to concepts of functional analysis. Topics include Banach and Hilbert spaces, contraction mappings, convergence, differentiation and integration, and Euclidean space. 1978 edition. 208pp. 5 3/8 x 8 1/2. 0-486-47882-3

FRACTALS, CHAOS, POWER LAWS: Minutes from an Infinite Paradise, Manfred Schroeder. A fascinating exploration of the connections between chaos theory, physics, biology, and mathematics, this book abounds in award-winning computer graphics, optical illusions, and games that clarify memorable insights into self-similarity. 1992 edition. 448pp. 6 1/8 x 9 1/4. 0-486-47204-3

SET THEORY AND THE CONTINUUM PROBLEM, Raymond M. Smullyan and Melvin Fitting. A lucid, elegant, and complete survey of set theory, this three-part treatment explores axiomatic set theory, the consistency of the continuum hypothesis, and forcing and independence results. 1996 edition. 336pp. 6 x 9. 0-486-47484-4

DYNAMICAL SYSTEMS, Shlomo Sternberg. A pioneer in the field of dynamical systems discusses one-dimensional dynamics, differential equations, random walks, iterated function systems, symbolic dynamics, and Markov chains. Supplementary materials include PowerPoint slides and MATLAB exercises. 2010 edition. 272pp. 6 1/8 x 9 1/4. 0-486-47705-3

ORDINARY DIFFERENTIAL EQUATIONS, Morris Tenenbaum and Harry Pollard. Skillfully organized introductory text examines origin of differential equations, then defines basic terms and outlines general solution of a differential equation. Explores integrating factors; dilution and accretion problems; Laplace Transforms; Newton's Interpolation Formulas, more. 818pp. 5 3/8 x 8 1/2. 0-486-64940-7

MATROID THEORY, D. J. A. Welsh. Text by a noted expert describes standard examples and investigation results, using elementary proofs to develop basic matroid properties before advancing to a more sophisticated treatment. Includes numerous exercises. 1976 edition. 448pp. 5 3/8 x 8 1/2. 0-486-47439-9

THE CONCEPT OF A RIEMANN SURFACE, Hermann Weyl. This classic on the general history of functions combines function theory and geometry, forming the basis of the modern approach to analysis, geometry, and topology. 1955 edition. 208pp. 5 3/8 x 8 1/2. 0-486-47004-0

THE LAPLACE TRANSFORM, David Vernon Widder. This volume focuses on the Laplace and Stieltjes transforms, offering a highly theoretical treatment. Topics include fundamental formulas, the moment problem, monotonic functions, and Tauberian theorems. 1941 edition. 416pp. 5 3/8 x 8 1/2. 0-486-47755-X

Mathematics–Logic and Problem Solving

PERPLEXING PUZZLES AND TANTALIZING TEASERS, Martin Gardner. Ninety-three riddles, mazes, illusions, tricky questions, word and picture puzzles, and other challenges offer hours of entertainment for youngsters. Filled with rib-tickling drawings. Solutions. 224pp. 5 3/8 x 8 1/2. 0-486-25637-5

MY BEST MATHEMATICAL AND LOGIC PUZZLES, Martin Gardner. The noted expert selects 70 of his favorite "short" puzzles. Includes The Returning Explorer, The Mutilated Chessboard, Scrambled Box Tops, and dozens more. Complete solutions included. 96pp. 5 3/8 x 8 1/2. 0-486-28152-3

THE LADY OR THE TIGER?: and Other Logic Puzzles, Raymond M. Smullyan. Created by a renowned puzzle master, these whimsically themed challenges involve paradoxes about probability, time, and change; metapuzzles; and self-referentiality. Nineteen chapters advance in difficulty from relatively simple to highly complex. 1982 edition. 240pp. 5 3/8 x 8 1/2. 0-486-47027-X

SATAN, CANTOR AND INFINITY: Mind-Boggling Puzzles, Raymond M. Smullyan. A renowned mathematician tells stories of knights and knaves in an entertaining look at the logical precepts behind infinity, probability, time, and change. Requires a strong background in mathematics. Complete solutions. 288pp. 5 3/8 x 8 1/2.

0-486-47036-9

THE RED BOOK OF MATHEMATICAL PROBLEMS, Kenneth S. Williams and Kenneth Hardy. Handy compilation of 100 practice problems, hints and solutions indispensable for students preparing for the William Lowell Putnam and other mathematical competitions. Preface to the First Edition. Sources. 1988 edition. 192pp. 5 3/8 x 8 1/2. 0-486-69415-1

KING ARTHUR IN SEARCH OF HIS DOG AND OTHER CURIOUS PUZZLES, Raymond M. Smullyan. This fanciful, original collection for readers of all ages features arithmetic puzzles, logic problems related to crime detection, and logic and arithmetic puzzles involving King Arthur and his Dogs of the Round Table. 160pp. 5 3/8 x 8 1/2. 0-486-47435-6

UNDECIDABLE THEORIES: Studies in Logic and the Foundation of Mathematics, Alfred Tarski in collaboration with Andrzej Mostowski and Raphael M. Robinson. This well-known book by the famed logician consists of three treatises: "A General Method in Proofs of Undecidability," "Undecidability and Essential Undecidability in Mathematics," and "Undecidability of the Elementary Theory of Groups." 1953 edition. 112pp. 5 3/8 x 8 1/2. 0-486-47703-7

LOGIC FOR MATHEMATICIANS, J. Barkley Rosser. Examination of essential topics and theorems assumes no background in logic. "Undoubtedly a major addition to the literature of mathematical logic." — *Bulletin of the American Mathematical Society.* 1978 edition. 592pp. 6 1/8 x 9 1/4. 0-486-46898-4

INTRODUCTION TO PROOF IN ABSTRACT MATHEMATICS, Andrew Wohlgemuth. This undergraduate text teaches students what constitutes an acceptable proof, and it develops their ability to do proofs of routine problems as well as those requiring creative insights. 1990 edition. 384pp. 6 1/2 x 9 1/4. 0-486-47854-8

FIRST COURSE IN MATHEMATICAL LOGIC, Patrick Suppes and Shirley Hill. Rigorous introduction is simple enough in presentation and context for wide range of students. Symbolizing sentences; logical inference; truth and validity; truth tables; terms, predicates, universal quantifiers; universal specification and laws of identity; more. 288pp. 5 3/8 x 8 1/2. 0-486-42259-3

Mathematics–Algebra and Calculus

VECTOR CALCULUS, Peter Baxandall and Hans Liebeck. This introductory text offers a rigorous, comprehensive treatment. Classical theorems of vector calculus are amply illustrated with figures, worked examples, physical applications, and exercises with hints and answers. 1986 edition. 560pp. 5 3/8 x 8 1/2. 0-486-46620-5

ADVANCED CALCULUS: An Introduction to Classical Analysis, Louis Brand. A course in analysis that focuses on the functions of a real variable, this text introduces the basic concepts in their simplest setting and illustrates its teachings with numerous examples, theorems, and proofs. 1955 edition. 592pp. 5 3/8 x 8 1/2. 0-486-44548-8

ADVANCED CALCULUS, Avner Friedman. Intended for students who have already completed a one-year course in elementary calculus, this two-part treatment advances from functions of one variable to those of several variables. Solutions. 1971 edition. 432pp. 5 3/8 x 8 1/2. 0-486-45795-8

METHODS OF MATHEMATICS APPLIED TO CALCULUS, PROBABILITY, AND STATISTICS, Richard W. Hamming. This 4-part treatment begins with algebra and analytic geometry and proceeds to an exploration of the calculus of algebraic functions and transcendental functions and applications. 1985 edition. Includes 310 figures and 18 tables. 880pp. 6 1/2 x 9 1/4. 0-486-43945-3

BASIC ALGEBRA I: Second Edition, Nathan Jacobson. A classic text and standard reference for a generation, this volume covers all undergraduate algebra topics, including groups, rings, modules, Galois theory, polynomials, linear algebra, and associative algebra. 1985 edition. 528pp. 6 1/8 x 9 1/4. 0-486-47189-6

BASIC ALGEBRA II: Second Edition, Nathan Jacobson. This classic text and standard reference comprises all subjects of a first-year graduate-level course, including in-depth coverage of groups and polynomials and extensive use of categories and functors. 1989 edition. 704pp. 6 1/8 x 9 1/4. 0-486-47187-X

CALCULUS: An Intuitive and Physical Approach (Second Edition), Morris Kline. Application-oriented introduction relates the subject as closely as possible to science with explorations of the derivative; differentiation and integration of the powers of x; theorems on differentiation, antidifferentiation; the chain rule; trigonometric functions; more. Examples. 1967 edition. 960pp. 6 1/2 x 9 1/4. 0-486-40453-6

ABSTRACT ALGEBRA AND SOLUTION BY RADICALS, John E. Maxfield and Margaret W. Maxfield. Accessible advanced undergraduate-level text starts with groups, rings, fields, and polynomials and advances to Galois theory, radicals and roots of unity, and solution by radicals. Numerous examples, illustrations, exercises, appendixes. 1971 edition. 224pp. 6 1/8 x 9 1/4. 0-486-47723-1

AN INTRODUCTION TO THE THEORY OF LINEAR SPACES, Georgi E. Shilov. Translated by Richard A. Silverman. Introductory treatment offers a clear exposition of algebra, geometry, and analysis as parts of an integrated whole rather than separate subjects. Numerous examples illustrate many different fields, and problems include hints or answers. 1961 edition. 320pp. 5 3/8 x 8 1/2. 0-486-63070-6

LINEAR ALGEBRA, Georgi E. Shilov. Covers determinants, linear spaces, systems of linear equations, linear functions of a vector argument, coordinate transformations, the canonical form of the matrix of a linear operator, bilinear and quadratic forms, and more. 387pp. 5 3/8 x 8 1/2. 0-486-63518-X

Browse over 9,000 books at www.doverpublications.com

Mathematics–Probability and Statistics

BASIC PROBABILITY THEORY, Robert B. Ash. This text emphasizes the probabilistic way of thinking, rather than measure-theoretic concepts. Geared toward advanced undergraduates and graduate students, it features solutions to some of the problems. 1970 edition. 352pp. 5 3/8 x 8 1/2. 0-486-46628-0

PRINCIPLES OF STATISTICS, M. G. Bulmer. Concise description of classical statistics, from basic dice probabilities to modern regression analysis. Equal stress on theory and applications. Moderate difficulty; only basic calculus required. Includes problems with answers. 252pp. 5 5/8 x 8 1/4. 0-486-63760-3

OUTLINE OF BASIC STATISTICS: Dictionary and Formulas, John E. Freund and Frank J. Williams. Handy guide includes a 70-page outline of essential statistical formulas covering grouped and ungrouped data, finite populations, probability, and more, plus over 1,000 clear, concise definitions of statistical terms. 1966 edition. 208pp. 5 3/8 x 8 1/2. 0-486-47769-X

GOOD THINKING: The Foundations of Probability and Its Applications, Irving J. Good. This in-depth treatment of probability theory by a famous British statistician explores Keynesian principles and surveys such topics as Bayesian rationality, corroboration, hypothesis testing, and mathematical tools for induction and simplicity. 1983 edition. 352pp. 5 3/8 x 8 1/2. 0-486-47438-0

INTRODUCTION TO PROBABILITY THEORY WITH CONTEMPORARY APPLICATIONS, Lester L. Helms. Extensive discussions and clear examples, written in plain language, expose students to the rules and methods of probability. Exercises foster problem-solving skills, and all problems feature step-by-step solutions. 1997 edition. 368pp. 6 1/2 x 9 1/4. 0-486-47418-6

CHANCE, LUCK, AND STATISTICS, Horace C. Levinson. In simple, non-technical language, this volume explores the fundamentals governing chance and applies them to sports, government, and business. "Clear and lively ... remarkably accurate." — *Scientific Monthly*. 384pp. 5 3/8 x 8 1/2. 0-486-41997-5

FIFTY CHALLENGING PROBLEMS IN PROBABILITY WITH SOLUTIONS, Frederick Mosteller. Remarkable puzzlers, graded in difficulty, illustrate elementary and advanced aspects of probability. These problems were selected for originality, general interest, or because they demonstrate valuable techniques. Also includes detailed solutions. 88pp. 5 3/8 x 8 1/2. 0-486-65355-2

EXPERIMENTAL STATISTICS, Mary Gibbons Natrella. A handbook for those seeking engineering information and quantitative data for designing, developing, constructing, and testing equipment. Covers the planning of experiments, the analyzing of extreme-value data; and more. 1966 edition. Index. Includes 52 figures and 76 tables. 560pp. 8 3/8 x 11. 0-486-43937-2

STOCHASTIC MODELING: Analysis and Simulation, Barry L. Nelson. Coherent introduction to techniques also offers a guide to the mathematical, numerical, and simulation tools of systems analysis. Includes formulation of models, analysis, and interpretation of results. 1995 edition. 336pp. 6 1/8 x 9 1/4. 0-486-47770-3

INTRODUCTION TO BIOSTATISTICS: Second Edition, Robert R. Sokal and F. James Rohlf. Suitable for undergraduates with a minimal background in mathematics, this introduction ranges from descriptive statistics to fundamental distributions and the testing of hypotheses. Includes numerous worked-out problems and examples. 1987 edition. 384pp. 6 1/8 x 9 1/4. 0-486-46961-1

Mathematics–Geometry and Topology

PROBLEMS AND SOLUTIONS IN EUCLIDEAN GEOMETRY, M. N. Aref and William Wernick. Based on classical principles, this book is intended for a second course in Euclidean geometry and can be used as a refresher. More than 200 problems include hints and solutions. 1968 edition. 272pp. 5 3/8 x 8 1/2. 0-486-47720-7

TOPOLOGY OF 3-MANIFOLDS AND RELATED TOPICS, Edited by M. K. Fort, Jr. With a New Introduction by Daniel Silver. Summaries and full reports from a 1961 conference discuss decompositions and subsets of 3-space; n-manifolds; knot theory; the Poincaré conjecture; and periodic maps and isotopies. Familiarity with algebraic topology required. 1962 edition. 272pp. 6 1/8 x 9 1/4. 0-486-47753-3

POINT SET TOPOLOGY, Steven A. Gaal. Suitable for a complete course in topology, this text also functions as a self-contained treatment for independent study. Additional enrichment materials make it equally valuable as a reference. 1964 edition. 336pp. 5 3/8 x 8 1/2. 0-486-47222-1

INVITATION TO GEOMETRY, Z. A. Melzak. Intended for students of many different backgrounds with only a modest knowledge of mathematics, this text features self-contained chapters that can be adapted to several types of geometry courses. 1983 edition. 240pp. 5 3/8 x 8 1/2. 0-486-46626-4

TOPOLOGY AND GEOMETRY FOR PHYSICISTS, Charles Nash and Siddhartha Sen. Written by physicists for physics students, this text assumes no detailed background in topology or geometry. Topics include differential forms, homotopy, homology, cohomology, fiber bundles, connection and covariant derivatives, and Morse theory. 1983 edition. 320pp. 5 3/8 x 8 1/2. 0-486-47852-1

BEYOND GEOMETRY: Classic Papers from Riemann to Einstein, Edited with an Introduction and Notes by Peter Pesic. This is the only English-language collection of these 8 accessible essays. They trace seminal ideas about the foundations of geometry that led to Einstein's general theory of relativity. 224pp. 6 1/8 x 9 1/4. 0-486-45350-2

GEOMETRY FROM EUCLID TO KNOTS, Saul Stahl. This text provides a historical perspective on plane geometry and covers non-neutral Euclidean geometry, circles and regular polygons, projective geometry, symmetries, inversions, informal topology, and more. Includes 1,000 practice problems. Solutions available. 2003 edition. 480pp. 6 1/8 x 9 1/4. 0-486-47459-3

TOPOLOGICAL VECTOR SPACES, DISTRIBUTIONS AND KERNELS, François Trèves. Extending beyond the boundaries of Hilbert and Banach space theory, this text focuses on key aspects of functional analysis, particularly in regard to solving partial differential equations. 1967 edition. 592pp. 5 3/8 x 8 1/2.
0-486-45352-9

INTRODUCTION TO PROJECTIVE GEOMETRY, C. R. Wylie, Jr. This introductory volume offers strong reinforcement for its teachings, with detailed examples and numerous theorems, proofs, and exercises, plus complete answers to all odd-numbered end-of-chapter problems. 1970 edition. 576pp. 6 1/8 x 9 1/4. 0-486-46895-X

FOUNDATIONS OF GEOMETRY, C. R. Wylie, Jr. Geared toward students preparing to teach high school mathematics, this text explores the principles of Euclidean and non-Euclidean geometry and covers both generalities and specifics of the axiomatic method. 1964 edition. 352pp. 6 x 9. 0-486-47214-0